d verniers
& Sharpe

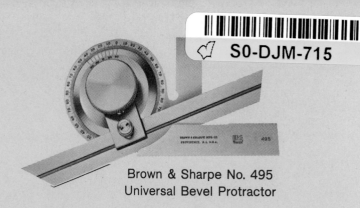

Brown & Sharpe No. 495
Universal Bevel Protractor

VERNIER PROTRACTORS
(Angular measurements)

DIAL:
Graduated in degrees (°)

VERNIER:
Each line represents 5 minutes (5′)

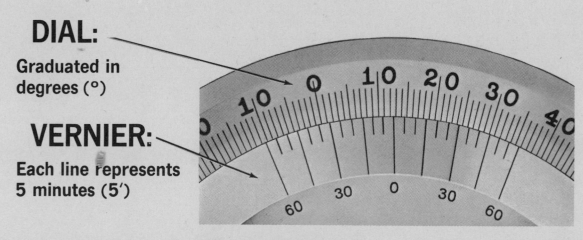

Example A

Number of whole degrees vernier
"0" has moved past scale "0" — 12 — **12°**

Number of vernier line that
coincides with a scale line
(reading in same direction) — 50 — **50′**

TOTAL READING 12°50′

NOTE: If vernier "0" and "60" lines coincide with two scale lines, reading is even, in degrees. You add no minutes.

Example B — **60°**

Example C — **73°5′**

ENGINEERING GRAPHICS

ENGINEERING GRAPHICS

FOURTH EDITION

COMMUNICATION • ANALYSIS • CREATIVE DESIGN

James S. Rising
Maurice W. Almfeldt
Paul S. DeJong

Iowa State University

WM. C. BROWN COMPANY PUBLISHERS
Dubuque, Iowa

Copyright © 1953, 1959, 1964 by
James S. Rising and Maurice W. Almfeldt

Copyright © 1970 by
James S. Rising, Maurice W. Almfeldt and
Paul S. DeJong

Library of Congress Catalog Card Number: 76-124851

ISBN 0-697-08601-1

All rights reserved. No part of this book may be reproduced in any form or by any process without permission from the Copyright owners.

Printed in the United States of America

Preface to the Fourth Edition

The preparation of a new edition is influenced primarily by a desire to improve the product and to bring it up to date in current practices. It is felt that these objectives have been accomplished by at least three major changes. These include the addition of new material on Creative Design and Application Drawing, a responsibility of the new coauthor; the work of two specialty assistants on Unit 24, Electrical and Electronic Diagrams, and on Unit 25, Computer Aided Design; and by rearranging and supplementing much of the previous editions to place increased emphasis upon industrial application.

Professor DeJong, the new coauthor, has an excellent experience background and an imaginative mind for developing the material on Creative Design, Manufacturing Processes, and Design Drawing.

When the decision was made that current practices in Electrical and Electronic Drawing, and in Computer Aided Design should be included in this edition, a search was made for the most knowledgeable source for this information. The greatest problem was to condense the profusion of important material into a brief, interesting, and beneficial compendium. This has been accomplished admirably both by Professor Granneman in Unit 24 and Mr. Aronson in Unit 25.

Other features which will improve the utility of this text include, perhaps of most importance, the addition of pictures, drawings, problems, and manuscript referring to the professional use of the many facets of graphics by all types of engineers. In addition, the several methods used to make pictorial drawings and sketches have been grouped into one unit called Pictorial Systems; Vector Geometry immediately follows the units on space geometry to illustrate its application; also the Differential and Integral Calculus are combined into one unit to better show their interdependence.

The authors are indebted for friendly criticism by the members of the Engineering Graphics department at Iowa State University; to the several Institutes, Societies, and Departments of Standards; and to a variety of industries (credits given) that have provided new material, drawings, and pictures to enhance the authenticity and thereby the value of the text. The manuscript typing and proof-reading by Mrs. LuOra Payne is also greatly appreciated.

New Problem Books, commonly called Work Books, keyed to the material of the text have been prepared by Professors Sanders, Arnbal, and Crawford of the department of engineering graphics at Iowa State University.

J. S. Rising
M. W. Almfeldt
P. S. DeJong

Contents

1 Introduction … 1
2 Creative Design … 5
3 Freehand Drawing … 14
4 Engineering Lettering … 22
5 Drawing Equipment … 31
6 Orthographic Projection—Points … 43
7 Orthographic Projection—Lines … 55
8 Orthographic Projection—Planes … 64
9 Orthographic Projection—Solids … 78
10 Pictorial Systems … 90
11 Sections and Conventional Practices … 105
12 Dimensioning—Basic Concepts … 117
13 Space Geometry—Points and Lines … 133
14 Space Geometry—Lines and Planes … 142
15 Vector Geometry … 155
16 Design Surfaces … 163
17 Surface Intersections … 174
18 Surface Developments … 185
19 Fasteners—Threaded … 198
20 Fasteners—Removable and Permanent … 206
21 Manufacturing Processes and Measurements … 217
22 Dimensioning for Production … 230
23 Piping and Structural Drawing … 244
24 Electrical and Electronic Diagrams … 261
25 Computer Aided Design … 267
26 Design Drawings … 277
27 Graphical Representation of Design Data … 292
28 Graphical Mathematics—Alignment Charts … 305
29 Graphical Mathematics—The Calculus … 323
   Appendix … 337
   Index … 377

unit 1

Introduction

1.1 Brief History of Drawing

The art of portrayal by means of lines dates back practically as far as the history of man. Archaeologists have discovered hieroglyphics on walls and stones as a mute witness of ancient man's inherent desire to draw. Gradually, the crude markings of prehistoric man evolved into symbols interspersed with drawings which were deciphered to tell a story. Even the Greeks must have considered writing and drawing as essentially the same since they used the same word for each. To this day, drawing is the universal tongue which is used when it becomes necessary to convey an idea to another who uses a strange written or spoken word.

Leonardo da Vinci, the great artist and engineer who lived in the last half of the fifteenth and early sixteenth centuries, has been called the father of modern drawing. He practiced and taught a method of graphical description which conveyed and recorded ideas about mechanical subjects.

Toward the end of the eighteenth century Gaspard Mongé, a French mathematician, introduced two planes of projection at right angles to each other for the graphical investigation of solid geometry problems. He generalized many of the methods heretofore employed into a code termed "la Geometrie Descriptive." This graphical method, known as orthographic projection, was extended to solve practical applications in the ship-building and military professions. Coincident with the development of the science of orthographic projection, came the beginnings of the interchangeable system of manufacturing and the establishment of technical education.

The period of industrial growth in this country during the late part of the nineteenth and early twentieth centuries witnessed the introduction of universal drawing standards and the gradual elimination of fancy shading, ornamental lettering and artistic decorations. It is said that the Art of Draughting had been lost, but the Business of Drafting had been discovered. Now, the Science of Graphics is the modern evolution.

1.2 Engineering Graphics

Definition. *Engineering Graphics is the combination of those arts and sciences of drawing applicable to the solution of engineering problems.*

Drawings may be made either freehand or with the aid of instruments, each method having specific merits and applications. Training in freehand work will emphasize form, speed, hand control, and appreciation of proportion. The ability to make freehand drawings is extremely valuable to the engineer because, by this means, he can quickly communicate ideas for design, explanations for problem solutions, or suggestions for changes in design, construction, or production methods.

Drawing instruments are used to create a high quality, realistically delineated, and accurately proportioned record of the design.

Engineering graphics has an important place in all types of engineering practice because it is the basis of all design, is necessary for the fabrication or manufacture of a product, and is the foundation for problem analysis and research. It is considered by some as the most important basic branch of study in the engineering curriculum with the possible exception of mathematics, the fundamental concepts of which are graphical.

The ability to make sketches, to use and interpret drawings, to direct their production, and to pass upon their correctness is inherent to the professional literacy of every engineer.

1.3 Divisions of Modern Engineering Graphics

Engineering Graphics may be considered to consist of three essential parts: (a) Communication, (b) Problem Analysis, and (c) Creative Design. Variable emphasis can be placed on each branch depending on the objectives of the course sequence and the time available.

1.4 Communication

It is no less true today than in the past that Enginering Graphics is the "Language of the Engineer." In order to write and speak this language, the engineer must have working knowledge of the alphabet, the vocabulary, the grammar, the idioms, and the composition.

As in any science such as mathematics, physics, or chemistry, there are many basic concepts upon which more advanced work is built. Certainly it would be impossible to master the intricacies of algebra or the calculus without a sound grounding in arithmetic. The student must possess the ability to form, understand, and use the numerals and symbols of which mathematics is composed. Likewise, the student in graphical expression must be able to understand graphical symbols and to read and write them legibly. Practice in the representation of points, lines, planes and solid objects in multiview projection will continue with instruments and sketches until the student becomes familiar with the symbols, conventions, and abbreviations of the language.

Pictorial drawing is another important part of graphical representation which the engineer often uses to convey his ideas to others. Pictures are also important for a quick interpretation of the intricacies of shape or construction to aid communication between the engineer and management, between the designer and production, and between sales and the layman or consumer.

The engineers' dependency upon *drawing as a medium of communication* gives documentary evidence to the accuracy of the Chinese proverb that, "A picture is worth ten thousand words."

1.5 Problem Analysis

When the student has mastered the fundamentals of the graphical language, he will be able to investigate more complex theory and ascertain the value of this tool for the analysis and solution of engineering problems. Much of this theory and application is ordinarily treated in a separate course called *Descriptive Geometry*. Considerable advantage is gained by the integration of geometrical theory and analysis with basic and application drawing, since the same systems of projection, nomenclature, and symbols are used throughout the course of study.

The analysis of trends, rates of change, and total changes from experimental and statistical data can be animated by graphical representation. Test data can be plotted on many types of coordinate graphs for analysis and if necessary, for derivation of empirical equations. Vector geometry, both coplanar and noncoplanar, is used in rocketry, guidance systems, electrical systems, and structural stress analysis. Alignment charts and other forms of graphical mathematics are advantageous to the engineer for the solution of a wide variety of simple and complex problems. Reoccuring computations of more consequence may require rapid solution by the computer.

No claim is made that graphical methods should replace or eliminate mathematics for the analyzing and solving of engineering problems. Rather, the engineer should be familiar with all methods in order that he may choose the easiest or most effective means of solution for a particular problem.

1.6 Creative Design

It has been said that the engineer's knowledge and skill should be used to create something new or improve something old for the benefit of mankind. For this reason, it is felt that the study of Engineering Graphics will be more interesting and more valuable to the beginning engineer by using *Creative Design* as the application climax.

The short time that the student studies Engineering Graphics allows only a brief introduction to some of the basic methods and potential uses of drawing for Communication, Analysis, and Creative Design. It is hoped that this introduction will provide a background which will make further development both pleasurable and profitable in subsequent related courses at the University and in future engineering careers.

Fig. 2-1. *Model of an Innovative Design, Lunar Excursion Module "Eagle."*

unit 2

Creative Design

> "Some men see things as they are and ask 'Why?'; I dream of things that never were and ask, 'Why not?'"
> —Robert Kennedy

2.1 Introduction

Most students enter the field of engineering because they are good at math, enjoy tinkering with real and abstract problems, like to make things, and fancy themselves somewhat inventive. They see themselves as builders or "do-ers." This is as it should be since we find most engineers are designers in one way or another. They design everything from lunar excursion modules, high thrust rocket engines and more powerful fuels to computer languages; from chemical processes and safer auto braking systems to dishwashers; and from television sets and printed circuits to bridges and tunnels. All these things are vastly different and require different scientific knowledge. However, all design problems have one objective: a satisfactory solution must be achieved by the designer. No magic is involved in the creation of a new product; the designer accepts the "raw" problem and finally arrives at a solution which must be as good or better than anything currently available. Occasionally some solution occurs by chance but generally there is a rational method of approach to any design problem.

Some people believe that tomorrow's products will be computer designed. Computers are powerful devices when used correctly and there is an important place for them as an aid to design, the subject of Unit 25. Changes in design parameters can be introduced with rapid computations and the result quickly plotted by a tape-controlled drafting machine. However, Man is the innovator and creator; the computer is a most desirable adjunct because of its tireless speed in solving and resolving problems thus freeing the designer for more creative endeavor.

2.2 Fundamental Concepts

It would be natural for the student to ask several questions: Just what is design, and specifically, creative design? How is graphics related to design? Why study creative design as a part of beginning graphics?

Design can be defined as the application of an ability to combine facts and relationships from the various fields of art or science to produce an acceptable system, mechanism, or process. In order to be aceptable, the design must satisfy some need, such as music that is pleasing to the ear, or devices which perform desired functions.

Creative Design is the application of an ability to combine *seemingly unrelated* facts and relationships in order to produce a *unique* or *novel* system or device. The design does not always involve the creation of a new device; sometimes it may be an improvement of a process or a product. The novel or unique design is conceived by the addition of *creativity*.

In both of the definitions of design the word *ability* has been used. Some people may immediately claim that they were not born with this ability. However, neither were they born with the ability to walk or to solve equations. Ability is gained not by birth or osmosis, but by learning and practice. This again implies that there must be some logical method used in the design process. Considering this, let us formulate another applicable and perhaps more useful definition: *Creative Design is a systematic plan or approach for combining seemingly unrelated facts in order to establish the form and details of an original, imaginative, and useful system or device.*

Iteration, a term frequently encountered in engineering, is a repetitive action in which successive trials are performed to continuously approach a desired goal or solution. Computers use iterative processes to converge on a solution to a problem, and instructors reiterate facts or concepts to improve comprehension by the student. Similarly, work done in a previous step of the creative process may be repeated if a later step shows an omission or new material which will improve the previous endeavor.

Engineering Design is an iterative, decision-making process. It involves the essential elements of creative design, augmented by knowledge and experience gained from other engineering disciplines and industrial experience.

Engineering design does not always lead to particularly novel results. The many different makes of automobiles or television sets are good examples of standard components assembled in essentially similar ways but resulting in a variety of products quite different in appearance and quality. A system can be well designed or fundamentally improved *only* if a systematic design process is followed.

2.3 How is Graphics Related to Design

Graphics is the *language* of creative design. It is the only medium through which form and substance can be given to abstract ideas, and is therefore one of the highest forms of communication. At the same time graphics is one of the most basic forms of communication because it is universally understood. It is the primary vehicle for conveying ideas to management, production, sales personnel, and to other engineers.

Sketches, graphs, and charts are frequently needed as an aid in visualizing the complex shape relationship and function of the component parts in a design. Complex mathematical solutions involving physical phenomena can often be quickly checked or avoided by graphical means. Many other examples could be cited.

2.4 Why Study Creative Design Now

Engineering Graphics must be used, not merely studied. It is important for the student to learn by practice how to convert his unique ideas into visible and understandable concepts. An appreciation for the relevance of engineering graphics will be gained by actual involvement in the communication of practical ideas rather than the negative effect of tedious solutions of purely academic problems.

Many applications will be encountered during this and subsequent engineering courses. The method outlined here can be applied to numerous projects successfully if it is applied *conscientiously*. As this text is studied, the potential application of the text material to the communication of a creative design shounld be kept in mind. Everything is applicable; one must merely recognize its potential.

Make no mistake, all design knowledge is *not* contained in this unit or this text. A tremendous amount of technical knowledge must be acquired in order to perform competent engineering design. Projects for creative design should be simple enough that day-to-day experience can be used to finalize the design. This unit is concerned with a more or less universal *design process* or approach rather than technical or theoretical knowledge.

2.5 The Design Process

This design procedure employs seven steps which have been found effective. It should be recognized that although the steps may not be as clearly defined in practice as they are here, each step of the design process should be performed in sequence. The emphasis on some design steps may vary for different types of problems. One may go back to a previous step, but must not skip the forward steps. To prevent "redesigning the wheel," problem solutions already available must be disregarded and attention focused on the fundamentals involved. The design will be initially nebulous but will become more concrete with each step. The process is arranged so that in Frank Lloyd Wright's words, "Form follows Function." This means that the best design evolves from a logical study of the characteristic uses of the product.

The seven steps listed and discussed in the following articles will be illustrated by an example of design.

2.6 Step 1. Analyze the Problem

Recognize Problem and Identify Needs. In this first step a problem is discovered and expressed as a human need in as general terms as possible. It should be recognized that *problems are caused by unsatisfied needs*, and that design begins with the awareness of these social or economic human needs. A need may be some fundamental requirement, such as a method of purifying urban air, or it may be an improvement of an existing product or situation like a more rapid method of mass transportation or a less costly process for converting sea water into palatable water. The need may call for a brand new product such as synthetic parts that are compatible with the human body. In an actual situation the Market Research Department of your company will probably determine the marketability of the product or process, which is really the ultimate test of need. The engineer's job is to create a product which will satisfy the need.

It is frequently necessary at this point to re-evaluate need in terms of its most general solution. For a simple example consider the need for "A better Rear View Automobile Mirror." This statement automatically limits the thinking of the designers. It would be better to state this need as: "A Means of Eliminating Blind Areas from Automobile Drivers." Mirrors are then considered as only one of the many avenues of approach. The question must be asked: "Is the need as stated inclusive enough for all possible solutions?" This isn't easily answered of course, but through experience one learns to recognize an all-inclusive statement of need.

Establish Goals. Upon identification of a specific human need, the desired goals should be clearly established. These goals should reflect the expectations of accomplishment of the design and give some desirable characteristics and limitations. They should be stated in terms that will not limit the design or define its form. The goals statement may include some general characteristics such as simplicity of operation, pleasing appearance, range of building and operating costs, safety of operation, simplicity of installation, and some idea of weight or size. Since projects vary in complexity, no fast rule can be made which states exactly what should be included in the statement of goals. It is primarily important for the designer to be able to define the purpose of his endeavor.

2.7 Step 2. Perform Research

Research includes an investigation of products presently available, what related patents exist, how existing complimentary devices operate, what materials are useable, what scientific knowledge is required before attempting the design, and what the market expects of the desired product. There are several journals which index and abstract specialized periodicals. Among these, the Engineering Index, received by most large university libraries, is valuable in this work. Such key words as *mirrors, optics, glass* or *safety* may lead to many related and informative articles regarding the problem cited in the previous article.

2.8 Step 3. Specify Design Parameters

The research done in step 2 will produce some real knowledge. This knowledge and limited experience make it possible to define some specific traits that the creation should possess. These traits are called design parameters since they are characteristics of the product and include such things as cost, size, simplicity, weight, etc. Their range of acceptability may be broad or narrow, depending on the subject but should not define in any way the *form* of the product.

Limited experience should not be considered a liability in design. Truly creative designers are too far ahead of the competition to make experience of primary importance.

As an example of specified parameters, consider the specifications the United States Air Force writes for new airplanes: maximum take-off distance, maximum and minimum air speeds, distance between refueling, minimum climb rate and payload, maximum landing distance, etc. These are known as performance specifications and do not describe the plane, but rather its design parameters.

2.9 Step 4. Ideate

During the ideation step it is important to list *every conceivable* method of problem solution without regard to feasibility and without criticism of what at first seem to be glaring faults or foolish suggestions. It is here that seemingly unrelated facts are often combined to obtain that "ideal" solution. Ideation is probably the most important single step in the creative design process. The probability of arriving at a really unique solution is proportional to the number of different ideas appearing on the ideation list. It is interesting to observe how easily ideas flow during a good "brainstorming" session when criticism is not permitted. "Thumbnail" or idea sketches can be employed to clarify ideas during these sessions. Thoughtful consideration of the less feasible ideas may result in modifications or combinations which *are* unique and workable. Frequently, after having pored over his ideas and given up in dejection, the designer suddenly experiences an inspiration for a better approach to a satisfactory solution. Apparently the subconscious mind has continued to consider various combinations, working tirelessly during this *incubation* period. *Real and serious thought and effort must be applied to ideation if one is ever to experience inspiration.*

2.10 Step 5. Conceptualize

Several of the apparently better ideas should be developed by good pictorial and orthographic sketches as discussed in Unit 3. These sketches, like Fig. 3-9, should include any notes which are necessary to adequately describe the outstanding features and operation of the design. Conceptual sketches are the only evidence of an abstract idea at this point and must be of high quality if the idea is to be understood or sold to management.

After examining the conceptual sketches, more ideas may occur. Remember that this is an iterative process and backtracking often progressively improves the design.

2.11 Step 6. Evaluate and Select

Each tentative design concept is analyzed for technical correctness and feasibility. Designs are compared with each other and with goals and specified parameters. Ultimately, by means of this comparison method one or two tentative designs are selected for possible prototype construction and testing.

One method of comparison is as follows: An Analysis and Decision Sheet is prepared similar to the sample in Fig. 2-2.

Column 1 lists typical design parameters which influence the rating of each proposed solution or design. The unavoidable overlapping of some of these factors must be taken into consideration while determining their relative importance to the overall design. Column 2 indicates the relative importance of each parameter to the design and must be determined by mutual agreement of the design team. The sum of the Relative Importance figures, which might be called a Figure of Merit, must be 100.

The concepts are compared *simultaneously* with respect to a particular parameter and rated as objectively as possible. The Concept Value Numbers are determined by multiplying each rating by the Importance Number for that parameter.

Since a perfect design would have ratings of 1.0 (100%), its Value Numbers would add up to 100%, its Figure of Merit. It follows that the best *available* concept will be the one having the largest value.

2.12 Step 7. Delineate and Finalize

Drawings and Sketches, whether made by the engineer or the draftsman, will be required for each Creative Project or Design. These drawings may include diagrams and layouts, pictorials, orthographic assemblies, bills of material, detailed descriptions, sections, and descriptive notes where needed. Many helpful ideas are found in Units 19-25, and the discussion of what constitutes a *set of working drawings* is found in Unit 26.

The drawings made here must be used to convince someone—an instructor, section leader, or employer—that the idea is sound and worth an investment of more time and money. Moreover, the appearance, correctness, and completeness of these drawings will often be equated with the competence of the designer and the likelihood of product success. Above all, if there is nothing specific except drawings to show, they must be

CREATIVE DESIGN

ANALYSIS AND DECISION MATRIX

DESIGN PARAMETERS (Select Those Appropriate)	RELATIVE IMPORTANCE	RATING / CONCEPT VALUE FOR CONCEPT				
		1	2	3	4	5
A. *Feasibility* (can solution be completed)	10	.5 5	.9 9	.8 8	.9 9	.5 5
B. *Reliability* (Degree of dependability)	15	.7 10.5	.8 12	.6 9	.9 13.5	.6 9
C. *Simplicity*	10	.5 5	.9 9	.7 7	1.0 10	.6 6
D. *Economy* (High rating for low cost)	10	.5 5	.8 8	.5 5	1.0 10	.5 5
E. *Convenience of Use*	20	.6 12	.8 16	.8 16	.9 18	.7 14
F. *Appearance*	15	.7 10.5	.8 12	.5 7.5	.8 12	.6 9
G. *Creativity*	20	.9 18	.7 14	.7 14	.9 18	.5 10
H. Other Considerations						
J.		*BEST*				
TOTAL (Figure of Merit)	100%	66.0	80	66.5	**(90.5)**	58

Decide on a RELATIVE IMPORTANCE (%) for each PARAMETER and enter in appropriate position. Values may vary considerably but must total 100%.

RATING SCALE

EXAMPLE

Excellent	90-100%
Good	80- 90%
Fair	70- 80%
Poor	60- 70%
Inferior	50- 60%

Design Parameter A, *Feasibility* is assigned a Relative Importance of 10%.

Concept #2 is rated *Good* (90%) with respect to A, *Feasibility*. Therefore its value is (.90) 10 = 9.

Record 9 under Concept #2, opposite A.

Fig. 2-2. *Analysis and Decision Sheet.*

clear. There must be no doubt as to the details of the design or its operation.

Written Reports are frequently required for a design proposal and are helpful to both reader and writer. Management can obtain a quick summary of the design features; the designer has an opportunity to review the proposal and verify that all features discussed are actually included.

The report should be brief but cover pertinent information not contained in the drawings. It should be complete, containing a justification for and purpose of the project, a description of the proposed design, and an emphasis on its principal features with references to the apppropriate drawings. Further elaboration should be considered as optional reading intended for those interested in details of the design.

Prototypes are hand made, accurate, and expensive pilot models of production designs.

Models are hand made approximations of designs and are often helpful in overcoming skepticism in management. They provide a physical concept of the design that can be checked before building a prototype. Models can be constructed without great expense by using ingenuity, time, and basic hand tools. The model need not be identical to the design in every way, but should be functionally correct and resasonably well executed. Poorly finished models will only act to destroy the chances of design acceptance. Some examples of good models are shown in Figs. 2-1, 6-1, and 7-1.

Presentations may be required to describe your design problems to groups for evaluation. Many good basic texts are available regarding techniques of public speaking, and no attempt will be made to replace them. A few reminders regarding technical presentations should be observed. The speaker should:

1. Dress appropriately. A business suit and good grooming is important. *He must sell himself to sell his idea.*
2. Speak *to* the audience. Eye contact and voice projection should be used. Reading notes should be avoided.
3. Avoid distracting mannerisms of speech and body motion.
4. Use graphical devices liberally to describe an idea. No complex device can be described verbally without confusion and loss of the audience's interest.
5. Use a direct approach and be brief. Get the audience's undivided attention and tell them immediately why and how the solution is best for the particular problem.
6. Hit only the important, novel points and ask for questions afterwards. If the interest of the audience has been generated, they will ask for more information when given the opportunity.

2.13 Case Study

To illustrate briefly the use of the 7-step technique previously outlined, an actual example of creative design has been summarized in the paragraphs that follow. Write down the seven steps on a piece of scratch paper and study this case. As it is recognized, identify each step of the process mentally or with a pencil. Note how the design grows naturally from a problem to a product when a systematic approach is used.

SPARE TIRE LIFTER

A student design team decided to undertake the task of improving the plight of some motorists. Owners, particularly women, of many recent model American-made station wagons encounter real difficulties in the event of a flat tire. Several factors contribute to this.

Frequently the storage area must be partially unloaded to get at the spare tire, which is located in a narrow side well. Its location and the 45-pound weight of the spare tire make removal not only clumsy but also dangerous to the back and abdomen. Measurements and inspections indicated the problem was common to most cars and the geometry was similar on each. It was therefore decided to design a device to correct this undesirable situation. The product, it was felt, would have to be simple, easily and quickly operated, and not require significant modification of the auto in order to use it. Dealers indicated that a price of less than ten dollars would probably result in a good sales rate.

It was suggested facetiously that buyers should get an accessory muscle man, or courses in body building or magic levitation. Guaranteed puncture proof tires or an aerosol can of tire patch to eliminate the need of a spare were rejected as a solution due to blowout possibilities. Aluminum rims would diminish weight, but cost would be prohibitive. A mechanical, hydraulic, or electrical jack might be installed under the spare, but room seemed to be at a premium in the well. A pneumatic tube or winch-operated flat belt under the tire would do the job and occupy negligible well space, but would require auxiliary equipment. A block and tackle using nylon cord would be quite compact and looked promising, even though a hook would be required in the roof panel. A tripod was considered to replace the hook. A simple lever was mentioned as a device not requiring installation. The lever could also replace the winch used in the belt lifting idea. A removable fender panel, with spare tire access from the outside was considered one of the best ideas but not compatible with the scope of the project.

The selected design was based on the lever idea since it best represented a simple, fast, economical, and marketable solution. The final configuration is shown in Figs. 2-3, 2-4, and 2-5.

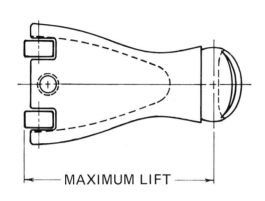

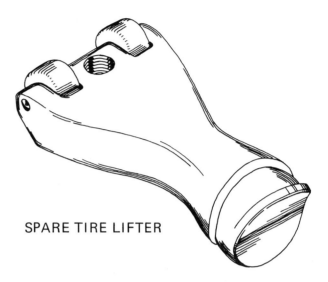

Fig. 2-4. *Creative Design, Pictorial Sketch.*

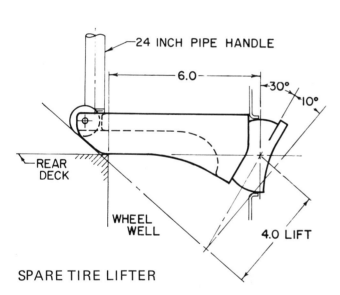

Fig. 2-3. *Creative Design, Orthographic Views.*

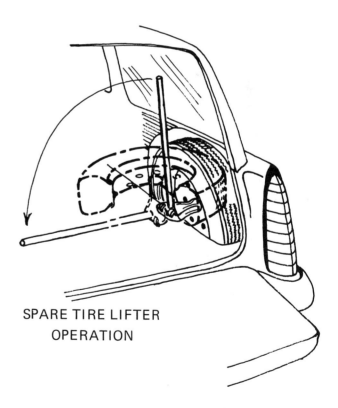

Fig. 2-5. *Creative Design, Explanatory Sketch.*

The tool head is contoured to fit into the wheel rim centering hole and grasp the hole flange. Rotating the lever first lifts the wheel vertically, then rotates it gradually to the horizontal position. With the spare in this position, the tool is rolled on its wheels to the tailgate where the spare tire is lifted off. The removable handle affords compactness and a large mechanical advantage.

The design team felt that this design, although untested, was fundamentally sound and would work to some degree of satisfaction. Prototype construction would no doubt improve the design; any errors in estimating angles, etc., would become apparent and could be corrected.

2.14 Conclusion

The design process discussed in this unit is by no means the only workable procedure; after some practice in its use individuals will find variations which adapt well to their own thinking and personality. However, this process represents an observed characteristic approach used and found effective by designers. The seven steps, Fig. 2-6, can be committed to memory as easily as a phone number for future reference.

This procedure *can* promote creative thought and produce positive results in a wide variety of problem-solving situations. It *will* do this only if it is studied and applied *conscientiously*. The opportunity to apply the process several ways is given in the problems which follow.

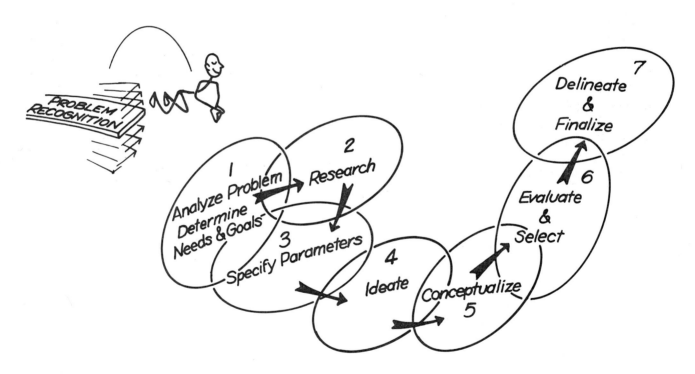

Fig. 2-6. *Steps in the Design Process. (They are not isolated but are related and somewhat over-lapping. However, it is important to apply the steps of design in their proper sequence.)*

PROBLEMS

Organizing and Executing A Creative Design Project

Interest, time for thought, and progressive learning are key factors in the success of an open-ended creative project. The latter two factors may be promoted by devoting approximately one class period every week to one step in the process, with two or three periods for finalizing. This program will allow the student to use his knowledge progressively and review it in the final delineation.

Interest can be generated by organizing design teams of four or five members who are all interested in the same design project. Project subjects may be suggested by students and drawn from the brief list which follows, or from the problem book.

Team work should be submitted frequently to the instructor for evaluation and suggestions. Individual outside assignments should be discussed and recorded before ending any team project session.

1. Creative Design Project Ideas.
   a. Extra-compact bait-casting fishing device.
   b. Paint roller cleaning device.
   c. Child's yard toy or game.
   d. Adult game.
   e. Styling design of a specific product such as desk caddy, lamp, scooter, trike trailer, safe wagon hitch.
   f. Compact and portable student furniture, such as bookcase, study table, and special chair for reading.
   g. Better bicycle seat.
   h. Urban snow disposal system.
   i. Transportation system for your campus.
   j. Auto coffee cup holder.
   k. Exterior frame for old-fashioned umbrella tents.

2. Can you improve on the design of the spare tire lifter, Art. 2?
   a. List any shortcomings of the design as proposed.
   b. Make needs and goals statements applicable to the spare tire lifter.
   c. List all possible ideas. Do any new ideas appear on the list?

3. Problems: Require analysis for need.
   a. Rubber stoppers for washroom sinks are stolen as fast as they are replaced.
   b. Very little storage space for luggage, etc., is provided in dormitory rooms.
   c. Most safety belts require two hands to fasten.
   d. Think of special problems with which you are personally concerned.

unit 3

Freehand Drawing

3.1 Objectives of Freehand Drawing

In this text the words *Freehand Drawing* and *Sketching* are used interchangeably because they refer to identical procedures. While some prefer to refer to sketching, others would rather call it freehand drawing.

The principal objective of a sketch is the *rapid rendition* of a *neat and legible* illustration suitable for its particular purpose. The final use determines the degree of precision or accuracy necessary for a freehand drawing.

An engineer often sketches preliminary designs or layouts in order to sell his idea to management, to the draftsman, or to laymen. If the preliminary design is patentable, the sketch should be dated and witnessed so that priority of conception can be established. Sketches are often used to clarify ambiguities which may arise in a design. Emergency jobs can usually be completed more rapidly when sketches replace instrument drawings because of the time delay involved in obtaining the usual drafting room print. Plans for temporary repairs, construction, or apparatus which do not require permanent drawing records may be sketched in detail with all necessary specifications.

The ability to rapidly make a readable freehand drawing may be used more frequently and will probably prove to be more valuable for the engineer than his ability to make a finished instrument drawing.

Pictorial sketches are valuable aids in the explanation of an idea or a problem to anyone unfamiliar with the operation involved. Pictures are used throughout this text to assist the student in analyzing problems and visualizing their solutions. The layman can understand a picture where a word description or a multiview drawing may be meaningless. Catalogs of industry "picture" their products and operations which they wish to sell to the customer. It is common for the engineer to make pictorial sketches from which the draftsman may make finished pictorial drawings for reproduction in service manuals, operation instructions, or catalog illustrations.

3.2 Sketching Equipment

Paper, pencil, eraser, and an observant eye are the only tools required for sketching. Cross-section paper, faintly ruled with a background grid, aids the beginner's control of straightness and alignment of a freehand drawing. However, when using a grid paper, the student should not try to construct details by counting squares. Instead, he should consider only the proportions of the sketch as a whole.

A well pointed pencil of grade HB or F is preferred for engineering sketches. When shading is to be applied, a softer grade such as the B or 2B may be used.

A soft plastic eraser is recommended for making changes, keeping the sketch clean, or eliminating any preliminary skeleton lines which detract from the finished sketch.

FREEHAND DRAWING

3.3 Types of Sketches

Single-View Sketches. Maps, areas, plats, circulation plans, charts, and graphs are delineated by one-view drawings since only the two dimensions, length and width, are necessary for complete descriptions. Objects of uniform depth or thickness, such as a gasket, can be fully described by a single contour view if a note indicating the thickness or third dimension accompanies the drawing.

Multiview Sketches. Details of more complicated solid objects or the solution of intricate space problems may require several related views, commonly called orthographic projections.

Pictorial Sketches. A pictorial sketch shows the three principal dimensions of length, width, and height of an object in a single view. It attempts to show what the eye would see when viewing the object from a particular position. To repeat from Art. 3.1, the ability to quickly make a picture sketch will be a valuable asset to the student and the engineer.

3.4 Technique for Freehand

Never losing sight of the fact that a sketch should be done *rapidly* if it is to accomplish its purpose, the steps for the preparation of a freehand drawing may include the following:

1. *Plan* the sketch on the paper.
2. *Outline* an overall skeleton.
3. *Check* proportions as details are added.
4. *Clean*, sharpen, and straighten the outlines for a finished product.
5. *Letter* information and titles.

Not only should a relatively soft pencil be used for sketching but it should always be kept well-sharpened with a slender, conical point. See Fig. 3-1.

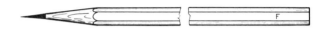

Fig. 3-1. *Well Pointed Pencil.*

The sketching pencil should be held lightly, with the fingers in a relaxed position at least 1 1/2 inches from the point. Long lines may be sketched more accurately by spotting the end points and making several trial strokes between these points. Sweeping strokes are made with a motion of the fingers and wrist rather than the arm. After necessary corrections for line straightness, the final lines can be sharpened. See Fig. 3-2.

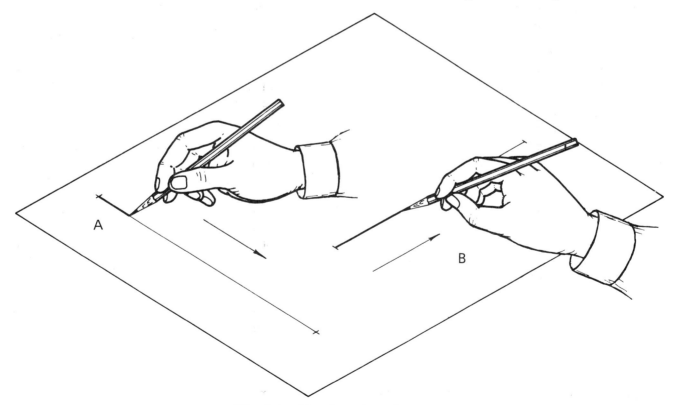

Fig. 3-2. *Sketching Straight Lines.*

Plan the Sketch. Even though a freehand drawing is not laid out to an exact size or scale, the finished drawing must show the relative proportions of the illustrated material. This can be achieved by first visualizing a drawing of correct proportions to fit the available working space of the drawing paper. If more than one outline is to be shown on the drawing, their proportions and positions must be carefully planned before starting the actual drawing.

Draw the Skeleton. Proportions of the details of any sketch can be controlled by the construction of skeleton boxes whose dimensions are equal to the overall dimensions of the proposed view. In the case of a one-view drawing the box is centered in the available working space; while for multiview drawings, the boxes are located so as to give correct alignment to the adjacent views. The basic outlines of the drawing are constructed within the boxes and are checked for proper proportion and position before adding smaller details. Note steps (a) and (b) shown in Fig. 3-3.

Add Finish. After all details have been shown and checked, the lines of the sketch are sharpened so as to provide good contrast for easy reading, Fig. 3-3c and d. The more important lines, such as the object outlines, are heavy or dark while others, including hidden lines, center lines, or construction lines are lighter for easier interpreta-

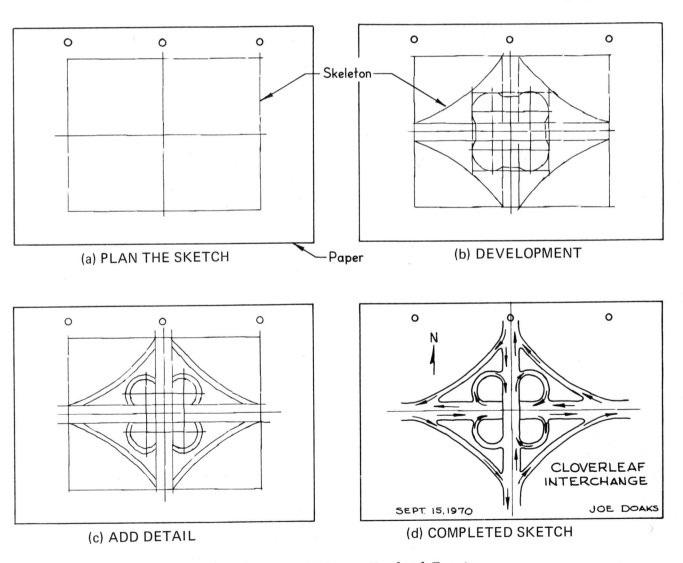

Fig. 3-3. *Steps in Making a Freehand Drawing.*

FREEHAND DRAWING

tion of the drawing. Fig. 3-4 illustrates an alphabet of sketch lines.

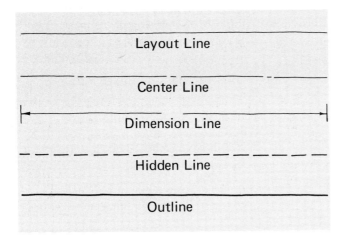

Fig. 3-4. *Alphabet of Sketch Lines.*

Lettering. Any lettered information pertinent to the sketch should be located to provide proper balance for the entire drawing. Such items as name and/or purpose of the illustration, the name or initials of the sketcher, and the date are essential for all sketches. Dependent upon the purpose of the drawing, other items which may be included are material specifications, size and finish, and name of school or company. Lettering should be of standard size and quality using very light, rapidly drawn, freehand guide lines. Standard styles of engineering letters are shown in Unit 4.

Freehand Circles and Ellipses. The first step in freehand construction of a circle, following the location of its center, is to sketch the center lines. These light, alternating long and short dashes are drawn at right angles to each other through the centerpoint as shown in Fig. 3-5a. Equal radial

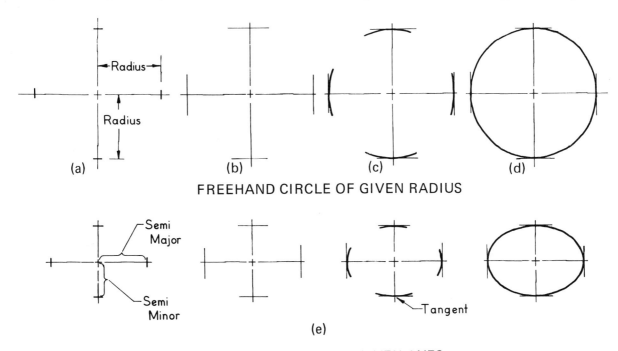

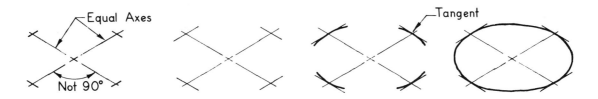

FREEHAND ELLIPSE OF EQUAL CONJUGATE AXES
(PICTORIAL OF A CIRCLE)

Fig. 3-5. *Sketching Steps for Circles and Ellipses.*

distances are marked off on the center lines in the four directions from the center point. At these four points, short perpendiculars to the center lines are drawn and arcs are then sketched tangent to these perpendiculars. The circle is completed by sketching a uniformly curved line closing the gaps between the tangent arcs. See Fig. 3-5b, c and d. All defects, such as lumps or flats, may be corrected before the circle outline is sharpened. Extremely large circles can be more accurately sketched by addition of more radial lines through the center of the circle.

Ellipses may be constructed by a modification of the method used for circles. On one center line lay off one half of the desired *major axis* on each side of the center; while on the other perpendicular center line lay off one half of the *minor axis* on each side of the center. The steps of procedure are illustrated in Fig. 3-5e. Practice, good judgment, and orderly procedure are necessary for the rapid construction of a smooth freehand circle or ellipse.

3.5 Pictorial Sketching

The most important function of a pictorial view is to illustrate a three-dimensional (length, width, height) object approximately as it would be seen by the eye. A picture view is valuable in the visualization of an object by a person who would be confused by a multiview drawing.

The three dimensions of an object in space, here referred to as length, width, and height, may be considered as its coordinate dimensions. If the coordinate dimensions are laid in three directions as shown in Fig. 3-6, a space figure or picture is formed.

Position of the Coordinate Axes. Before taking a picture, the photographer selects the best position for viewing the important details of his subject. In sketching a picture the same considerations must be observed in selecting the best coordinate axes. Fig. 3-7 shows several positions of the axes which give different pictures of the rectangular block. One of the most common positions, shown at (a), is known as an ISOMETRIC. This type of pictorial gives an equal amount of distortion to each of the three visible planes. A more natural view of the block can be shown by using a larger angle between the length and width axes,

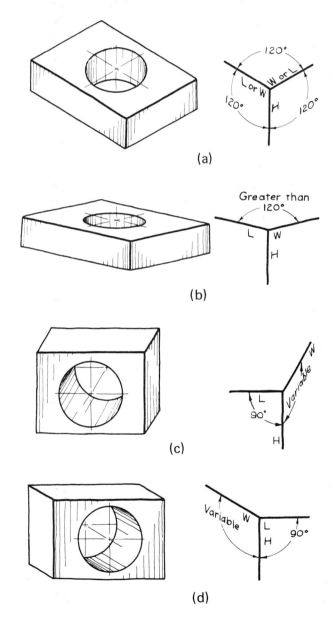

Fig. 3-7. *Positions of Coordinate Axes in Pictorial.*

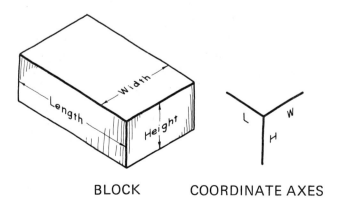

Fig. 3-6. *A Pictorial Sketch.*

FREEHAND DRAWING 19

as illustrated in (b). The position of the axis shown in (c) and (d) gives a drawing known as OBLIQUE. Oblique drawings are advantageous for illustrating objects which have circular contours in the front face (L & H) and a *relatively short receding axis* (W). Otherwise, placing the length and height axes at an angle greater than 90° gives the pictorial a more pleasing appearance.

Circles and curves are shown in pictorial by using coordinate boxes and following the sketching steps shown in the illustration of Fig. 3-5. If the center lines for a circle, when drawn in pictorial, are other than 90° to each other, the pictorial view will be an ellipse. Note the circular hole in the block shown in Fig. 3-7a and b.

A more detailed description of pictorial drawing will be found in Unit 10. It is recommended that the student leaf through the pages of this text and observe the number of pictorial drawings used. Study these pictorials and try to identify the axes of length, width, and height.

Fig. 3-8 shows a sketch of a desk. Some shading is used to improve the effectiveness of the picture.

Practice in making sketches, either in two or three dimensions, will help the hand to obey the mind and will develop an appreciation of proportion and scale. One of the objectives of the study of Graphics is the development of the ability to visualize. This can only be achieved by continued practice in the reading and writing of this language.

3.6 Application of Sketches to Design

As the student has learned in Unit 2 and will find repeated in nearly all units of study, the end result of either freehand or instrumental drawing is to accomplish the several parts of an Engineering Design. Fig. 3-9 has been included to show how pictorial, single view, and multiview sketches are used to develop ideas into a form that can be communicated to others. Many idea sketches were needed before making these more complete conceptualization sketches. Following the decision as to which is the best design for its purpose, more complete and accurate pictorials may be drawn with instruments following one or more of the pictorial systems explained in Unit 10. In addition, detail and assembly drawings with material lists will be essential to consummate the design. It wil be necessary to assimilate the information contained in Units 9, 11, 12, and many others, especially Unit 26, before designs of consequence can be concluded.

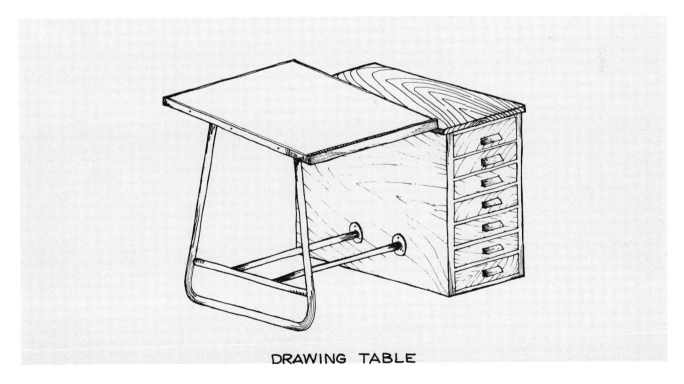

Fig. 3-8. *Pictorial Sketch.*

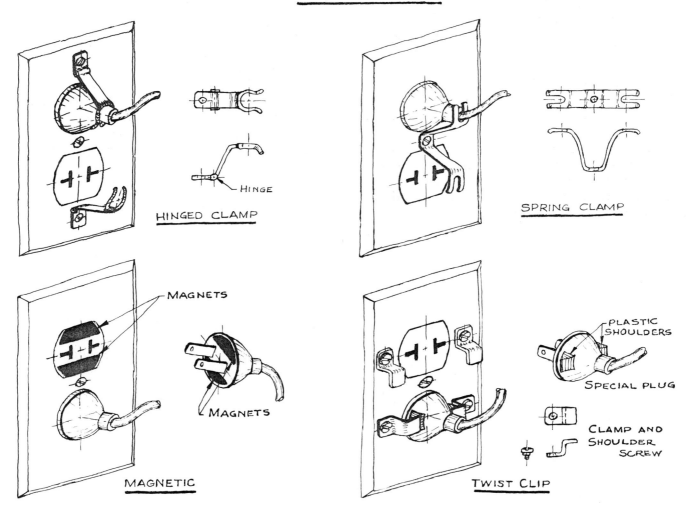

Fig. 3-9. *Conceptual Sketches for Product Design.*

PROBLEMS

Problems one and two are intended to give the student practice in the fundamenals of freehand sketching. Problems three through six have been designed to facilitate the development of the student's ability to formulate, clarify and communicate his ideas with freehand sketches. All of the problems are to be drawn freehand (without instruments), on suitable paper, employing the principles outlined in the text.

1. **Single-View Sketches.** Make a single-view sketch of:
    a. the floor plan of your room
    b. a map of your town
    c. the front of your desk or drawing table
    d. the front of your house
    e. a bull's-eye with three concentric rings
    f. an electric light bulb
    g. a hammer
    h. a hacksaw
    i. a screwdriver
    j. a carpenter's level
2. **Pictorial Sketches.** Make a pictorial sketch of:
    a. a stairway with four steps
    b. a triangular scale
    c. a T-square
    d. a waste basket
    e. a cup and saucer
    f. a hammer

g. a pencil file
   h. a screwdriver
   i. a roll of masking tape
   j. a mechanical pencil
   k. a composition—scale, triangle and tape
   l. a living room interior

Design Problems

3. Given a board 10 x 3/4 x 4'-0 and an unlimited supply of strap iron 1 1/2 x 1/8. Using these materials and any standard fasteners, design a book shelf which can be attached to a wall. Make a pictorial sketch of your design.
4. Given the material of Problem 3 above,, design a book rack for your desk.
5. Given a 6 inch cube of steel. Make a step pulley with the following specifications. Each step of the pulley must be two inches wide to accommodate the drive belt. Use the following diameters for the steps: six inches, four inches, and two inches. An inch hole is drilled through the center of the pulley. Make a pictorial sketch of the pulley.
6. You have just requested a raise from your boss and he has countered with a request of his own. He needs a simple, inexpensive pencil-holder paperweight for his desk and suggests that you design one. Knowing that your raise depends upon how well you do, design a solution to his problem and make a pictorial sketch of your design.

unit 4

Engineering Lettering

4.1 Historical

Our modern alphabet is considered to have its beginning with the ancient Egyptian hieroglyphics. The symbols were formed into a 22 letter alphabet by the Phoenicians and later adopted by the Greeks. This alphabet was passed on to the Romans as the old-world culture progressed westward. Little change has taken place over a period of 2000 years except additions to form our present alphabet of 26 letters.

Previous to the invention of printing in the fifteenth century, all writing was done by hand. In many cases the letters were embellished and decorated by the artisans of that time until their scrolls were works of art. Since the advent of the printing press there has been a tendency toward a more uniform and standardized letter form. The great majority of present-day commercial printing is done in Modern Roman letters. Block or Gothic letters came from the Germans as a modification of the Roman style.

It is important that the lettering on an engineering drawing should be easy to read and, as well, easily and rapidly executed.

4.2 The Use of Engineering Lettering

A completed drawing, whether it be in the form of a sketch, a working drawing, or a graphical solution of a problem, requires a certain amount of lettering for notes, dimensions, and titles.

There is variation with respect to the style of lettering, although the single-stroke commercial Gothic letter form has almost universal adoption. A single-stroke letter is one where single lines are used to make the outlines of the letters without heavier stems, serifs, or terminating embellishments like those used in these printed words. A reference to Fig. 4-4 will show an alphabet of single-stroke Gothic letters and numerals. The principal differences in industrial lettering standards lie in the variation of Vertical or Slope letters, and Capitals or Lower Case. It might be noted that over the last decade or two there has been a trend, especially noticeable in the aircraft and automotive fields, toward the use of vertical letters and in most cases to capital letters only. On the other hand, many older firms prefer to use capital and lower case slope letters. The ANSI Y14.2 standard shows both the capitals and lower case in the slope and vertical lettering styles. The SAE Automotive Drafting Standards, Section A3, illustrates both the slope and vertical letters but only in capitals. The lettering chart in Fig. 4-4 shows complete alphabets of letters and numerals, both capitals and lower case, vertical and slope.

The engineering departments of the larger industries usually have a set of drafting standards which are printed and revised periodically for the benefit of their design and drafting rooms. These standards, evolved to meet the needs of the company, include lettering, dimensioning, specification writing, and delineation.

ENGINEERING LETTERING

The important consideration for the student is that he must conform to the standard of the organization with which he becomes associated, and be able to adapt himself quickly and consistently to that style.

4.3 Lettering Equipment

Pencil lettering should be done with a relatively soft pencil preferably an F or H grade, depending on the roughness of the paper. It is important that the pencil should be well sharpened with at least 1/4 inch of lead protruding. After sharpening so that sufficient lead is exposed, a long conical point should be formed by a fine file, special lead pointer, or sandpaper pad. See Fig. 4-1.

Some draftsmen and students prefer to use mechanical drafting pencils because of the ease of maintaining a uniform long lead. Leads of desired hardness can be obtained from any supply store.

Guide Lines. Guide lines must be used if letters are to be of uniform height. The first lettering practice may be done on paper upon which faint guide lines have been printed. Subsequent work will require construction of guide lines before the lettering can be applied. They may be drawn by first marking correctly spaced points, and then with the aid of the T-square, drawing *very faint* parallel lines with a relatively hard (4H or 6H) pencil.

Lettering Device. Guide lines can be spaced and drawn more uniformly, and usually more rapidly, by means of a lettering device. The Ames Lettering Guide is used as illustrated in Fig. 4-2; for more details study the directions which accompany each new instrument.

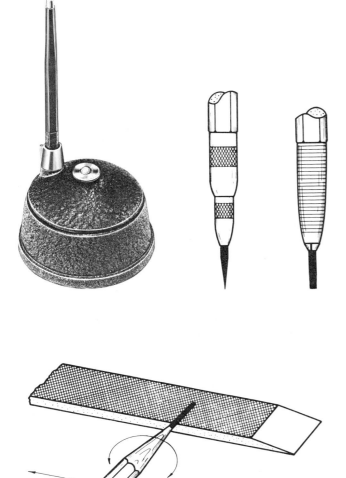

Fig. 4-1. *Pointing the Pencil Lead.*

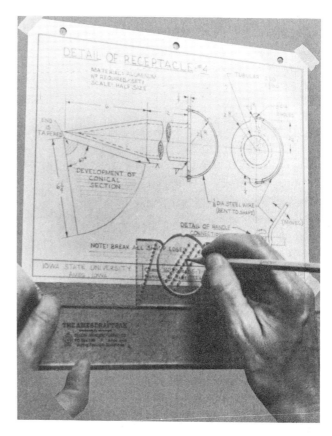

Fig. 4-2. *Ames Lettering Device. (Courtesy O. A. Olson Co.)*

Paper. Different qualities of paper have a "pencil feel" which depends upon the smoothness and hardness of the surface. To produce good quality work on paper with a relatively soft surface, a softer grade of lead should be used than on a smooth, hard surface. Whatever quality of paper, the objective is to produce a clean-cut, distinct line or letter using a pencil that is soft enough not to groove the paper yet firm enough to maintain the desired sharp quality.

Eraser. The eraser is one of the draftsman's most valuable tools. Care and diligence in the use of the soft rubber or plastic eraser will eliminate pencil lines and leave a clean area in which new lettering can be applied without smudging.

4.4 Operation Lettering

Good lettering is produced by about 90% headwork and 10% handwork. The student should observe two rules whenever lettering is to be done.

1. The pencil should be grasped lightly with a normal position of the thumb and fingers. See Fig. 4-3. Too tight a grasp will tend to cramp the fingers and arm thus making lettering tedious and tiring.

Fig. 4-3. *Position for Lettering.*

2. Place the paper in such a position that the forearm is supported by the table or board. The arm should rest in a comfortable position to relax the muscles.

A firm stroke should be used, as opposed to a sketch stroke, without exerting excessive pressure which will break the pencil point or groove the paper. Particular care should be taken that the strokes *cover* the guide lines, not simply touch them. *The control of lettering height is within the power of anyone and its violation is inexcusable.*

After several strokes are made with the pencil, the lead will become flattened at the contact point. Continued use of a worn point will produce fuzzy letters with too broad outlines. Rotate the pencil a small amount after each letter in order to produce uniform wear on the point. Repoint or resharpen the pencil often enough to maintain a sharp, uniform line quality.

4.5 The Vertical Capitals

The student should bear in mind the statement that lettering is a great deal more a function of the *mind* than of the *hand*. Your mind must tell the hand where to start, in what direction to proceed, and where to stop. Your hand must be controlled with respect to the shape, width, slope (or lack of), and spacing of the letters. Unless these details are firmly fixed, the letters will not look regular and uniform. The natural tendency for the student to make too narrow letters should be diligently avoided. Refer to the chart of Engineering Letters, Fig. 4-4.

Capital letters are nearly always used in titles and are the only style found on many drawings. All vertical capital letters consist of vertical lines, horizontal lines, oblique lines, circles and curves. One of the most difficult parts of vertical lettering is to keep the vertical strokes perpendicular to the guide lines. Slight variation, either forward or backward, produces a slope letter which when mixed with vertical ones destroys the symmetry of the lettering. For beginners, light vertical lines should be drawn frequently on the guide lines to assist the regularity of the letter outlines.

The student should become familiar with a natural order of stroking. As a general rule, it is most natural and faster to draw first the verticals from top to bottom, and then the horizontals from left to right, for example, the H: ¹|⟶³|². Also note that the single stroke-letters I and J have no serifs in engineering letters, thus: |, J.

The curved letters are usually considered the hardest to make. The O is best made in two strokes, each starting from the top guide line,

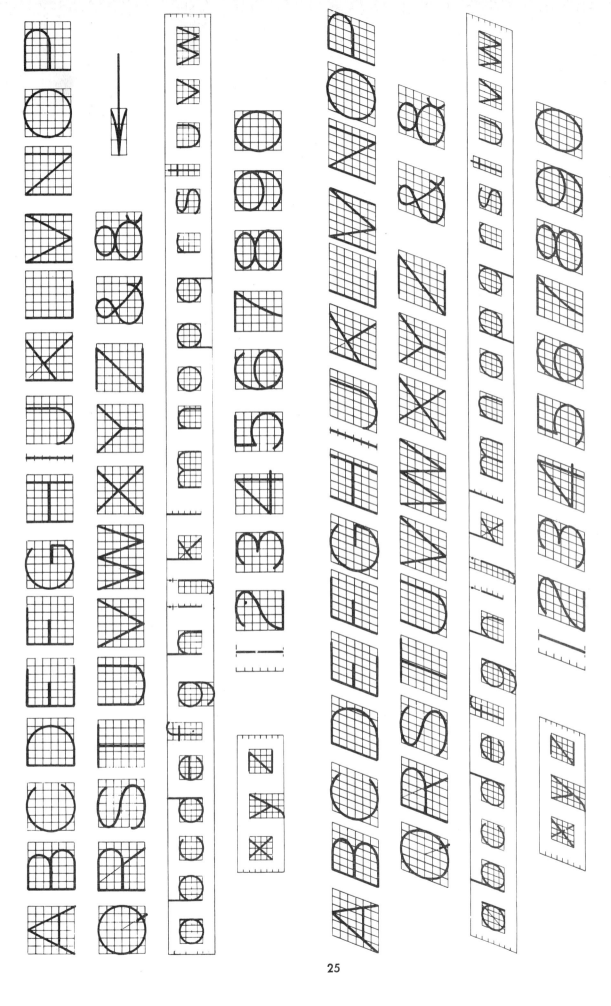

Fig. 4-4. *Standard Letters and Numerals.*

thus: ◯. The S is perhaps the most difficult letter to form. It is composed of portions of two ellipses, one above the other, the lower one being larger to give stability. Avoid a narrow letter. A suggested order of strokes is as follows: S. Numerals are used on drawings to supply dimensions and specifications. Study their shapes and proportions on the lettering chart, Fig. 4-4. Care must be exercised in forming the 6 and 9: 6, 9. The 3 and 8 have similiar shapes. Notice the engineering style 4 and watch that its horizontal stroke is located at least two-thirds of the way from top to bottom. For a drawing to be legible, the numerals must be well made, distinct, and uniform.

Dimensions, many of which include fractions, are very important since only one unreadable number renders the entire drawing worthless to the user. *Fractions are twice as high as the whole numerals.* In other words, if the spacing of guide lines is 1/8 inch apart for the whole numbers, the overall height of the fraction will be 1/4 inch. The numerator and denominator of a fraction are separated by a *horizontal bar,* spaced midway between the top and bottom guide lines. Care should be taken that the numerator and the denominator do not touch the horizontal dividing bar. Study Fig. 4-5. The Ames lettering device may be used to draw special guide lines for numerals and fractions.

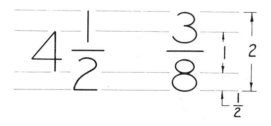

Fig. 4-5. *Whole Numbers and Fractions.*

4.6 Lower Case Vertical Letters

Three guide lines are required for lower case letters with the top and bottom lines spaced for capitals. The third line is located two-thirds the distance from the bottom to the top guide line as shown in Fig. 4-6.

Fig. 4-6. *Proportions of Lower Case Letters.*

A study of the lettering chart, Fig. 4-4, reveals that the circle is the foundation of each of the following letters: a, b, c, d, e, g, o, p, q. The circle should be made first to cover the guide lines and the proper vertical line added for the completion of any of these letters.

The letters c, o, s, u, v, w, x, z, have shapes and proportions that are nearly identical to the capitals.

Take particular notice of the letters i, l, t, which consist of only a vertical line with t having a lesser height than l. There is no "tail" on these letters. The capital I and the lower case l are identical when made in single-stroke commercial Gothic letters, for example: Illinois.

4.7 Letter Spacing

Good appearance and readability depend upon spacing as well as letter shapes. As a general rule, it might be said that *area* between successive letters should be equal. In Fig. 4-7 the actual clearance is greater between some letters than others due to their peculiarities of shape. It is a good rule to keep the letters close to each other with a much wider space between words. Do not forget: *The letters should be nearly as wide as they are high.*

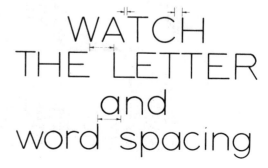

Fig. 4-7. *Letter and Word Spacing.*

4.8 Composition

The primary objective in lettering should be readability and appearance of the work. This is promoted by full bodied letters, close together, and by good word spacing. The best general rule is to make the space between words about equal to the total letter height. See Fig. 4-7.

Titles or words may be spaced by lettering them rapidly, at approximately the proper height and spacing, on a piece of scratch paper. If this template is placed immediately above the space to be lettered, the proper starting and stopping points for the series of words are predetermined. See Fig. 4-8. If the space is too short for the necessary words, the above described method will determine that the height of the letters must be decreased, or that more compressed letters must be used. Sometimes more than one template will be necessary. In any case, *avoid off-center lettering* or crowding.

4.9 Slope Capitals and Lower Case

Styles of lettering vary in different companies; some prefer only vertical capitals, while some

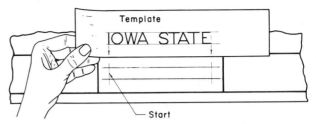

Fig. 4-8. *Use Template to Center Lettering.*

combine the capitals and lower case. If the student can do one style well, he should be able to adapt himself quickly to the standard of the company with which he is associated. The Civil Engineer will be called on more often than others to use slope lettering. The standard slope angle is about 68° formed by a ratio of 5 vertical to 2 horizontal. Slope guides may be drawn by laying out a 5 on 2 ratio line or by using the slope line on a lettering device.

4.10 Mechanical Lettering Instruments, Guides, and Templates

Most lettering and dimensions on drawings are freehand. However, a number of devices for mechanically producing letters are available.

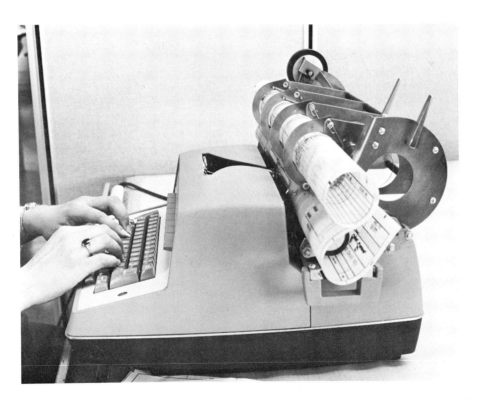

Fig. 4-9. *Open Carriage Typewriter. (Courtesy IBM and Cooper Bessmer Co.)*

Fig. 4-10. *Gritzner Lettering Typewriter. (Courtesy Milmanco Corp., Seattle)*

The Vari Typer, a machine similar to a typewriter, has several sizes and styles of types available. Many companies are able to save a great deal of time by having notes and specifications typed on the drawings. There are also open carriage drafting typewriters which will type on drawings of any length. See illustration in Fig. 4-9.

The Gritzner Lettering Typewriter is a small machine which operates on the drawing board. The machine's horizontal guide rail replaces the horizontal scale in any drafting machine. See Fig. 4-10. Not only lettering but numerals and arrowheads are produced rapidly by this machine.

Most lettering guides on the market consist of a template through which a pencil, special pen, or scriber is inserted and guided as the operator spaces the letters and words by sliding the template along a straight edge. One of the most widely used is the Leroy Lettering Instrument shown in Fig. 4-11. The Wrico Lettering Guide, Fig. 4-12, is also successfully used by many people. With practice, rapid and perfect letters can be executed by either of these instruments. Perfect uniformity is obtained and lettering time is saved since it is unnecessary to draw guide lines.

Fig. 4-11. *Leroy Lettering Instrument. (Courtesy Keuffel and Esser Co.)*

4.11 Ink Lettering

The Leroy or Wrico offers particular advantages for *large ink letters* where freehand work is apt to be ragged and irregular unless rendered by an expert.

Lettering with ink is often accomplished by using a steel pen and ordinary pen holder. The

ENGINEERING LETTERING

Fig. 4-12. *Wicro Lettering Guide. (Courtesy Wood-Regan Instrument Co.)*

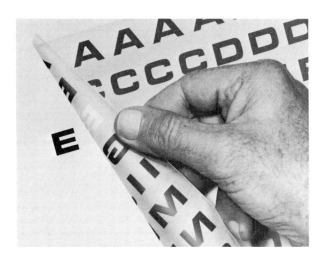

Fig. 4-13. *Transfer Letters. (Courtesy Prestype, Inc.)*

choice of pen depends upon the style of the letter, the size and boldness with which the strokes are made, and the preference of the draftsman. There are also many patented lettering pens in a variety of sizes and shapes.

Transfer letters as shown in Fig. 4-13 are available in many sizes and styles. Their most important use is for the large lettering found on titles, charts, and display drawings where high quality lettering is essential. Prestyle and Letraset are two well known trade names of transfer letters.

4.12 Special Alphabets

Architectural drawings, maps, large titles, charts, notices, plaques, and other decorative lettering jobs sometimes need a special style or shape of letter chosen from the many available alphabets. Those in greatest use are the Modern Roman, Old Roman, and Old English, each with many variations. Many modern alphabets have been devised with open faced, shaded, italicized, or embellished letters. Architectural drawings often contain attractive lettering with shading and special effects. Samples of these many styles can be found in a library reference.

4.13 Conclusions

This Unit is concluded by a statement of the *Law of the Five S Words*:

"Size, Shape, Slope, Space, and Speed adhibit good lettering." Also, repeating the *Headwork Law*:

"Lettering is 90% headwork and 10% handwork."

LETTERING EXERCISES

Exercises one through five are intended to allow careful study of the size, shape and proportions of the individual letters. Exercises six through eight stress word spacing and composition, and exercises nine and ten provide opportunity to develop the esthetics and logic of title design. All of the problems should be done on suitable paper using a properly sharpened F or H grade pencil. *Single-stroke vertical letters should be used throughout.*

Letter Forms and Spacing

1. Using very light guide lines 3/8" apart, letter the following straightline capitals: I H L T F E Z A M N V X Y K W. Determine the size, shape, and proportion of the letters by referring to Fig. 4-4. Take special care to space letters evenly. Repeat as necessary for satisfactory results.

2. Repeat the above procedure with the following curved-line capitals: J U D B P R S O Q C G &.

3. Repeat the above procedure with numerals in the following order: 1 4 7 0 2 3 8 5 6 9. Space numerals evenly and pay careful attention to Fig. 4-4.

4. Construct proper guide lines for 1/4" *numerals* (see Fig. 4-5) and letter the following

fractions 1 3/4, 5 9/16, 17/32. Repeat as necessary for satisfactory results.
5. Using properly spaced guide lines, construct 3/16″ lower-case letters in the order given: o c e a d b p q g i l t z w x a k y j f r h m n u s. Refer to Fig. 4-4.

Word Spacing and Composition

6. a. Using 3/16″ high capitals letter the five important "s" words of lettering: SIZE, SHAPE, SLOPE, SPACE, SPEED.
   b. Repeat (a) using 3/16″ lower-case letters.
7. a. Using 1/8″ capitals, letter the following sentence: "GOOD LETTERING IS PRODUCED BY ABOUT 90% HEADWORK AND 10% HANDWORK."
   b. Repeat above sentence using 1/8″ lower case letters.
8. Draw the guide lines for four lines of 3/16″ capitals, making each line about 6″ long. Taking care to center the words on each line, letter as follows: Line 1, Your School Name; Line 2, Your Town and State; Line 3, Your Name—Class of 19...; Line 4, Your Home Address.
9. Design a title block for an assembly drawing of a cam shaft stabilizer as made by the aircraft engine division of Aerodyne Corporation, San Diego, California. Scale of the drawing is 6″ = 1′-0 and its number is E4134. Space allowed is 3 x 5 inches.

unit 5

Drawing Equipment

5.1 Introduction

The purchase of drawing instruments, equipment, paper and books needed by the engineering student represents a considerable investment. It is recommended that the student engineer purchase the best quality he can afford, since he will probably use the materials for many years. When buying drawing equipment from a reputable dealer, it is wise to consider that the quality of the product parallels the cost.

Not only should the drawing equipment be carefully selected, it should be cared for and used professionally. The following articles list the drawing instruments necessary for the work in the field of graphics and go into some detail as to the proper use of this equipment. The ability of the student to complete his assignments rapidly and correctly depends to a great degree upon a knowledge of the proper use of his drawing tools and the efficient application of that knowledge.

5.2 Equipment Needs

Equipment manufactured by reliable companies, such as Gramercy Import Company, Keuffel & Esser, Frederick Post Company, and others, is available at the local supply stores.

The following drawing equipment, exclusive of paper, texts and problem books, is considered necessary for the graphics courses of an engineering student.

Drawing Equipment Needed by an Engineer
Drawing board (not required if drafting tables are available and in good condition)
T-square (length depends upon size of board or drawings to be made)
Triangles: 45° and 30°-60°
Architects or Mechanical Engineers scale
Engineers scale
Ames lettering guide
Irregular curves: 1 large, 1 small
Protractor
Drawing pencils, grades: 4H, 2H, H, F (or mechanical drafting pencils with corresponding hardness of leads)
Pencil pointer: file, sandpaper, mechanical
Erasers: hard and soft rubber or plastic
Erasing shield
Drafting tape

The student should mark each piece of his equipment, including the individual drawing instruments, by scratching initials or other identification upon them as soon as they are purchased. This will provide a means of positive identification in case any item is lost or stolen.

5.3 Care of Instruments and Equipment

Cleanliness is a prerequisite to good drafting. The drawing equipment must be clean if the drawing upon which it is used is to be kept clean. At frequent intervals the triangles, T-squares, scales, etc., should be washed in soap and water and *thoroughly* dried. Frequent wiping of the drawing instruments will prevent the formation of rust and tarnish due to handling. It is obvious that the drafting table and the draftsman's hands must also be kept clean. Carbon shavings deposited on the hands, tools and table because of careless sharpening or pointing of a pencil lead are one of the most common sources of dirty or smudgy drawings. The pencil pointer should be kept in an envelope or case when not in use so as to prevent the transfer of carbon particles to other tools.

5.4 Drawing Instruments

There is some difference of opinion as to what should be included in a set of drawing instruments. For many years standard sets generally have contained a relatively large number of individual items. Sets with a lesser number of instruments have recently gained in popularity. Such a set is satisfactory for engineering graphics courses and should include items pictured in Fig. 5-1. In addition, some like a small bow compass with their instruments for small circle work. It is reasonable to assume that a set which includes a lesser number of instruments can be purchased at a lower cost. It is more practical to own a high quality, small set than one of lesser quality with more instruments.

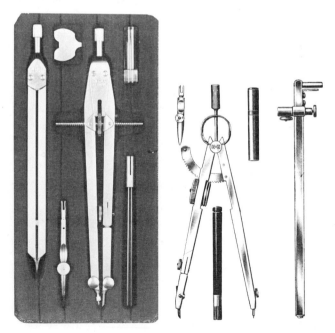

Fig. 5-1. *Drawing Instruments. (Courtesy Grammercy Import Co. and Keuffel and Esser Co.)*

Compass. The compass, the most frequently used instrument, should be equipped with lead which will provide the correct line quality. In general, the compass lead should be as soft or softer than the grade of lead in the pencil used for other line work.

Good drafting practice requires that the compass and pencil leads be kept well sharpened in order to insure uniform line quality. Fig. 5-2 illustrates the method of pointing a compass lead.

One leg of the compass is equipped with a needle point, one end of which is long and tapered; the other end formed with a shoulder. *The shoulder end* should be used for general compass work, especially when drawing concentric circles because it will absorb the pressure applied to the compass and prevent enlargement of the center hole in the paper. The shoulder point should be adjusted so that it extends slightly beyond the compass lead when the legs of the compass are closed. Note Fig. 5-2.

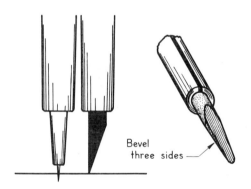

Fig. 5-2. *Sharpening the Compass Lead.*

The advantage of a bow compass lies in its rigidity due to a screw adjustment. As pressure is exerted to produce a "bright" line, the compass legs will not spread.

Dividers. Dividers are used to divide given curved or straight line distances into any number of equal parts, to transfer measurements, or to set off a series of equal distances by the trial method. If the instrument is held correctly, all adjustments can be made by one hand. Fig. 5-3 shows the correct method of holding a pair of dividers. The functions of a divider may be accomplished with a compass. In fact, many prefer the compass for measuring distances thereby avoiding prick holes in the paper.

DRAWING EQUIPMENT

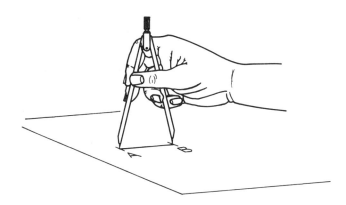

Fig. 5-3. *Using the Dividers.*

Ruling Pens. Ruling pens are designed for inking lines by following a guide, either straight or curved. They are not used for freehand work or for lettering.

The Rapidograph pen has become popular for both ink line work and lettering. The pen points are available in 8 sizes for variable line weights. See Fig. 5-4.

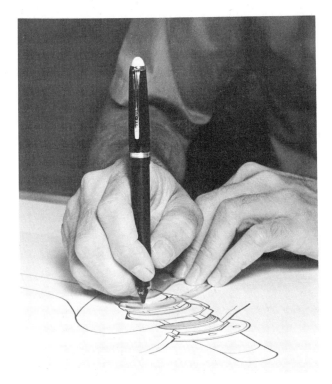

Fig. 5-4. *Rapidograph Pen. (Courtesy Kohi-noor, Inc.)*

5.5 Drawing Board and T-Square

Drawing boards are available in sizes from 10" x 120" to 60" x 120", with 20" x 26" as a common size, satisfactory for student use.

The T-square, available in a variety of sizes and qualities, consist of a hard wood head rigidly attached to a blade edged with transparent celluloid or made completely from transparent plastic. In order to insure efficient operation of the T-square, the drawing paper should be placed toward the top of the drawing board in a convenient position for the draftsman. Align the paper so that the

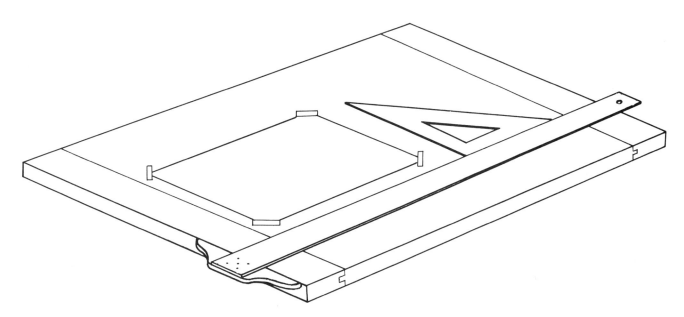

Fig. 5-5. *Drawing Board, T-Square, Triangle, and Paper.*

top border line is parallel to the edge of the T-square blade and the *left-hand edge of the paper is fairly close to the T-square head*. When the paper is in proper position, fasten it to the drawing board by placing strips of drafting tape across the corners. See Fig. 5-5.

5.6 Triangles

Varying sizes of triangles, made of plastic or metal, are on the market. The most popular triangles are the transparent 45° and the 30°-60° combination shown in Fig. 5-6. It is, of course, essential that the triangle edges be straight and free of any nicks and that the angles be accurate. Many patented triangles are also available with variable angles and other built-in features.

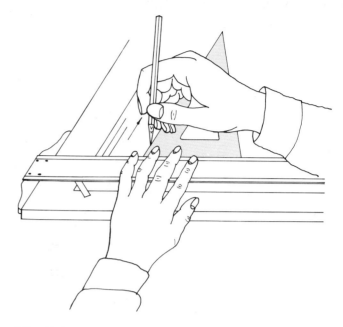

Fig. 5-7. *Position for Drawing Lines Perpendicular to T-Square.*

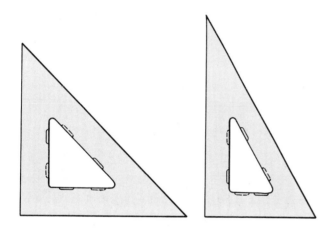

Fig. 5-6. *Standard 45° and 30°-60° Triangles.*

Horizontal and Vertical Lines. It is important that the beginner develop good habits for efficient use of his tools. When drawing lines along any guiding edge, *the hand should be above the guiding instrument* with the pencil or pen held at about 75° to the paper measured in a vertical plane along the straight edge. The right-handed draftsman constructs lines perpendicular to the T-square blade by placing one of the edges of a 90° angle in contact with the T-square while the other edge serves as the guide for the pencil or pen. He positions the triangle so that his right or ruling hand is resting on the triangle rather than the paper. His left hand moves the T-square, steadies the triangle, and keeps the T-square head constantly snug against its guiding edge. See Fig. 5-7.

The same procedure will apply to a left-handed draftsman if the words right and left are interchanged.

Inclined Lines. Lines inclined at 45°, 30° or 60° to the horizontal or vertical are drawn by guiding the proper triangle along the T-square. Keep the *pencil hand above the triangle* at all times even if it is necessary to shift your body position towards either end of the board. Angles of 15° and 75° to the horizontal, sloping either to the left or right, can be formed by combining the angles of the 30°-60° and 45° triangles guided by the T-square. See Fig. 5-8a.

Parallel Lines. Any set of parallel lines can be drawn by aligning the edge of one triangle with the desired direction and sliding its adjacent edge along another triangle or the T-square. See Fig. 5-8b.

Perpendicular Lines. Since a right triangle consists of a 90° angle and two other angles whose sum is 90°, two methods of this angle relationship are available for construction of perpendicular lines.

1. In *The Sliding Method* a leg of a triangle is aligned with the given line to which a perpendicular is desired. A second triangle is placed with its hypotenuse in contact with the hypotenuse of the first triangle. Slide the first triangle along the guiding hypotenuse until its

DRAWING EQUIPMENT

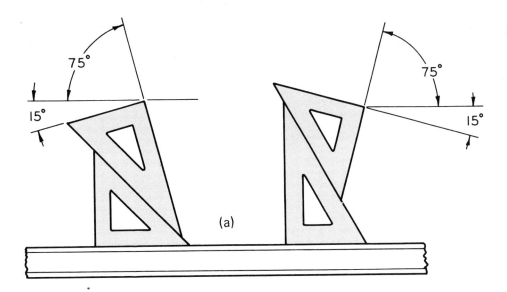

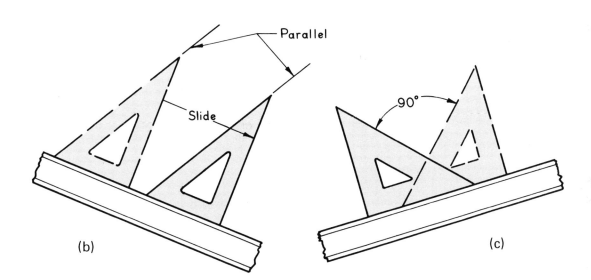

Fig. 5-8. *Drawing 15°, 75°, Parallel, and Perpendicular Lines.*

other leg is in the desired position and perpendicular to the given line.

2. *The Turning Method* consists of aligning the hypotenuse of a triangle with a given line and placing a second guiding triangle or T-square in contact with either leg of the first triangle. Turn the first triangle until its *other leg* is in contact with the guiding edge which shifts the hypotenuse to a position perpendicular to the given line.

Study Fig. 5-8c to become familiar with this procedure which will often be used to draw perpendicular lines.

5.7 Scales

Four classifications of scales, the *Mechanical Engineers,* the *Architects,* the *(Civil) Engineers,* and the *Metric* scales are available for use in different types of engineering design. Fig. 5-9 shows the Architects, the Engineers, and the Mechanical Engineers scales, which the student will use in graphics courses. The Architects and the Engineers scales are examples of "open-divided" and "full-divided" respectively. An open-divided scale has only the end unit sub-divided; while a full-divided scale is one whose face is sub-divided throughout its entire length.

Objects represented on machine drawings vary in size from small fractions of an inch to machines with large dimensions. A common practice in machine and related drawing is to specify dimensions in inches up to 72 or greater. However, this is a flexible figure.

5.8 The Architects Scale

The triangular Architects scale in common use has 11 different scales, each representing one foot and subdivided (except for the 16 face) into a multiple of 12 parts to represent inches and fractions.

With your Architects scale in hand study Fig. 5-10. Note the form in which scales are expressed on a drawing. One popular form is to express the scale by words. Example: Full Size, Half Size, Quarter Size, Twice Size, etc. Also, and perhaps more commonly, the scale is expressed as an equation of drawing size in inches or fractions of an inch equated to one foot dash (—) zero inches. Examples: $3'' = 1'-0$; $1/2'' = 1'-0$. (Note that inch marks are omitted on the 0). A ratio of the drawing size to its actual size is also used. Examples: 1/1, 1/8, 10/1.

5.9 The Engineers Scale

On a triangular engineers scale, inches are fully divided into ten, twenty, thirty, forty, fifty, and sixty parts. When using the ten scale, full-size values with decimal parts can be laid out conveniently. This face is also used for $1'' = 1'$ (not used if dimensions are given in feet and inches); $1'' = 10'$; $1' = 100'$; $1'' = 1000$ yds.; etc. In like manner the twenty scale is used for $1'' = 2''$; $1'' = 20'$; $1'' = 200$ mi.; etc. The thirty, forty, fifty, and sixty scales are adapted for similar use. See Fig. 5-11. In addition to linear dimensions, other quantities can also be expressed graphically with the engineers scale. Examples: 1 inch equals

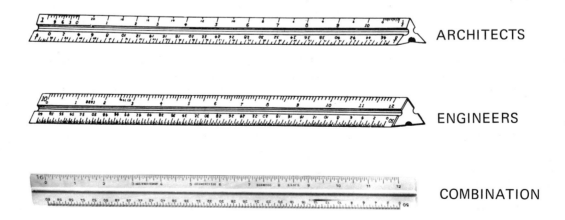

ARCHITECTS

ENGINEERS

COMBINATION

Fig. 5-9. *Architects (above), Engineers (middle), and Special Combination Scales. (Courtesy Keuffel and Esser Co. and Grammercy Import Co.)*

DRAWING EQUIPMENT

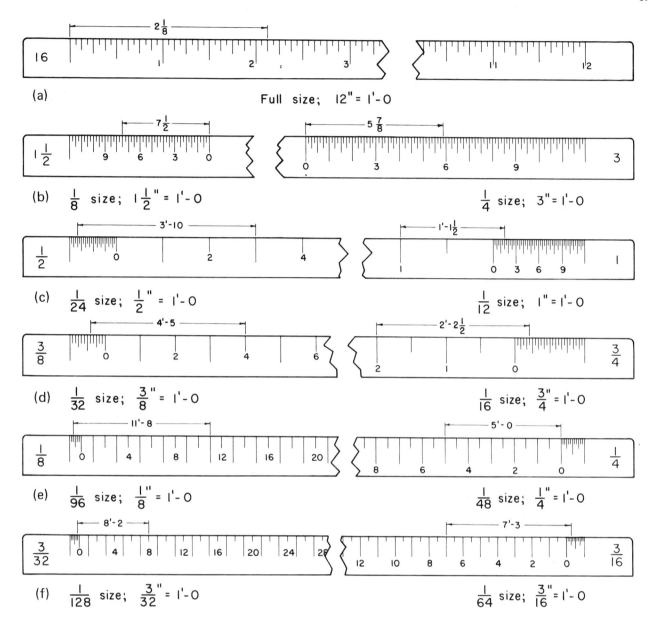

Fig. 5-10. *Faces on Architects Scale.*

500 pounds; or 1 inch equals 40 degrees fahrenheit. This scale is indispensable for laying out surveys, plotting graphs, etc.

5.10 The Mechanical Engineers Scale

The Mechanical Engineers or Draftsman scale is similar in appearance to the Architects scale. It is available in fully divided as well as the more common openly divided style. Because the Mechanical Engineers scale is concerned with *inches only* as opposed to feet and inches, the subdivisions are in halves, quarters, eighths, sixteenths, etc. The full-size scale is the same as the 16 face on the architects scale. Other usual faces are 3/4, 1/2, 3/8, 1/4 size. Most Mechanical Engineers scales also include a decimal scale with fifty divisions to the inch from which decimal inches can be read to .02 and estimated to .01. See illustration of the Combination scale in Fig. 5-9.

5.11 Other Scales

The *Metric Scale* is used in many countries where the meter is the standard unit of linear measurement. Many movements have been started

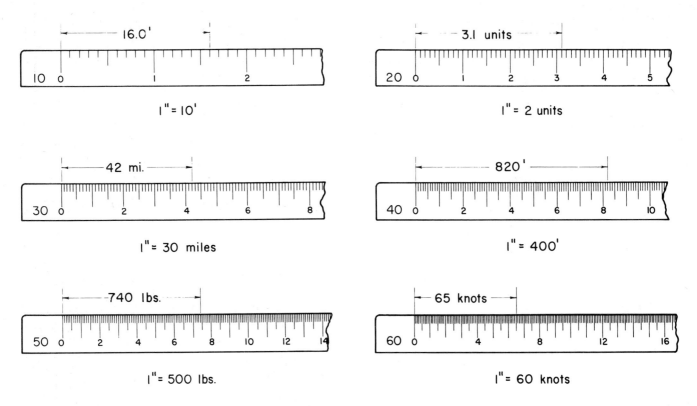

Fig. 5-11. *Faces on Engineers Scale.*

in this country to make a complete change to the c.g.s. (cm., gram, sec.) system with the only consistent progress being made by scientists. The principal objection to a change to the metric system is the gigantic cost (estimated at $20 to $30 billion) and the necessary re-education of our people to the new system. It is predictable that a complete change to the metric system will not come in the immediate future. However, the decimalized inch is rapidly gaining popularity in the mechanical industry and is used throughout this text.

Decimalized Inch. The Society of Automotive Engineers (SAE) and the American National Standards Institute (ANSI) recommend the use of linear dimensions in decimal values. Two variations include the two-place decimal system and three-place system with no tolerance implied. Dimensions depend upon the size of commercial tools, gages, or stock sizes will reflect the size identity of the tool, gage, or stock size. This is more completely explained in Unit 12.

In the two-place system, dimensions are expressed as multiples of .02 where practicable. All dimensions have at least two decimal places rather than the decimal equivalents of fractions. See Fig. 5-12.

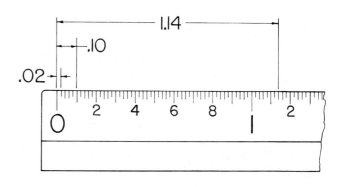

Fig. 5-12. *The Decimal Scale.*

In the three-place system, dimensions are expressed as multiples of .010, if practical, rather than exact equivalents of fractions. No fewer than three decimal places are used for any dimension on the drawing when using this system.

DRAWING EQUIPMENT

5.12 Irregular Curves

Irregular curves, or French curves, are used as a guide for drawing curved outlines which cannot be made with the compass. A wide variety of sizes and shapes are available ranging from very small templates to the large ship curves used for fairing long radius curvatures. Fig. 5-13 shows two popular irregular curves for student's use.

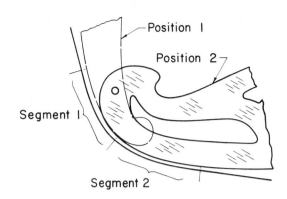

Fig. 5-14. *To Draw Curve Through Plotted Points.*

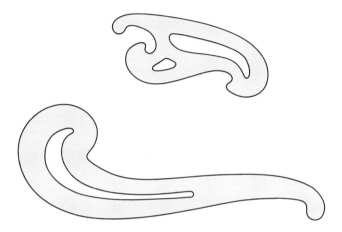

Fig. 5-13. *Large and Small Irregular Curves.*

Most drafting rooms are equipped with curves with a flexible metal ruling edge backed with rubber, or a spring band with an enclosed lead core. This type of curve is very convenient because of the innumerable variations of curvature which can be formed.

Curve Fitting. Fitting a smooth curve to a series of plotted points is a job that requires patience, care, skill, and judgment. The French curve is applied so that its ruling edge will pass through a minimum of three points, following the "trend" of curvature. A satisfactory "trend" curve can be determined by lightly freehanding a smooth curve through the plotted points. In this way, a greater portion of the curve's guiding edge can be fitted, thereby lessening the possibility of "humps" or "scallops" on the finished curve. Be sure that the trend of curvature of the plotted points is matched by the same trend on the French curve. In other words, if the desired curve is going from a longer to shorter radius, be sure that the French curve is applied with its radius of curvature changing from longer to shorter. See Fig. 5-14. Hold the pencil nearly perpendicular to the paper when tracing around the French curve.

5.13 Protractor

The protractor is an instrument for measuring angles. A very satisfactory model for student use is one made of celluloid or plastic. It is usually semicircular in shape, 3 to 6 inches in diameter, divided in degrees from 0 to 180, reading from either end of the diameter. Fig. 5-15 shows a satisfactory protractor for engineering students. A precision protractor is larger, usually made of metal and equipped with a vernier.

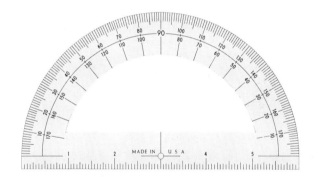

Fig. 5-15. *Protractor.*

5.14 Pencils

All pencils must be kept *well pointed* and ready for use. The choice of proper hardness of lead to be used on a drawing will depend on a combination of 1) the type and quality of paper; 2) the "brightness" or quality of line desired; and 3) the individual using the pencil. A harder grade of lead should be used to draw clean, sharp, readable lines on a smooth surfaced paper. This same quality of line may be duplicated on a rough surfaced paper by using the softer grades of lead. A 2H grade is recommended for general line

Hardness Label	Hardness	Use
9H, 8H, 7H, 6H	Very Hard	Fine layout work or precision graphical computations.
5H, 4H, 3H	Medium Hard	Light weight lines.
2H, H	Medium	Bulk or sharp detail work.
F, HB	Medium Soft	Lettering and Sketching.
B, 2B, 3B, 4B, 5B, 6B	Very Soft	Sketching and Shading.

Fig. 5-16. *Lead Hardness Table.*

work on a hard surface detail paper, while some prefer a softer lead on vellum to bring out black, printable lines. For lettering or sketching, a softer lead, F, HB, or even B should be used.

Guide lines, construction lines or the fine skeleton used for instrument layout requires a harder grade pencil (4H) than used for line drawings.

Since some people have the bad habit of pressing too hard with the pencil, they use this as an excuse for using a grade of lead harder than the recommended standard. Remember that changing the hardness of the lead is not the proper method of correcting excessive pressure. Likewise, laziness in keeping the pencil sharp is not a valid excuse for the selection of a hard grade of lead. Dirty, smudgy drawings are the result of carelessness rather than the use of a 2H pencil instead of a 4H. Study carefully the hardness table shown in Fig. 5-16 and follow these recommendations for the selection of proper grades of lead needed to produce satisfactory line quality.

Many draftsmen prefer to use automatic pencils, one for each grade of lead desired, thus eliminating the time consumed in sharpening a wooden pencil. Draftsman's automatic pencil lead may be obtained in any of the seventeen grades available in ordinary pencils.

Sharpening. The drafting pencil should *always* have at least 1/4" of exposed lead. The method of sharpening was discussed in Art. 4.3 with an illustration in Fig. 4-1 showing the "conical" pointed lead.

5.15 Erasers

The student should have several good erasers available and should know the proper use for each of them. A firm rubber eraser is satisfactory for removing ink and heavy lines, while the soft plastic or rubber eraser is best for sketches, lettering, and light pencil work. A *Dry-Clean Pad*, made of finely pulverized eraser crumbs in a mesh bag, is handy to clean a drawing. A *very light* sprinkling of eraser crumbs will also help keep the drawing clean. Some find a brush indispensible.

Erasers containing abrasive material should be avoided. Although their action is more rapid, they leave a damaged area over which it is difficult to add clean, new work.

Drafting rooms are often equipped with electric erasing machines, but care must be exercised because their action can quickly damage or burn a hole in the drawing.

If it is necessary to remove a portion of a line in a congested area without disturbing other lines, an appropriate slot in an erasing shield, Fig. 5-17,

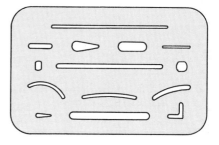

Fig. 5-17. *Erasing Shield.*

is placed over the line and the erasure is completed. A hard, smooth surface like a triangle placed under the paper at the point of erasure is sometimes helpful. Make sure that the paper is held taut to prevent wrinkles or slipping during the erasing operation.

5.16 Special Equipment

The following list of special tools and instruments for the draftsman includes some of the items already mentioned. Many of these instruments are too expensive for the student to purchase for his individual use, but will be found in the well equipped drafting room or graphics department. Pictures and complete description can be found in catalogs of drawing instruments companies.

Special Drafting Equipment

Border Pen	Special Triangles
Railroad Pen	Adjustable Curves and
Contour Pen	Splines
Drop Bow Pen	Ship and Railroad Curves
Proportional Divider	Electric Eraser
Beam Compass	Ellipse Templates
Planimeter	Assorted Templates
Pantograph	Wrico Lettering Set
Section Liner	Leroy Lettering Set
Specially Divided Scales	Drafting Machine

5.17 Drafting Media

The purpose of the drawing determines the type of paper to be used. If multiple copies of a drawing are needed, the original is often made on either translucent paper or cloth from which the necessary number of copies can be easily reproduced on sensitized paper by several methods. In the past it was common practice to make a pencil drawing on cream or buff detail paper and then ink trace the drawing upon tracing cloth. Presently, most drawings are made directly upon tracing paper with pencil, or upon pencil tracing cloth or drafting film. Well made, bright pencil lines produce good reproductions.

Detail Paper. A buff, cream, or light green paper, which comes in a variety of qualities and sizes, is used for layout pencil drawings which do not need to be reproduced. Although this paper is available in rolls, schools or industry may provide sheets cut to standard sizes. Borders and standard title blocks are often printed on the sheets to save drafting time.

White Paper. A white paper is appropriate for inked or colored display drawings. The best quality Whatman's hot pressed paper has a smooth surface which is excellent for fine line pencil drawings or ink work. Cold pressed paper is rougher and more appropriate for topographical drawings or architectural renderings. Industry uses a very light bond paper for many drawings upon which specifications can be typed and from which prints can be made.

Bristol Board. A stiff, white stock which comes in several thicknesses, 2-ply, 3-ply, 4-ply, etc. may be used when a self-supporting, heavy paper with good surface is needed. Patent Office drawings are made on 2- or 3-ply bristol board.

Tracing Paper. Two types of tracing paper, natural and treated, are available in a variety of qualities and sizes. The natural paper derives its transparency from the raw material and the method of manufacture. Generally, the treated paper, commonly called Vellum, is more transparent but may not be as durable as the natural paper. The great majority of industrial production drawings are made with pencil directly on tracing paper. However, ink can be used on tracing paper for better reproduction as is the case for all illustrations in this text. Like detail paper, tracing cloth, and drafting film, tracing paper is available in various sized rolls and cut sheets.

Tracing Cloth. Tracing cloth is a thin, finely woven cotton fabric which has been treated by a solution of starch or plastic compound to become translucent and to provide a good working surface which will "take" ink lines. Tracing cloth usually has a dull and shiny side. It is common practice to work on the dull side which, due to its "tooth," provides a more satisfactory working surface, especially for pencil guide or construction lines. Inked drawings on cloth are durable and can be handled and stored for long periods without deterioration. A white tracing cloth designed for pencil use is also available. A fixative is sometimes applied to the pencil drawing to prevent smudging.

Drafting Film. Mylar polyester film is the base material for a translucent, dimensionally stable, waterproof, and nearly indestructible drafting media. It is marketed by many companies under different trade names. Film is more expensive than drafting paper but its other qualities make it popular in the drafting room.

5.18 A Drawing Philosophy

One of the best ways to learn is by doing. This means intelligent practice using good habits and correct standards of drawing technique. Bad habits of procedure, once established, are inefficient, time-consuming, and hard to overcome.

We repeat again, drawing is not an automatic skill strengthened only by practice, but rather is the exercise of the mind over the hand. To accomplish a good drawing of any kind the hand must be directed *how*, *what*, and *where* to draw.

PROBLEMS

The purpose of the problems in this unit is twofold: (1) to give the student practice in the use of instruments and equipment and (2) to familiarize him with some of the more common geometrical constructions used in Engineering. The Geometry section of the Appendix should be referred to for the constructions of problems 3 through 10. If additional problems are desired, assignment of other geometrical constructions is recommended because they provide excellent practice in the use of instruments and equipment. The problems should be done on suitable paper combining, if desired, several problems on a sheet. Care should be taken that the constructions are correct and that good draftsmanship is achieved.

1. Architects Scale:
   Lay off the following dimensions to the scale indicated. Space dimension lines 3/8 apart, insert dimensions and terminate both ends of each line with an arrow head. See Fig. 5-12 for sample dimension lines.

Value	Scale
a. 7 7/16	$12'' = 1'-0$
b. 32'-5	$1/4'' = 1'-0$
c. 10'-7 1/2	$3/3'' = 1'-0$
d. 2'- 1 7/8	$3'' = 1'-0$
e. 2'-9 3/4	$1 1/2'' = 1'-0$
f. 30'-8	$1/8'' = 1'-0$
g. 7'-1 1/2	$1/2'' = 1'-0$
h. 8'-7	$3/16'' = 1'-0$
i. 3'-5 3/4	$1'' = 1'-0$
j. 1 7/8	$24'' = 1'-0$

2. Engineers Scale:
   Lay off the following dimensions using instructions in problem 1.

Value	Scale
a. 4.7'	$1'' = 1'$
b. 605'	$1'' = 60'$
c. 310'	$1'' = 50'$
d. .29''	$1'' = .1''$
e. 148 yds.	$1'' = 40$ yds.
f. 2140'	$1'' = 300'$
g. .008''	$1'' = .002''$
h. 660 mi.	$1'' = 200$ mi.
i. 660 mi.	$1'' = 600$ mi.
j. 225 mm.	$1'' = 60$ mm.
k. 225 yds.	$1'' = 50$ yds.
l. 18,600 rods	$1'' = 4000$ rods

3. Draw a straight line 2.50 long and divide it into eleven equal parts.
4. Construct a triangle whose sides measure 1.88, 3.19, 2.44.
5. Construct a regular hexagon circumscribing a 2'' dia. circle. What is the length of each side of the hexagon?
6. Construct a regular pentagon inscribed in a 3'' dia. circle.
7. Construct a regular polygon of nine sides. Make each side 1.50.
8. Illustrate the following constructions using suitable circles, lines and arcs:
   a. two lines tangent to a circle
   b. four lines tangent to two circles
   c. an arc tangent to two non-parallel, non-perpendicular lines
   d. an arc tangent to two perpendicular lines
   e. an arc tangent to a circle and a straight line
   f. an arc tangent to two circles
9. a. Construct an approximate involute (use compass) to a 1 1/4 dia. circle.
   b. Construct a true involute (use irregular curve) to a 1 1/4 dia. circle.
10. Use the trammel method to plot an ellipse whose major and minor axes equal 2 1/2 in. and 1 1/2 in. respectively. Use your irregular curve to draw the curve formed by the plotted points.

unit 6

Orthographic Projection-Points

6.1 Introduction

A photograph or a person's normal vision gives an image or picture of an object which represents the three dimensions of space sometimes called width, height, and depth, or "3D." The artist depicts three dimensional space by shades and shadows, and size and proportion by comparisons to common objects. The pictures by a good artist may be said to have excellent depth. Maps, which are basically two dimensional drawings, sometimes represent the third dimension of elevation by shading or color. However, at best none of these methods are accurate enough to give definitive values for size specifications.

Some segments of industry are using scale models from which the three dimensions of space may be measured. Many process operation units or structures of the Proctor and Gamble Company are constructed directly from models similar to the one shown in Fig. 6-1 which was used in the construction of the Ivorydale boiler house.

The majority of drawings necessary for the design of any product or system must accurately show the three space dimensions upon a two dimensional piece of paper. Multiview drawing is used for this more accurate delineation rather than pictorial representation. Multiview drawing is simply a logical grouping of the necessary orthographic views with each different view of the object at 90° to its adjacent view.

Most of us think of an object as a solid having volume. Without considering the details of method at this time, perhaps the student can visualize a set of six principal views of the familiar object shown in Fig. 6-2. A pictorial drawing is also shown in order to more quickly identify the faces of the object. Attention is directed on the drawing of the DIE to a particular "Spot A" one of many points which are joined to produce both the picture and the multiview images of the object. Because the multiview drawing of *One Point* is more easily visualized than are the many points on the visible and hidden outline of a solid object, the standards of *Multiview* or *Orthographic Projection* are introduced by a study of single points. The next step will be to consider the geometry of a straight line which is simply the shortest connector between two points in space. Subsequently, a plane may be represented by a point and line, or two parallel or intersecting lines. Therefore, the progression from the simple to the more complex, from points to lines, to planes, to solids, seems an efficient and correct progression while learning orthographic projection. The first important objective for graphical *Communication, Analysis* and *Design* is to understand the *theory and standards* of this system of drawing.

6.2 Orthographic Projection

Definition: *A system of drawing composed of images of an object formed by projectors from the*

Fig. 6-1. *Use of a Model for Plant Layout.*
(Courtesy Proctor and Gamble)

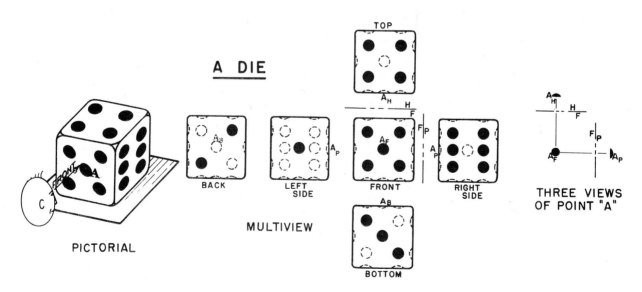

Fig. 6-2. *Pictorial and Multiview Drawing.*

ORTHOGRAPHIC PROJECTION—POINTS

object perpendicular to one or more desired planes of projection is called Orthographic Projection.

There are four significant features of orthographic projection as defined above, all of which are necessary for a complete definition.

1. *The Object* whose image is to be projected (point, line, plane, or solid).
2. *The Plane of Projection* or picture plane upon which the image of the object is formed.
3. *Projectors* or imaginary lines from the object perpendicular to the image plane and therefore parallel to each other.
4. *The Image* or projection of the object formed on the plane of projection.

In the construction of orthographic views, the object remains in a fixed position, the viewer orienting himself so that his line of sight is perpendicular to the image plane upon which the desired orthographic view is to be shown.

6.3 Spatial Location of a Point

Three coordinate dimensions are necessary to fix the location or position of a point in space. In mathematics these three dimensions of location are usually known as the X, Y, and Z coordinates. The exact location of a point, called A, inside a room, can be fixed if its distance X from one of the side walls, its distance Y from the front wall, and its distance Z below the ceiling are known; providing side and front walls and ceiling are properly identified.

The orthographic projection of point A upon the front wall, F, shows two coordinate dimensions X and Z. The projection on the ceiling, H, shows the coordinate dimensions X and Y. The side wall or P projection gives the Y and Z coordinate dimensions. This is illustrated in the pictorial, Fig. 6-3. The projections of point A are also shown upon the rear wall and the floor as well as the front, top, and side walls.

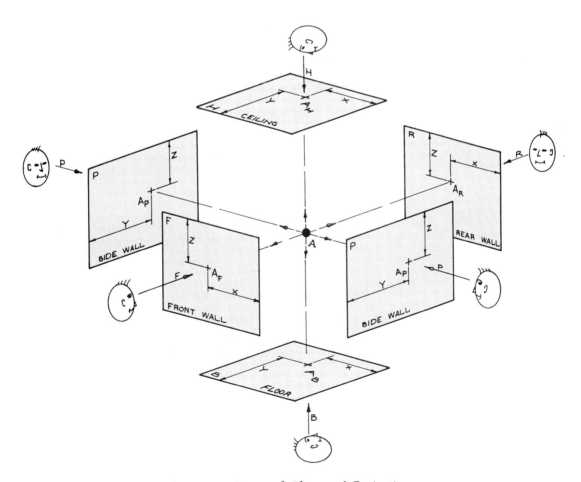

Fig. 6-3. *Principal Planes of Projection.*

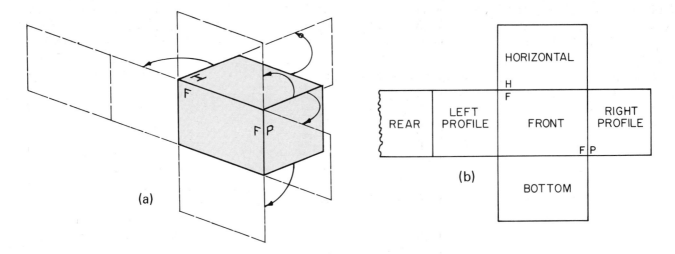

Fig. 6-4. *Principal Planes of Projection.*

6.4 The Box of Projection

The perpendicular walls of the room can be likened to a "box of projection" consisting of six transparent, perpendicular planes. These six planes, Front, Horizontal (top), Right Profile (right side), Left Profile (left side), Rear, and Bottom are known as the *Principal Planes of Projection*. If we imagine these planes to be hinged, this box can be opened to lie in the plane of a piece of paper as shown in Fig. 6-4.

6.5 Principal Projections of a Point

Since adjoining principal planes of projection are perpendicular to each other in space and the lines of sight for an orthographic projection are perpendicular to the image plane, the orthographic view will be shown in true size and the adjoining planes of projection will appear as edges. This is illustrated in Fig. 6-5.

When the *horizontal plane* is viewed from the direction shown by the line of sight in Fig. 6-5a the front and profile planes appear as edges. In Fig. 6-5b, the horizontal plane is in true size, therefore, the true distance X to the left of the profile plane and the distance Y behind the front plane is shown and labeled.

Fig. 6-6a and b shows the front elevation projection of point A. The *front plane* of projection, Fig.

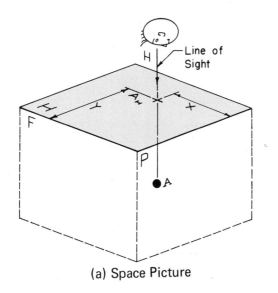

(a) Space Picture

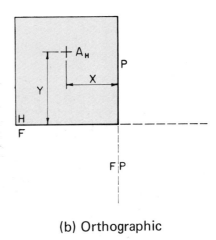

(b) Orthographic

Fig 6-5. *Horizontal Projection.*

ORTHOGRAPHIC PROJECTION—POINTS

6-6b, is in true size in the orthographic view; consequently, the distance X to the left of the profile plane and the distance Z below the horizontal plane are shown in true magnitude.

The views in Fig. 6-7a and b illustrate the projections on the *right profile plane* of projection. In the orthographic view of point A in Fig. 6-7b, the profile plane of projection is in true size and the true distance Z below the horizontal plane and the distance Y behind the front plane can be measured.

6.6 Relationship and Alignment of Principal Orthographic Views

In Art. 6.4 it was mentioned that the planes of the Projection Box can be opened until they lie in the plane of a two-dimensional piece of drawing paper (Fig. 6-4a and b). In the most common orientation of the planes of the Projection Box, the Horizontal, Right Profile, Left Profile, and Bottom planes are hinged to the Front plane, and the Rear plane is hinged to the Left Profile plane of projection. Thus the Horizontal plane and the Bottom plane have the same width as the Front plane and lie respectively above and below the Front plane. The Right Profile, Left Profile, and Rear planes of projection have the same height as the Front plane and are respectively located to the right and left of the Front plane. This standard alignment and relationship of views is important for universal understanding of orthographic projection.

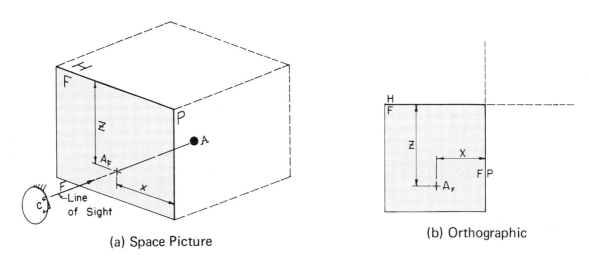

(a) Space Picture   (b) Orthographic

Fig. 6-6. *Front Projection.*

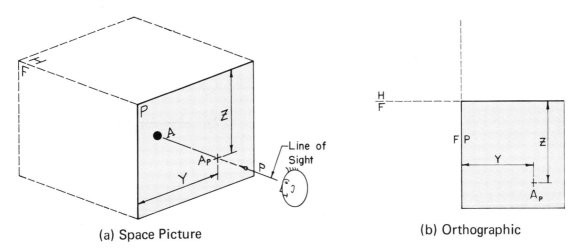

(a) Space Picture   (b) Orthographic

Fig. 6-7. *Profile Projection.*

One exception frequently occurs in the arrangement described in the previous paragraph. Here, the Profile view is adjacent to and hinged from the Horizontal plane instead of the Front plane. This alternate position of the Profile plane is used for reasons of space limitation and to provide a more advantageous solution of some particular problems. Such a solution is illustrated in Unit 7 when the true length and inclination of a Profile line are determined.

In Art. 6.5 (Figs. 6-5b, 6-6b, 6-7b) individual Horizontal, Front, and Profile views were constructed to illustrate the meaning of each view. These projection planes can be considered to be three mutually perpendicular planes of the Projection Box, Fig. 6-8a, and can then be rotated so as to lie in the plane of the drawing paper, as shown in Fig. 6-8b. It follows that any projection lying upon a plane before rotation will retain the same coordinate position after rotation into the plane of the paper.

6.7 Standards of Projection

Principal Views. The image on any of the six planes of the box of projection is called a principal projection or principal view. Each principal view should occupy a standard position with respect to an adjacent principal view. Refer again to Fig. 6-4.

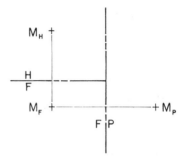

Fig. 6-9. *Hinge Lines.*

Adjacent Views. Any two adjoining orthographic projections that are properly aligned are said to be adjacent views. In Fig. 6-9, M_P and M_F are adjacent views. Likewise, M_F and M_H are adjacent.

Related Views. Any two views that are *adjacent to the same intermediate view* are called related views. In Fig. 6-9, M_H and M_P are related views. It should be noted that *the distance from an image point on related views to the hinge lines between the related and the intermediate view is always identical.* Or in Fig. 6-9, the distance from M_H to H/F and M_P to P/F is the same.

Hinge Lines. The line of intersection or hinge line between two adjacent planes of projection may have two additional interpretations when viewed in orthographic projection. For example, a line labeled H/F is:

1. The line of intersection between the Horizontal and Front planes of projection.
2. The edge view of the Horizontal plane when the line of sight is perpendicular to the Front plane of projection.
3. The edge view of the Front plane when the line of sight is perpendicular to the Horizontal plane of projection.

The latter two interpretations are used in plotting the space locations of a point, line, or plane.

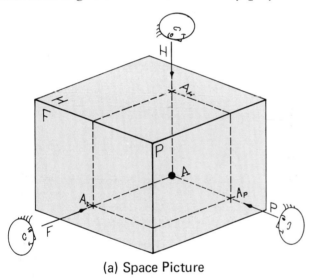

(a) Space Picture

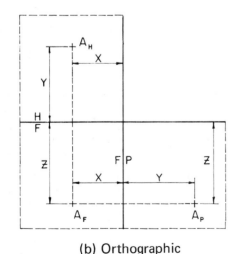

(b) Orthographic

Fig. 6-8. *H, F, and P Projection.*

ORTHOGRAPHIC PROJECTION—POINTS

Hinge lines are represented symbolically, Fig. 6-9, by a series of long and short double-dash lines. It is important that hinge lines be properly labeled to facilitate the plotting of point projections. A *letter or numeral* representing the *Name* of the plane of projection is placed on the side of the hinge line which is adjacent to the area of the designated plane shown in true size. Examples: F/P, H/1, 1/2, 6/13.

In Fig. 6-9 each projection of point M bears a subscript which signifies the plane of projection for that particular view of the point. The projection of the point is indicated by a cross, +, at the correct location on the plane of projection. Very light projection lines (always perpendicular to the hinge lines) should be used to relate or align the projections in adjacent views.

Coordinate Dimensions. The true distance X to the *right* or *left* of a Profile plane can be seen in either the Front or Horizontal view since the Profile plane appears as an edge in both. See Fig. 6-10a.

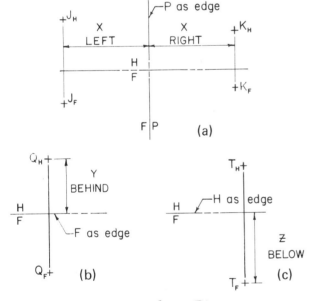

Fig. 6-10. *Coordinate Distances.*

The Horizontal view shows the Front plane of projection as an edge. Therefore, the desired distance Y *behind* the Front plane can be measured in the Horizontal view, Fig. 6-10b. By similar reasoning the Profile view also shows the distance behind F.

The Horizontal plane appears as an edge in the Front view and therefore the distance *below* the Horizontal plane appears in this view, Fig. 6-10c.

It now becomes apparent that *the distance of a point from a particular plane can be seen in a view where that plane is an edge.*

6.8 Plotting of Internal Points (Within Box of Projection)

Problem: Locate the Horizontal, Front, and Profile views of a point M which lies 3/4 of an inch to the left of a Profile plane, 1/2 inch behind the Front plane and 1 inch below the Horizontal plane. See Fig. 6-11.

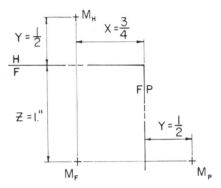

Fig. 6-11. *Point Location.*

Procedure

1. Construct the reference planes (hinge lines) labeled H/F and F/P.
2. Since the *Profile* plane appears as an edge in the H or F views, lay off the 3/4 inch distance X to the left of the P plane in the H or F view.
3. In the *Horizontal* view, the F plane of projection appears as an edge and the 1/2 inch coordinate Y behind F must be located in the H view. The projection of point M in the Horizontal view is labeled with the subscript H.
4. Since in the *Front* view the H plane of projection appears as an edge, the 1 inch distance Z below H is shown in the Front view. The resulting projection of Point M is given the subscript F which indicates that it is the projection of point M on the Front plane.
5. Because the *Profile* view shows elevation Z below H and also distance Y behind F, M with the subscript P is located in orthographic projection in the left-right alignment with the adjacent Front view. Also, the distance Y behind F will be located on this alignment 1/2 inch to the right (behind) of hinge line F/P. This projection is M with subscript P.

6.9 Auxiliary Views

In addition to the Principal Views, there are other orthographic views called Auxiliary views. These views may be classified as *Auxiliary Elevation*, *Auxiliary Inclined*, and *Auxiliary Oblique Views*.

Auxiliary views are used in the solution of problems involving true length and slope, slope angle, or per cent grade of oblique lines. They are also useful in finding the edge view and true size and shape of oblique planes. Units 7, 8, 14 and 15 will deal with many applications of Auxiliary Views.

6.10 Auxiliary Elevation Views

Definition: *An Elevation view is any view which shows the Horizontal plane as an edge.* Any Elevation view which is not parallel to any of the four principal elevation views, Front, Rear, Right Profile, Left Profile, is called an *Auxiliary Elevation View*.

The theory of auxiliary elevation views can be clearly perceived by folding a stiff piece of paper so as to form the projection planes shown in the pictorial illustration, Fig. 6-12a, and then observing, in the relationship of the three planes, that the elevation distances Z (below H) shown on planes F and 1 are identical.

Fig. 6-13 shows several auxiliary elevation views of point K. The given principal views are above the solution views. Note that the elevation of the point, that is, its position below the horizontal plane of projection is the same in each view. The edge views of the auxiliary elevation planes 1 and 2, may be located at any distance from the horizontal projection, K_H. The student should visualize a point held before him and picture the location of this point in the principal elevation views and the innumerable auxiliary elevations. He should also visualize how these views are represented on paper in Fig. 6-13.

6.11 Auxiliary Inclined Views

Definition: *Any non-vertical Auxiliary view in which either the Front plane or the Profile plane of projection appears as an edge is called an Auxiliary Inclined View*. Fig. 6-14 shows an auxiliary inclined view of point "O" which is Y distance behind the Front plane. Again note that in orthographic projection the planes for two adjacent

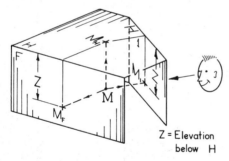

(a) SPACE ANALYSIS

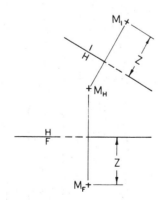

(b) ORTHOGRAPHIC

Fig. 6-12. *Auxiliary Elevation View.*

GIVEN
(2 PRINCIPAL VIEWS)

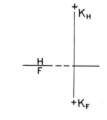

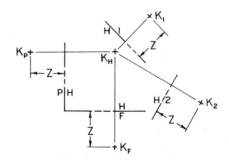

Fig. 6-13. *Auxiliary Elevation Views.*

ORTHOGRAPHIC PROJECTION—POINTS

views must always be perpendicular to each other. Therefore, an auxiliary inclined view will be "hinged" from (perpendicular to) the Front or Profile projections. For example, Fig. 6-14a shows the H and F planes with plane 1 perpendicular to F but not perpendicular to H.

The auxiliary inclined plane can be vividly illustrated in space by folding a piece of paper, following the pictorial in Fig. 6-14a. The same paper model that was used for the auxiliary elevation, Art. 6.10, can be adapted to the auxiliary inclined view if the H and F planes are interchanged.

Several auxiliary *elevation* and *inclined* views of point Q have been plotted from the given principal views in Fig. 6-15. Try to determine how each view is located and note why it is an elevation or an inclined view. As well, visualize point Q in space and identify the several views by name.

6.12 Oblique Auxiliary Views or Successive Auxiliaries

Definition: *A view that is neither parallel nor perpendicular to a principal view is known as an* Oblique Auxiliary View. Such a view is adjacent to an auxiliary view or another oblique view. A number of successive oblique views may be necessary to determine desired relationship between points, lines or planes. Any number of views with true orthographic relationship can be plotted if the principles of orthogonal projection are followed.

If two adjacent orthographic views are given, any other desired view adjacent to either given view can be plotted. In other words, a third view may be determined from any adjacent pair of views by means of orthographic principles. See Fig. 6-16. The relationship of these views will be similar to that which exists between the H, F, and an auxiliary elevation or an auxiliary inclined view. The paper model suggested in Arts. 6.10 and 6.11 can again be used for space visualization of the orthographic relationship for any three adjacent views by renaming each plane as desired.

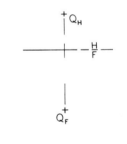

GIVEN
(2 PRINCIPAL VIEWS)

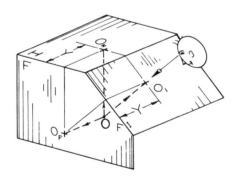

(a) SPACE ANALYSIS

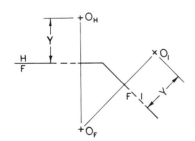

(b) ORTHOGRAPHIC

Fig. 6-14. *Auxiliary Inclined Views.*

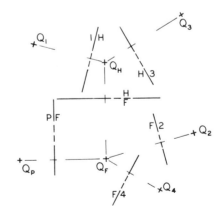

IDENTIFICATION OF VIEWS
PRINCIPAL: Q_H, Q_F, Q_P
AUX. EL.: Q_1, Q_3
AUX. INCL.: Q_2, Q_4

Fig. 6-15. *Auxiliary Elevation and Inclined Views.*

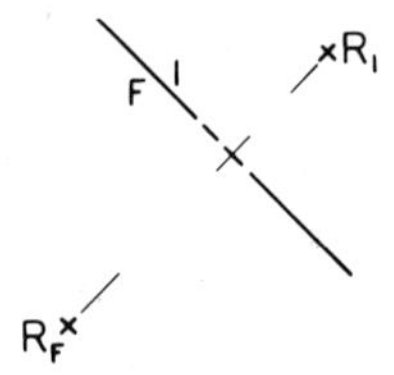

GIVEN
F AND I VIEWS

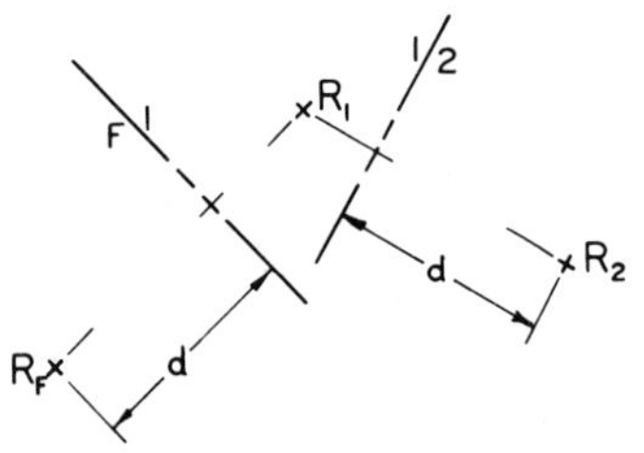

TO FIND VIEW 2

Fig. 6-16. *Adjacent and Related Views.*

Note the given and the solution diagrams in Figs. 6-16 and 6-17. Also note the common distance (d) for related views R_F and R_2 using the principle developed for distance Z in Fig. 6-12 and distance Y in Fig. 6-14. This principle also can be

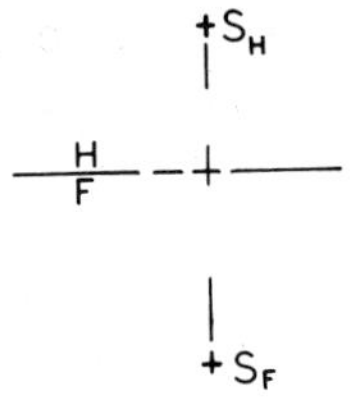

GIVEN
2 PRINCIPAL VIEWS

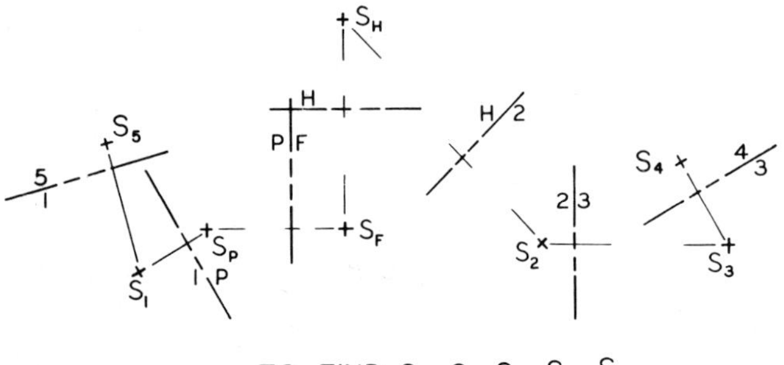

TO FIND S_P, S_1, S_3, S_4, S_5

Fig. 6-17. *Successive Projections.*

clearly demonstrated with the previously mentioned folded paper model. In Fig. 6-17 the eight successive views include principal, auxiliary elevation, auxiliary inclined, and auxiliary oblique views. Identify the coordinate distances used to plot each projection.

6.13 External Points (Outside the Box of Projection)

In addition to the orthographic views of points that are inside the "room" such as all projections shown in Figs. 6-8 through 6-17, it may be necessary to locate points that are in front of, rather than behind, the Front planes; above, instead of below, the Horizontal plane; or any combination of these positions. Illustrations should be self-evident because they follow the same principles of adjacent and related views discussed in Arts. 6.6–6.13 for points "inside the box." Both *Principal* as well as *Auxiliary Views* of two "outside" points are shown in Fig. 6-18. Study the construction carefully.

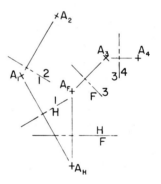

POINT A: ABOVE H
IN FRONT OF F

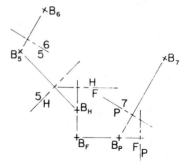

POINT B: BELOW H
IN FRONT OF F

Fig. 6-18. *Principal and Auxiliary Views (outside box).*

ORTHOGRAPHIC PROJECTION—POINTS

PROBLEMS

This unit studies orthographic projection of points in space. The relationship of principal views and standards illustrated in Fig. 6-8 should be used for the following problems. Horizontal plane, Front plane, and Profile plane have been abbreviated to H, F, and P respectively. The student should consider the spatial characteristics of the problems as well as their orthographic solutions. All points should be properly labeled.

1. Locate the following points in the Horizontal, Frontal, and Profile projections:

Point A: 1.74 in. left of P; 1.74 in. behind F; 1.74 in. below H.

Point B: 1.24 in. left of P; .62 in. behind F; .88 in. below H.

Point C: .50 in. left of P. 2.24 in. behind F; 1.38 in. above H.

Point D: 1 in. right of P; on F; 1/2 in. below H.

2. a. The ceiling, the front wall, and the side wall of a room are represented by planes. Locate point D, the center of a heating duct which lies: on the side wall; 1'-4 behind the front wall; 2'-0 below the ceiling. Scale: $1'' = 1'-0$.

   b. Locate point E, the center of an electrical outlet which lies: 2'-6 to the left of the side wall; on the front wall; 2'-6 below the ceiling. Scale: $1'' = 1'-0$.

3. Locate the following points in the horizontal frontal and profile projections and draw an auxiliary elevation view. Scale: $1'' = 5$ miles.

Point F: 13.0 mi. left of P; 9.6 mi. below H; 11.3 mi. behind F.

Point G: 6.9 mi. left of P; 17.7 mi. below H; 3.0 mi. in front of F.

Point H: 8.8 mi. left of P; 4.2 mi. above H; 12.4 mi. behind F.

4. Locate the following points in the horizontal and frontal projections and draw an auxiliary inclined view.

Point J: 3 in. right of P; 2 in. above H; 2 in. behind F.

Point K: 1 in. left of P; 2 in. above H; 3 in. in front of F.

5. See Fig. 6-19. Locate point Z in all views indicated. Write a word description similar to Problem 4, for the location of Point Z. Scale: $1'' = 1.0'$.

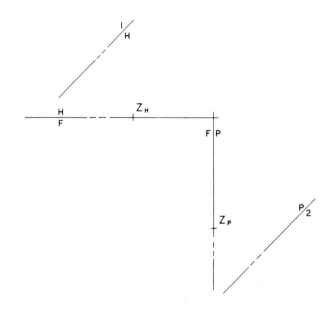

Fig. 6-19.

6. See Fig. 6-20. Locate point X in all views indicated. Write a word description of the location of point X. Scale: $1'' = 10'$.

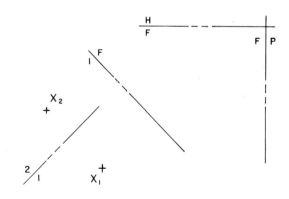

Fig. 6-20.

7. See Fig. 6-21. Locate in all views indicated:
Point M: 1 1/4 in. above H; 3/4 in. left of P; 1/2 in. behind F.
Point W: 3/4 in. right of P; 1/2 in. in front of F; 1 in. above H.

8. Fill in the table to give the coordinate locations of points M, K, Q, S, Fig. 6-22.

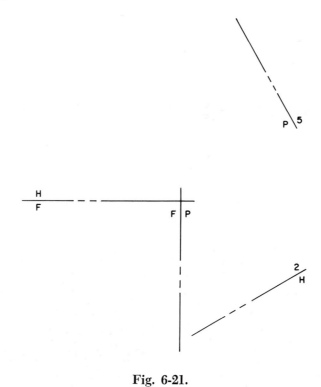

Fig. 6-21.

Fig.	POINT	DIST.- PROFILE	DIST.- FRONT	DIST.- HORIZONTAL
6-11	M			
6-13	K			
6-15	Q			
6-17	S			

Fig. 6-22.

unit 7

Orthographic Projection – Lines

7.1 Definition

A line may be defined as the path or locus of a point moving through space. This locus may be straight, curved upon a plane, or curved in three dimensions like a spiral. This unit will discuss the orthographic description and delineation of straight lines.

The graphical representation of a straight line segment is the shortest connection between its end points. By definition, a line may be considered to be continuous beyond its location points although the engineer usually considers only a finite segment.

The principal reason for studying lines is to be able to understand and write specifications for many industrial applications, for example, pipe lines and guy wires. The picture in Fig. 7-1 shows a model of part of a Chemical Process Plant containing many pipe lines. The size, location, length, and in some cases inclination of each pipe has been carefully designed for the desired operation. The resulting specifications were used in fabricating the system and erecting the structure.

7.2 Classification of Lines

In general, lines are classified or named according to the plane or planes of projection to which they are parallel. These classifications include Horizontal, Front, and Profile lines, as well as Vertical lines which are parallel to both the Front

Fig. 7-1. *Model of a Chemical Process Plant. (Courtesy Proctor and Gamble)*

and Profile planes. These lines are illustrated by orthographic projections in Fig. 7-2a, b, c, and d.

A line that is *not* parallel to any principal plane or projection is called an Oblique line. See Fig. 7-2e.

7.3 Line Specifications

Either of two types of specifications are used to convey necessary information about a straight line segment:
1. Coordinates of the end points. The orthographic projection may be delineated or the mathematical equation written from the given end point coordinates of the line.
2. Coordinates of a point on the line and:
   a. True length
   b. Bearing
   c. Inclination or declination of the line to a horizontal reference plane

If only the type 1 specification is known, the values of true length, bearing, and inclination can be found on the drawing.

Each of these specifications will be developed and illustrated in the following articles.

7.4 Space Analyses

Before solving space problems it is important to plan the steps of solution to be followed from the given position of the points, lines, planes, and/or solids involved, to the position or view where the desired answer is found. In other words, it is important that the Space Analysis of a problem be visualized so that the steps of procedure for the orthographic solution can be readily determined. The statement of a space analysis should contain: a) a statement of the problem; and b) a conclusion describing the view, condition, or

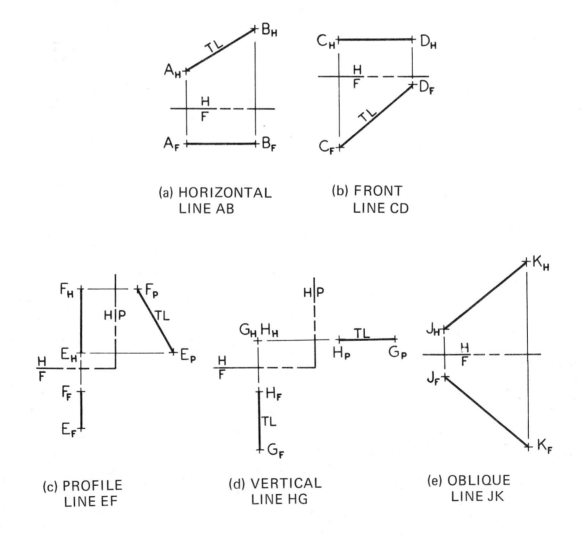

Fig. 7-2. *Classification of Lines.*

ORTHOGRAPHIC PROJECTION—LINES

position of the component parts of the problem to show the desired answer. Hereafter, the space analysis will be stated in italics for each new basic problem as it is encountered.

7.5 True Length of a Line

SPACE ANALYSIS. *The true length of any line in space* (statment of problem) *is found only upon a plane of projection parallel to the line* (condition of solution).

In Fig. 7-2a, b, c, d, each of the lines is parallel to a principal plane and therefore the projection is in true length on that plane. Note the label T.L. on the projection showing true length.

In Fig. 7-3, line RS is not parallel to a principal plane and is therefore known as an oblique line.

The true length of RS is found in an auxiliary elevation view, Fig. 7-3b, or an auxiliary inclined view, Fig. 7-3c.

7.6 True Length by Rotation

If an oblique line is rotated until it is parallel to one of the principal planes, the criteria for true length will be satisfied. The projection on that principal plane will then show the line in true length. Rotation is sometimes used to avoid drawing auxiliary views, although the principal views may often become congested by the rotated positions which can add confusion to the solution. The use of colored pencils will help keep the projections clear.

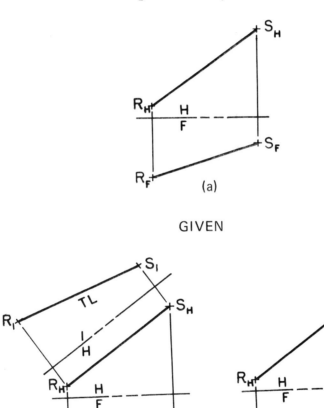

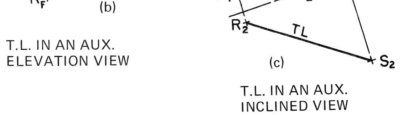

Fig. 7-3. *True Length of an Oblique Line.*

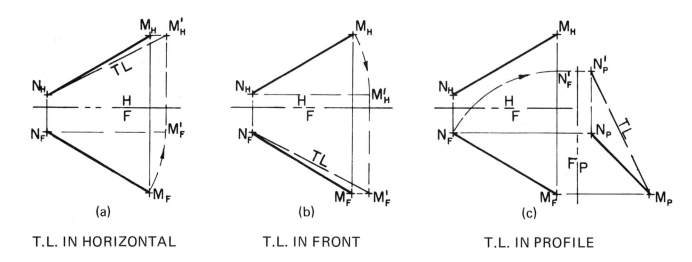

Fig. 7-4. *True Length by Rotation.*

Fig. 7-4a shows the rotation of line MN until it is parallel to the *Horizontal* plane and thereby will be in true length on this H plane. In Fig. 7-4b the line MN is rotated until parallel to the *Front* plane upon which it will show in true length. Line MN is rotated in Fig. 7-4c until parallel to and therefore in true length on the *Profile* plane. Note that the revolved projection is labeled with prime letters and proper subscripts.

Two operations must be observed when rotating a line about an axis to find its true length.

1. Determine the Point View of the axis of rotation in the view where the motion of rotation will appear circular. (View F in Fig. 7-4a, View H in Fig. 7-4b, and View F in Fig. 7-4c). Rotate the projection of the line about the axis (point) until the line is parallel to an adjacent plane of projection (line's projection parallel to the hinge line) upon which it will show in true length.
2. In the view adjacent to the circular motion, the ends of the line will move along a path parallel to the hinge line between views. (M_H to M'_H in Fig. 7-4a).

The student should visualize the steps of rotation by holding a pencil in space to represent the line MN or any other line whose true length is desired. Follow the steps of rotation in Figs. 7-4a, b, and c by turning the pencil about the appropriate axis until parallel to either H, F, or P.

7.7 Bearing of a Line

Definition: *Bearing is the deviation of a line from the North-South direction.*

SPACE ANALYSIS. *The bearing of a straight line is found as the acute angle between the horizontal projection of the line and a north-south reference drawn through the origin of the given line.* Examples: $N\beta°W$, $S\beta°E$.

The surveyor and the navigator find the designation of bearing important in specifying a line or course. Anyone reading a map uses bearing to locate one point with respect to another; for instance, a given highway may have a north-south direction. However, if you are at a particular point of origin, say Minneapolis, and wish to go to Dallas, Texas, you would proceed approximately due south; but, if you were in Dallas and wished to go to Minneapolis, the direction would be due north with respect to the origin of the line. In problems where the terminal points of the lines are lettered, it is common, unless otherwise indicated, to use the alphabetical order in determining which terminal is the origin point of the line. For example, point K is the origin for the line KM.

It is standard practice in graphical representation on maps to locate north toward the top of the sheet. Fig. 7-5 shows the location views of several lines with the bearing properly indicated.

What is the bearing of line LM in Fig. 7-6?

ORTHOGRAPHIC PROJECTION—LINES

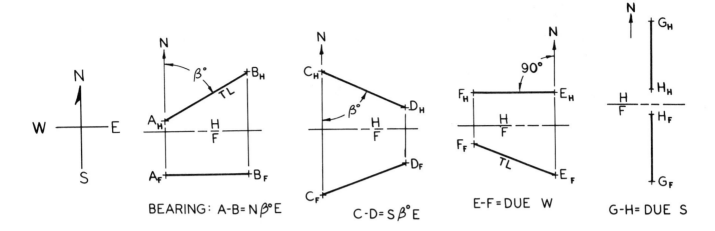

Fig. 7-5. *Bearing of Lines.*

7.8 Inclination or Declination—Slope Angle, Slope, Percent Grade

Definitions: The SLOPE ANGLE of a line is the acute angle formed between the line and a horizontal reference plane through the line's origin. The angle is positive (+) if inclined upward and negative (−) if downward from the origin.

The SLOPE of a line is the rise or fall of a line divided by its run. This is equivalent to the tangent of the slope angle. Likewise, slope is expressed as positive or negative respectively if it rises or falls from the origin.

The PERCENT GRADE of a line is equal to its slope multiplied by 100. It also must be prefixed by a + or − sign to indicate uphill or downhill from the origin. Note that the + for inclination or − for declination can be easily determined in any *elevation* view but it is most obvious in the front elevation.

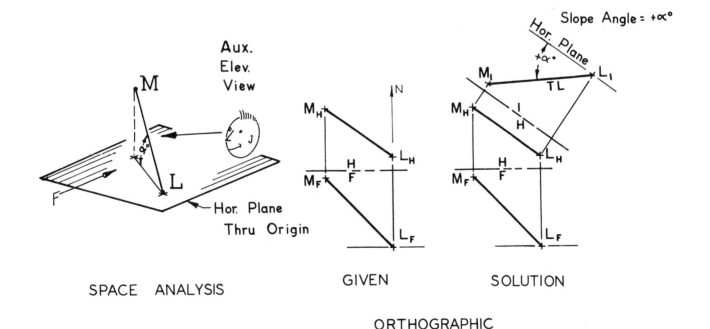

Fig. 7-6. *Slope Angle of a Line.*

SPACE ANALYSIS. *The value of the slope angle, slope, or percent grade is found ONLY in an elevation view which shows the line in true length.*

An elevation view which shows the line in true length will show a horizontal reference plane through the origin of the line as an edge. The origin is commonly considered to be the earlier letter in the alphabet such as L in the line LM, Fig. 7-6. The line could have been defined as ML in which case the inclination would have been negative (−) or downhill from M to L.

When slope, rise or fall divided by the run, is to be expressed as a positive or negative quantity, or the grade is to be expressed in percent, it is convenient to determine these values graphically using the run (denominator of the fraction) as 100. Use the 40, 50, or 60 face on the engineer's scale to lay off 100 units parallel to the H plane (H/1 hinge line) from the origin of the line in its true length elevation view. Using the same scale, the RISE (+) or FALL (−) is the length of a line perpendicular to the 100 unit run at its end point and extending to the true length projection of the given line or its extension. The *slope* is simply the value of this rise (+) or fall (−) divided by 100. The *percent grade* will equal the rise or fall with the + or − prefix. See examples in Fig. 7-7 where the values of slope angle, slope, and percent grade are labeled and tabulated. Refer to the Roof Problem, Fig. 8-12 and the Lunar Module, Fig. 2-1 for several interesting applications of True Length, Bearing, and Inclination.

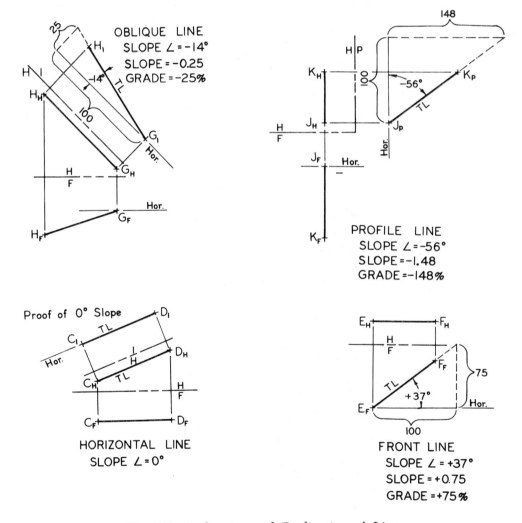

Fig. 7-7. *Inclination and Declination of Lines.*

7.9 Inclination by Rotation

Since inclination is measured in a true length elevation view, the horizontal projection of the line may be rotated about a vertical axis (perpendicular to H) until this projection is parallel to the front plane. The resulting front view of this rotated line will be a true length elevation view showing inclination. See Fig. 7-8.

7.10 Point View of a Line

SPACE ANALYSIS. *The view or projection of a line as a point is found ONLY upon a plane which is perpendicular to a true length projection of the line.* Fig. 7-9 shows the true lengths and point views of the center lines of three portions of the exhaust tail pipe for an automobile. Point views are used to test for clearances, true lengths, and bearings; slope angles are necessary to construct the jigs for bending the pipe. Note that true length views are necessary before finding the point views. Other applications of the point view in addition to clearance for pipes include the determination of distance between wires and edges of planes and solid objects. The analysis of such problems will be found in Unit 14.

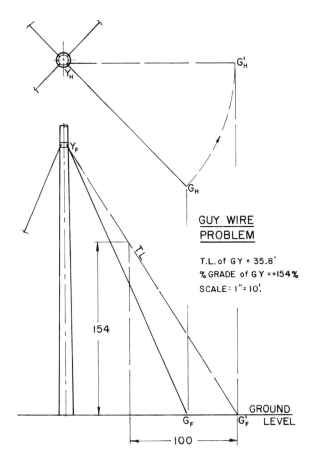

Fig. 7-8. *Inclination by Rotation.*

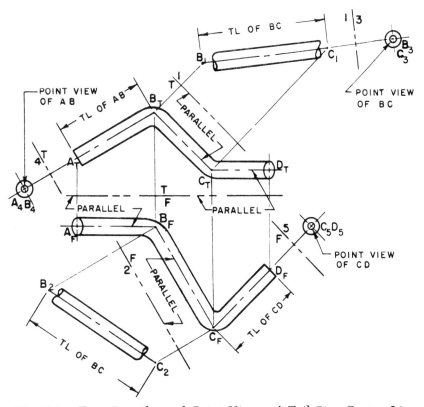

Fig. 7-9. *True Lengths and Point Views of Tail Pipe Center Lines.*

Rotation can be used to satisfy the criteria for finding the point view of a line by first revolving the line to give its true length on a principal plane. A second rotation will be necessary to turn this true length view perpendicular to an adjacent principal plane upon which the point view will be found. Fig. 7-10 shows an example of a point view of a given line found by rotation.

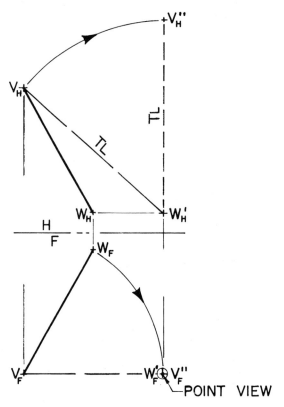

Fig. 7-10. *Point View by Rotation.*

PROBLEMS

Theoretical and practical problems involving the specifications of lines are included in this unit of study. The problems should be done on suitable paper and the student should take care to properly label and identify all work.

1. Locate point A which lies: .50″ below H; .50″ left of P; .50″ behind F.
   Locate Point B which lies: 2″ below H; 2″ left of P; 1″ behind F.
   Connect A_H to B_H, A_F to B_F, and A_P to B_P. What is the classification, true length, bearing, and slope angle of the resulting line AB?

2. Make freehand sketches and label all points on the orthographic views necessary for complete description of the lines specified (estimate lengths and angles).

   a. horizontal line AB, 2″ long, bearing N 30° E
   b. frontal line CD, 1 1/2″ long, slope angle of +60°
   c. oblique line EF, 2 1/2″ long, bearing S 40°W, slope angle of +30°
   d. profile line JK, 1 3/4″ long, with a slope of −0.7
   e. Find the point view of line EF.

3. Show the H and F projections and the point views of the following lines:

Line	Bearing	Inclination	True Length
AB	S 30° W	−30°	2 in.
MN	N 15° W	+.6	1.74 in.
EF	N 75° E	−42%	2.50 in.

4. A rocket leaves the ground at point A and bears N 30° E climbing at a constant rate for six minutes until it reaches point B. If its average air speed was 3000 MPH and the slope of its climb was 0.58, how many ground miles has it covered and how many miles is it from the surface of the earth when it reaches point B. Assume that the surface of the earth is flat. Scale 1" = 50 miles.

5. Two tunnels start from the common point A in a large shaft.
   Tunnel AB bears N 40° E, falls 20% and is 150 feet long.
   Tunnel AC bears S 75° E, falls 30% and is 200 feet long.
   The two ends are to be connected by a tunnel BC. Find the bearing, grade, and true length of tunnel BC. Scale: 1" = 50'.

6. Scale: 1" = 100 ft. Given a pipe line located as follows:

Line	Bearing	Inclination	True Length (ft.)
AB	N 60° E	+80%	200
BC	Due N	−20°	110
CD	S 60° E	−0.4	270

   a. Draw the H and F views of the pipe line.
   b. Find the bearing, grade, and true length of a single pipe from A directly to D. Which run of pipe would be the most economical from the standpoint of pipe cost?
   c. Solve by auxiliary views; check by rotation.

7. See Fig. 7-11. A guy wire is attached to each of the tower corners as indicated in the two given views. Each wire has the same slope angle. Find the total length of wire needed to support the tower. Allow 18" at each end for fastening. Scale: 1" = 30'.
   Solve only by rotation.

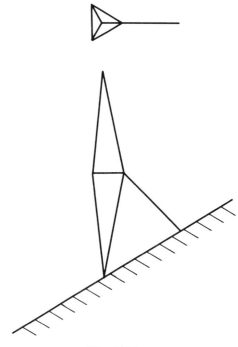

Fig. 7-11.

8. List the 7 possible classifications of lines in space.

unit 8

Orthographic Projection - Planes

8.1 Uses for Plane Areas

It is interesting to contemplate the many engineering applications of areas or planes. The following is a list of some common and some uncommon types of plane applications.

Ground planes, level and sloping, on land and underwater, especially in ocean bed exploration.
Underground strata used in mining and excavation.
Sloping and level surfaces on machines and structures.
Roof planes.
Vertical and sloping walls and mirrors.
Abutments and retaining walls.
Any area encountered which can be delineated by one of the four methods given in Art. 8.2.

8.2 Definition and Representation of a Plane

Definition: *A plane is a flat area on which any two points may be connected by a straight line lying wholly on the surface.*

Any of the following combinations can be used to graphically delineate a plane:

1. Three points, not in a straight line.
2. A straight line and a point not on the line.
3. Two intersecting lines.
4. Two parallel lines.

Some plane surfaces are limited by definite boundaries while others extend infinitely. Fig. 8-1 illustrates a space picture and orthographic projections of planes limited by a triangular area, intersecting lines, and parallel lines.

PRINCIPLES.

1. *If two lines intersect thus forming a plane, the projected images of their intersection will align in any pair of adjacent views.* Note Fig. 8-1.
2. *If two lines are parallel in space, the projection of the lines will be parallel in every orthographic view. Also, if one of the two lines is projected to a point view, the other will also appear as a point in the same projection.*

8.3 Classification of Planes

Planes can be classified in the same manner as lines, namely: a) horizontal, b) front, c) vertical, d) profile, e) oblique. The student should be able to identify each plane from its orthographic projections. See Fig. 8-2.

8.4 Graphical Specifications for a Plane

Comparable to lines, planes are specified by combinations of the following information:

1. Coordinates of three non-linear points lying on the plane.
2. True size and shape of the plane area.
3. Bearing or Strike of the plane.
4. Slope or Dip of the plane.

ORTHOGRAPHIC PROJECTION—PLANES

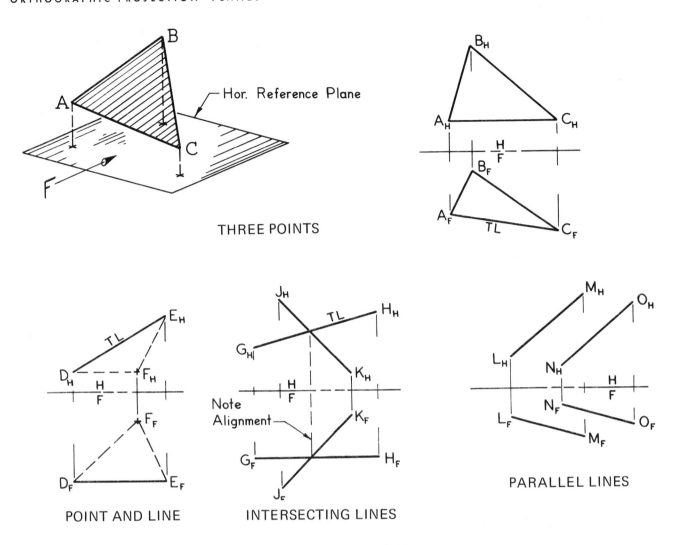

Fig. 8-1. Definition of Planes.

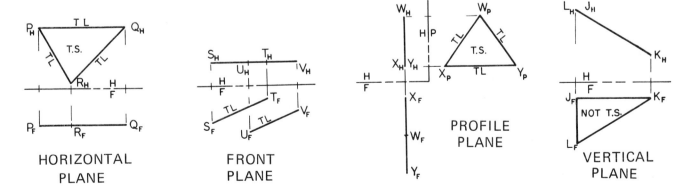

Fig. 8-2. Classification of Planes.

8.5 Points and Lines on a Plane

SPACE ANALYSIS. *A line is located upon a plane by aligning its intersections in adjacent views with two or more lines on the plane.* Reference: Principles 1 and 2, Art. 8.2.

To locate a horizontal line on a plane, first locate its projection in an elevation view of the plane, as in Fig. 8-3. The horizontal projection of the line, which will be in true length, can be readily located by orthographic means. See line DX in Fig. 8-3a.

To locate a front line on a plane, first locate its horizontal projection which is always a constant distance behind the front plane. The intersection of the front line with the given plane's outline in the horizontal view can be projected to any desired view as in Fig. 8-3b. Profile lines may be located in a like manner, Fig. 8-3c.

8.6 Location of a Point on a Plane

SPACE ANALYSIS. *A point is located upon a given plane by fixing a line on the plane which contains the point.*

The method of solution for a point location is to pass a straight line, lying on the plane, through a given projection of the point. By application of the first principle stated in Art. 8.2, any adjacent

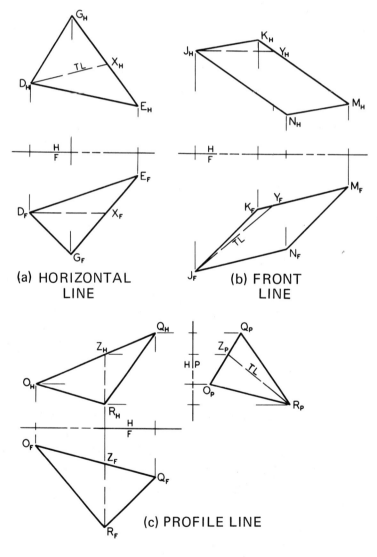

Fig. 8-3. *Lines on a Plane.*

ORTHOGRAPHIC PROJECTION—PLANES

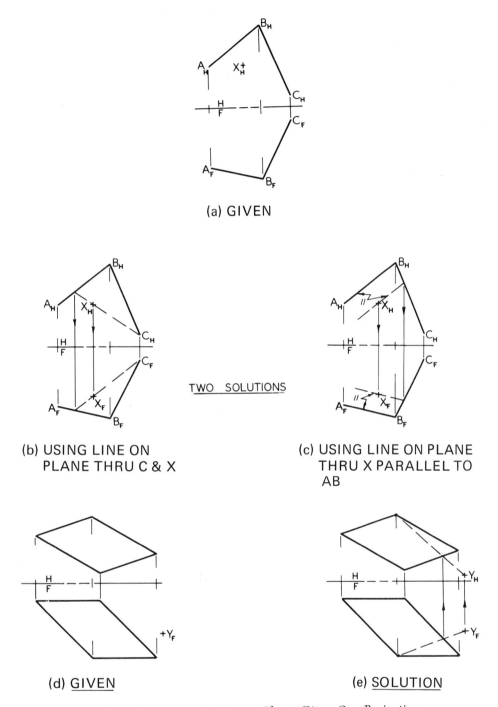

Fig. 8-4. *Locating a Point on a Plane, Given One Projection.*

view of the line on the plane can be located and therefore the corresponding projection of the point can be found. This is shown in Fig. 8-4; study carefully.

8.7 Edge View of a Plane

SPACE ANALYSIS. *The edge view of a plane is found in a view where any line in the plane appears as a point.*

A plane may be considered to consist of a series of parallel lines which will appear as a straight row of points if the lines are viewed from the end. This straight line of points will be the edge view of the plane. The method for viewing a line as a point is described in Art. 7.10. Therefore, to obtain an edge view of a plane, the following steps are required. See Fig. 8-5.

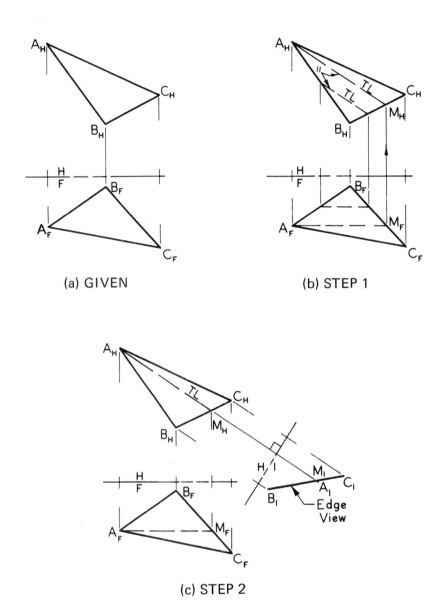

Fig. 8-5. *Edge View of a Plane.*

Step 1. Establish a line on the given plane that is parallel to one of the planes of projection (H, F, or P). Find the true length of this line. (Line AM in Fig. 8-5b).

Step 2. Project the given plane onto a plane of projection that is perpendicular to the true length projection of the line, view 1 in Fig. 8-5c.

Caution. Less confusion will result if points and lines are carefully labeled *as they are drawn*.

8.8 True Size and Shape of a Plane

SPACE ANALYSIS. *The true size and shape of a plane is found in a view whose line of sight is perpendicular to any edge view of the plane. (A view whose hinge line is parallel to edge view.)*

ORTHOGRAPHIC PROJECTION—PLANES

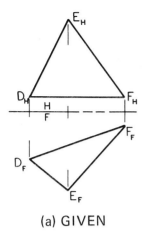

(a) GIVEN

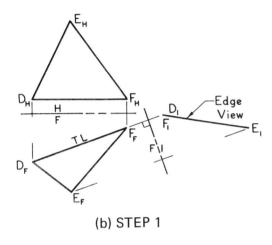

(b) STEP 1

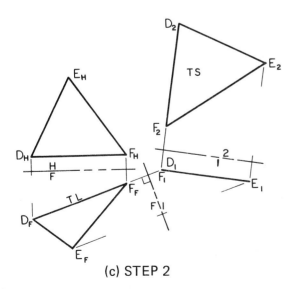

(c) STEP 2

Fig. 8-6. *True Size and Shape of a Plane.*

Two steps of procedure are necessary to locate the true size view of a plane.

Step 1. Find the edge view of the plane, Fig. 8-6b.

Step 2. Project the given plane onto a plane of projection (plane 2) that is parallel to the edge view of the given plane, Fig. 8-6c.

8.9 Edge View and True Size of a Plane by Rotation

The edge and true size view of a plane can also be obtained by rotating the given plane to the positions for viewing described in Arts. 8.7 and 8.8. The procedure necessary for edge and true size by rotation is demonstrated in Fig. 8-7.

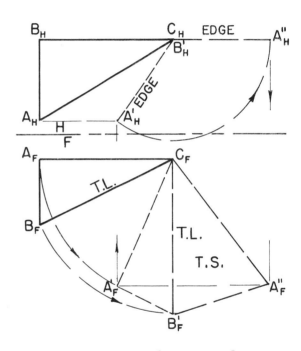

Fig. 8-7. *Edge View and True Size by Rotation.*

8.10 Bearing or Strike of a Plane

Definition: *A Bearing or Strike Line is a Horizontal line lying in the plane.*

SPACE ANALYSIS. *The Bearing or Strike of a plane is found in the horizontal projection as the ACUTE angle of bearing of the STRIKE LINE measured from the North.*

As explained in Art. 7.7, bearing of a line is always measured in the horizontal view. Because a horizontal line lying on a plane does not have a specific point of origin, the North-South reference can be taken on the horizontal projection of the strike line at any convenient point for measuring the acute angle with North. Fig. 8-8 shows the strike specifications for two different planes.

8.11 Slope or Dip of a Plane

Definitions: *The Slope Line of a plane is the steepest downhill line (line of maximum declination) lying on the plane.* It is perpendicular to the *Strike Line* (Art. 8.10) and directed toward the side of lower elevation.

The SLOPE DIRECTION of a plane is the compass quadrant toward which the Slope Line points.

DIP is the Slope Angle of the plane expressed with the Slope Direction. Example: 32°SE.

SPACE ANALYSIS. *The Slope or Dip Angle of a plane is found in the true length elevation view of the slope line. The Slope or Dip Direction is found in the horizontal projection as the quadrant direction toward which the slope line points.*

In Fig. 8-9a the strike and slope lines are located and labeled. It is important for the student to understand not only how but why the strike and slope lines are located and as well be able to use them to determine the slope or dip of a plane. Refer to Figs. 8-9 and 8-11.

The plane's slope or dip can likewise be found by rotation. The student should carefully follow the steps in Fig. 8-10.

8.12 Points, Lines, and Planes Compared

At this point the student should have a firm grasp of the orthographic relationship of points, lines, and planes in space. He should be able to find the projection of any desired series of points and label each in a uniform manner consistent with standard practice. In addition, he should be able to locate, recognize, or solve for the following items of information by orthographic views or by rotation.

The Point	*The Line*
Coordinate location	Location
Principal views	True length
Auxiliary elevations	Bearing
Auxiliary inclined views	Inclination
Oblique views	Point view

The Plane

Point or line in plane
Edge view
True size and shape
Bearing or Strike
Slope or Dip

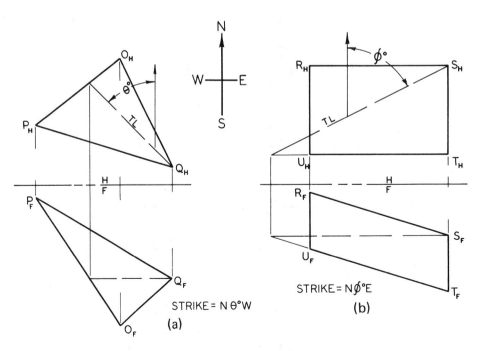

Fig. 8-8. *Strike of a Plane.*

ORTHOGRAPHIC PROJECTION—PLANES

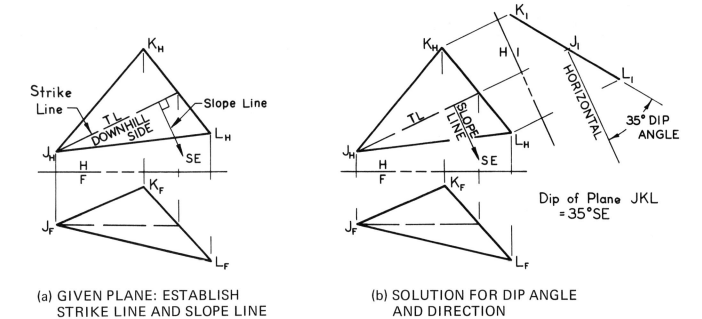

(a) GIVEN PLANE: ESTABLISH STRIKE LINE AND SLOPE LINE

(b) SOLUTION FOR DIP ANGLE AND DIRECTION

Fig. 8-9. *Dip Angle and Direction.*

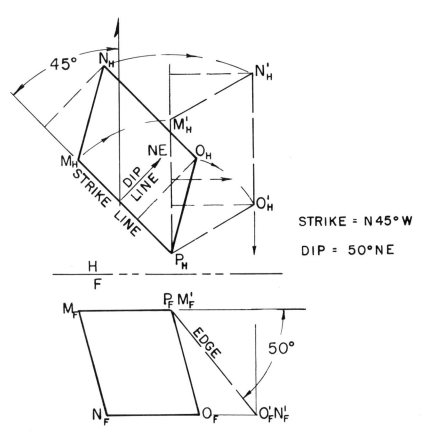

Fig. 8-10. *Dip Angle by Rotation.*

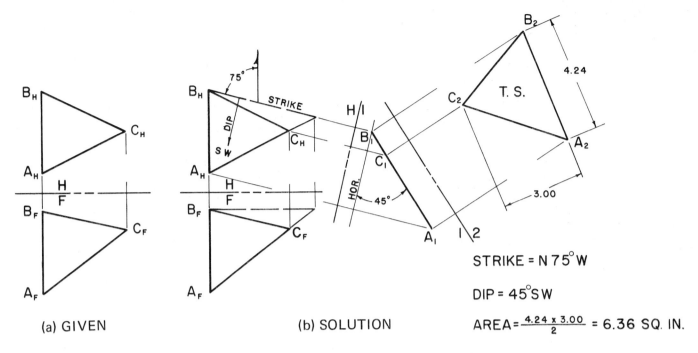

Fig. 8-11. *Strike, Dip, and True Size of a Plane.*

8.13 Application Problem—Points, Lines, and Planes

Problem. The following measurements have been taken to determine the necessary data for a roof replacement. The pictorial drawing in Fig. 8-12 identifies the points.

Location of A: 20 ft. east of a N-S lot line, 44.5 ft. north of curb line, and 25 ft. above the level ground.

Location of C: 65 ft. east of the lot line, 30.5 ft. north of the curb, and at the same elevation as A.

Points D, E, J, K, L, M are all 10 ft. above the level ground.

Ridge line AB has a bearing of N 60° E and CB is N 30° W.

End of roof ADE is vertical and has a strike of N 30° W and span of 40 ft.

End of roof CJK is vertical and strikes N 60° E with a 50 ft. span.

Adjacent roofs to a ridge line have equal pitch.

Pitch of a roof is the ratio of rise over *total span*, expressed for a rise of 1 unit. Example: 1/4 pitch.

Information needed for the roof replacement:

1. Area of each roof plane, reduced to the total number of squares of roofing (100 sq. ft.). Add 10% for waste.
2. Pitch of all roof planes.
3. Bearing, slope angle, and length of the Valley, BM, and Hip, BL.

The following list shows the order of procedure to solve this problem from the given data.

Steps:
1. Locate the H project of the lot line and curb line.
2. Locate H and F projections of A and C from the given coordinates using an appropriate scale.
3. Locate Ridge lines AB and CB in H and F.
4. Locate H position of end planes ADE and CJK.
5. Find D, E, J, M, K, L in the front elevation.
6. Using the given span, find the true size views of roof ends ADE and CJK.
7. Complete and letter all points on the plan and elevation views of the roof.
8. Find the four roof planes in true size.
9. Find the bearing, slope, and true length of the valley, BM, and hip, BL.
10. Measure and compute as necessary. Tabulate results.

ORTHOGRAPHIC PROJECTION—PLANES

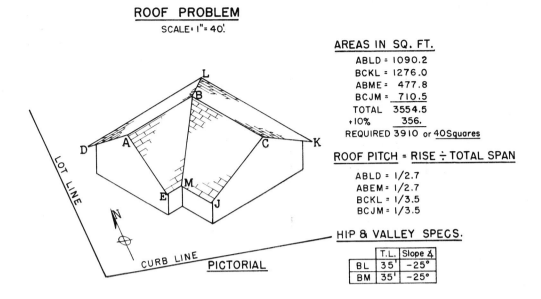

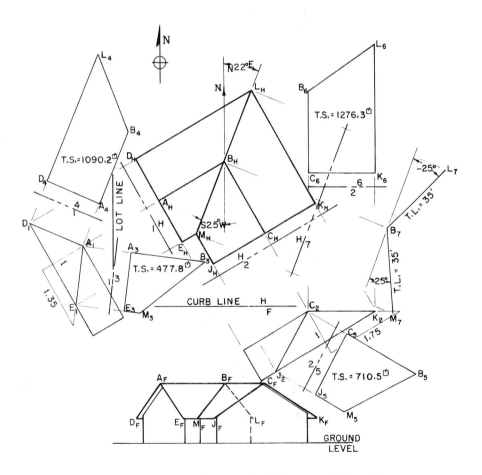

Fig. 8-12. *Application Problem—Roof Specifications.*

PROBLEMS

The problems of this unit may be redrawn at double scale or the student may wish to render a freehand facsimile of them on suitable paper. In either case, space should be provided for the views necessary to determine the required information. Problems 1, 2, and 3 may be worked directly on the page if desired. It is strongly recommended that all work be properly labeled and that important characteristics be identified.

1. Which of the three figures define planes? Sketch construction necessary to prove your answers. Fig. 8-13.

2. Sketch the construction used to locate the missing view of the lines or points which are in the given planes. Fig. 8-14.

3. Indicate freehand the strike and dip of the given planes. Identify the angles used and estimate their values. Fig. 8-15.

4. Find the strike and dip of the given planes. Fig. 8-16. Find dip angle by rotation.

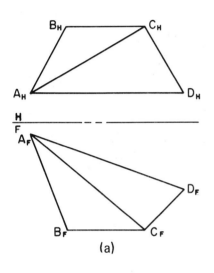
(a)

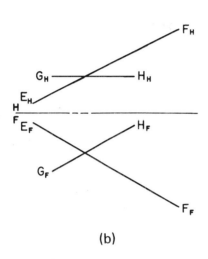
(b)

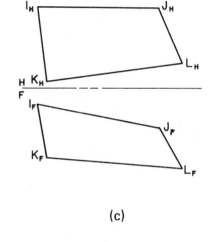
(c)

Fig. 8-13.

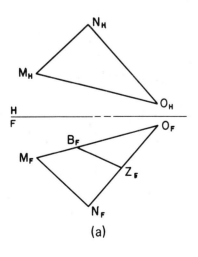
(a)

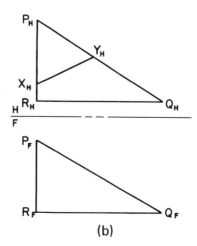
(b)

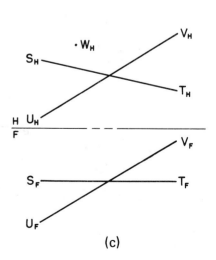
(c)

Fig. 8-14.

ORTHOGRAPHIC PROJECTION—PLANES

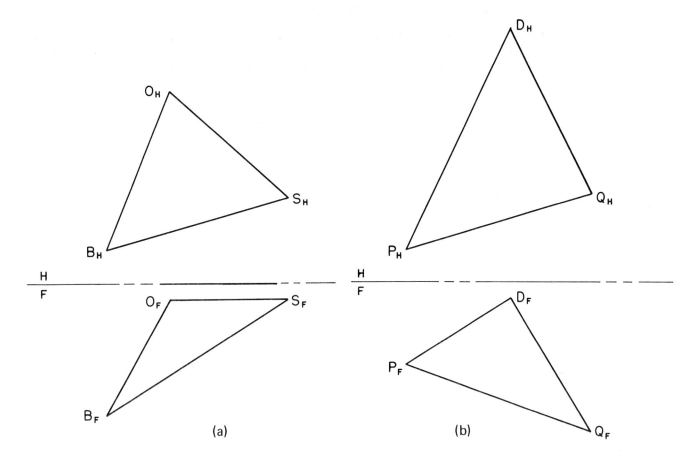

Fig. 8-15.

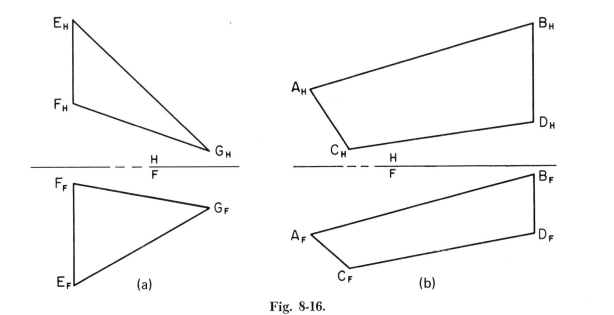

Fig. 8-16.

5. a. Find the strike, dip, and true size of the plane JKL.
   b. Locate in all views point M which lies in the plane, within the boundaries of JKL, one inch from points J and K. Fig. 8-17.

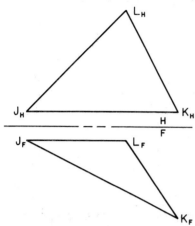

Fig. 8-17.

6. Point R lies in plane XYZ. Find the true angle between RY and RZ. Fig. 8-18. Use only the two given views with rotation.

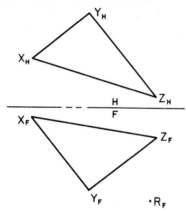

Fig. 8-18.

7. Find the H and F projections and the strike and dip of plane MNOP. Fig. 8-19.

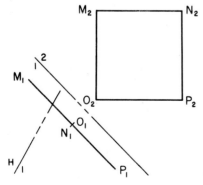

Fig. 8-19.

8. Given plane ABC. Find the true length, bearing, and slope angle of a line CD perpendicular to AB. Fig. 8-20.

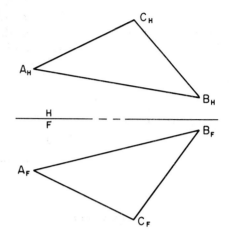

Fig. 8-20.

9. Point S is 74 in. from a plane whose strike is N 30° E and dip is 45° SE. Line SA has a bearing of S 60° E and is a horizontal line. Point A is in the given plane. How long is line SA? Solve using only rotation.

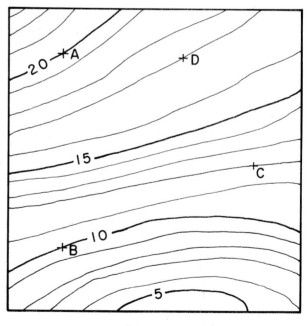

Fig. 8-21.

10. Locate the H and F views of a plane whose strike is N 60° W and dip is 40° NE and

whose shape is an equilateral triangle 2 in. on a side.

11. The contour map shows the location of bore-holes at points A, B, and C. Ore is found at depths of 5, 15, and 20 feet respectively below ground level. If the ore is a smooth strata, at what depth will a bore-hole strike ore at D? What is the strike and dip of the ore vein? Fig. 8-21.

12. The roof plan, Fig. 8-22, shows five distinct areas. The slope of roof planes A and B is 0.5. Find the slope of planes C, D, and E. Find the total roof area. Refer to the solution of Fig. 8-12.

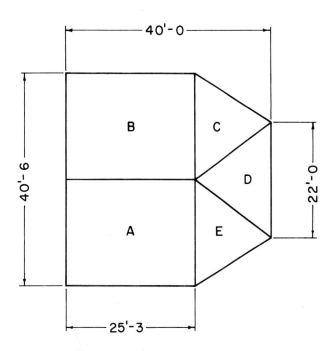

Fig. 8-22.

unit 9

Orthographic Projection – Solids

9.1 Principal Views of Solids

Orthographic projections of *solid objects* are delineated in a manner similar to that used for the point, line, or plane. The student should have no difficulty with the projection of solids if he understands the principles of orthographic projection.

In the previous units, the top, front, and profile or side were the only principal views discussed in detail. However, it must be emphasized that the projections upon any of the six planes, Top, Front, Right Side, Left Side, Bottom, or Back, are called the principal views. See Figs. 9-1 and 6-2.

One or more of these six principal views, supplemented by auxiliary views, are used to portray complete shape description of a solid object. Two or three of the principal views are usually sufficient to define the shape of simple objects.

9.2 Position of Views

As indicated in Art. 6.6, the profile or side view may be either adjacent to the front or top views. Reasons were advanced for having a true length view of a line adjacent to the top view when the inclination of the line was needed. However, with solids the most conventional position for a side view is adjacent to the front because the reader of the drawing can better observe the height dimensions in profile when they are parallel to the corresponding heights in the front view.

All six principal views of the Block are shown in Fig. 9-1b to illustrate the alignment position of each view, should more than two or three views be needed. For this problem the top, front, and right side views show complete shape description. If sufficient word specifications were given, two views, top and front or front and right side, would be sufficient.

9.3 Choice of Views

The purpose of an orthographic drawing of a solid object is to completely describe the shape of the object in a *minimum number of views*. The number of essential views depends upon the intricacy of the object and may vary from one to three on simple objects while complicated designs may require more views in order to show complete detail. In particular, it may be necessary to show several sectional views to fully describe the interior of an object. Sectional views and their uses are described in Unit 10.

ORTHOGRAPHIC PROJECTION—SOLIDS

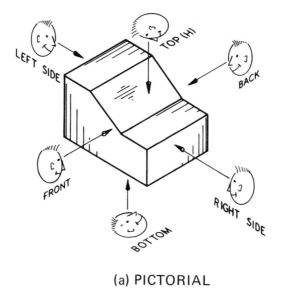

(a) PICTORIAL

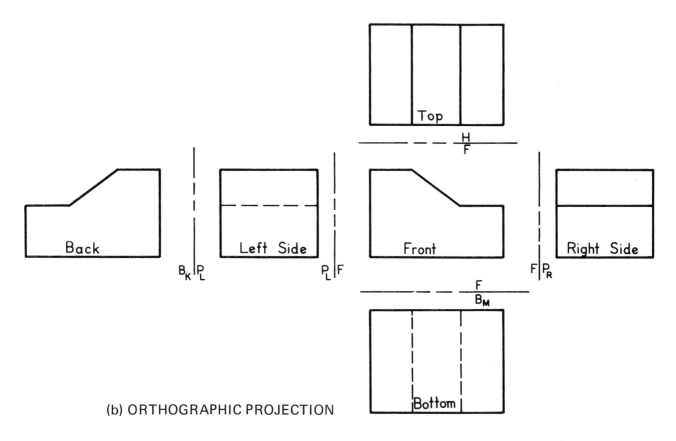

(b) ORTHOGRAPHIC PROJECTION

Fig. 9-1. *Position of Orthographic Views.*

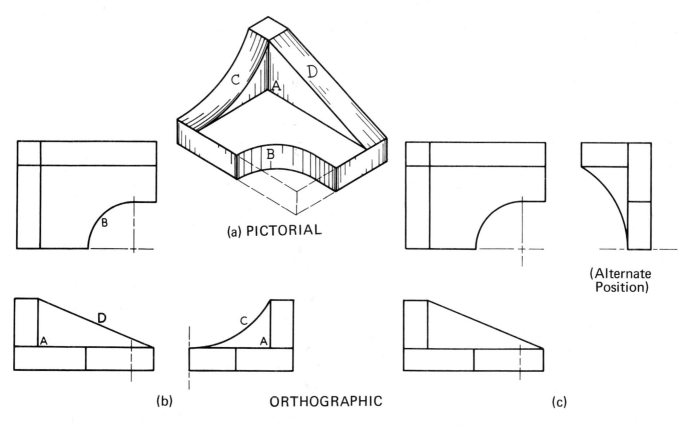

Fig. 9-2. *Shape Description—Wedge Block.*

Referring to Fig. 9-2b and Fig. 9-2c, several important points should be noted:

1. The relationship of aligned views, one to another, is identical to that previously studied.
2. Three views are used to fully describe the object because of *critical contour* in each of the three planes.
    a. The front elevation view shows critical contour of the object with respect to the corner A and the sloping wall D.
    b. The horizontal projection or top view is needed not only to show depth but also to describe the curved detail at B.
    c. The right side elevation or profile view shows the curved surface at C. Note that the detail of C is not accurately portrayed in the top and front views.
3. A right side view is preferred to a left side in this case because more detail is visible from the right side.
4. Hinge lines are omitted.

9.4 Necessary Views

As previously stated, the object should be completely described in a *minimum number of views*. The three dimensions of length, width, and height can only be completely described in a minimum of two orthographic views because any single projection will show no more than two dimensions. For example, the front view shows length and height; the top view shows length and width; the side view shows height and width.

Industry is anxious, wherever possible, to reduce drafting cost by using what is often referred to as minimum line delineation. In other words, never waste time to make an unnecessary view.

PRINCIPLE. *An object requires only the views necessary to show the shape of each intricate contour.*

Examples:
1. There are many simple objects whose basic geometrical shapes can be completely described by a short word description with no ortho-

ORTHOGRAPHIC PROJECTION—SOLIDS

graphic views necessary. For example, a sphere of given diameter, a cylinder of given diameter and length, etc. Fig. 9-3 shows an object with critical contour in only one plane and therefore can be described by one view and a supplementary note stating its thickness.

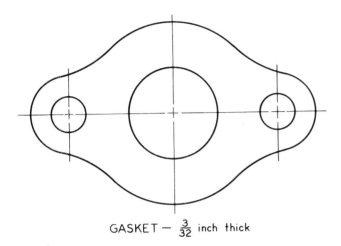

GASKET — $\frac{3}{32}$ inch thick

Fig. 9-3. *Single-View Drawing.*

2. An object which can be described satisfactorily by two views is shown in Fig. 9-4. Note the critical contour in each view.

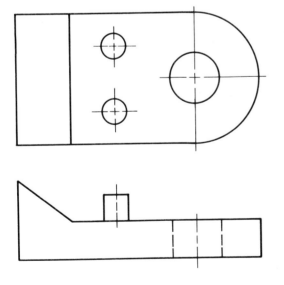

Fig. 9-4. *Two-View Drawing.*

3. Fig. 9-2 described the Wedge Block which has critical contour showing in three principal projections.

4. Examples will be encountered later that require more than three views for complete shape description. Fig. 9-11 shows an object needing five partial views.

9.5 Line Conventions

Visible Lines or outlines represent visible edges or contours of objects. Visible lines, in contrast with dimension and other secondary lines, should be drawn so that the views will stand out boldly on the drawing.

Hidden Lines are used to show the hidden features of an object. They are slightly finer than visible lines and consist of a series of dashes that have a minimum length of .12 inch, with short spaces of about .04 inch. A larger drawing will have longer dash lengths and spacings. Fig. 9-5a, b, c, and d show standards of hidden lines intersecting visible and other hidden lines. Hidden lines are sometimes omitted if their use is not required for the clarity of the drawing.

Center Lines. Center lines are drawn for all arcs and circles to register the axes of symmetry, to fix the center points for the circles and curves, and for dimensioning purposes. Center lines are represented by fine lines consisting of alternating long and short dashes. The long dashes vary in length from .75 inch to 1.50 inches or more, depending on the size of the drawing. The short lines should be about .12 inch long with spaces at a minimum of .06 inch. They should extend uniformly, a minimum of .25 inch beyond the object or feature of the drawing. Nearly all of the illustrations in this unit show center lines.

The principal views of positive or negative cylinders are the circle and the rectangle. Note in Fig. 9-5e that two center lines *always* intersect at 90° in the circular view, extending through the circle, and one center line *always* extends along the axis of the cylinder in the rectangular view.

Line Precedence. Center lines, hidden lines, and outlines are sometimes superimposed in an orthographic view. Outlines take precedence over center lines and hidden lines. Hidden lines take precedence over center lines. See Fig. 9-5f.

Fig. 9-5 should be studied carefully to become familiar with the line conventions.

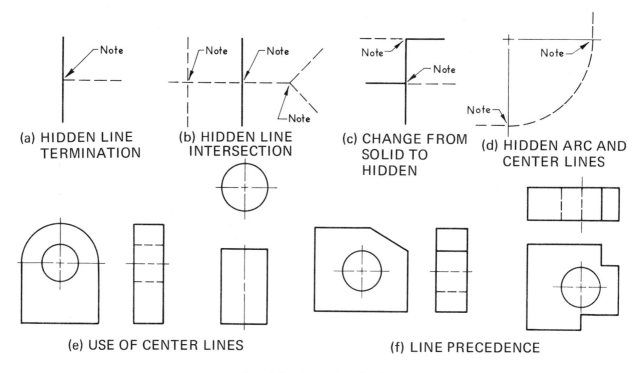

Fig. 9-5. *Line Standards.*

9.6 Order of Drawing or Sketching

Not only are accuracy, proper spacing, neatness and line quality essential to a drawing, but equally important are the rapidity and efficiency of its construction. Accuracy, appearance, and speed are accomplished by continuous attention to the following:

1. *Knowledge* of standards and basic procedure.
2. *Advance thinking,* visualization, and planning. In other words it is necessary to plan in advance a logical order of steps to be followed in drawing or sketching.
3. *Efficient handling of the drawing tools.* This includes habits of neatness, organization and placing of papers, books, triangles, sharp pencils, erasers, etc., to avoid loss of time in their manipulation.

In making multiview drawings of solid objects, or for that matter, any type of drawing or sketch, certain helpful procedures can be followed.

1. *Plan the sheet.*
   a. Draw a border and title strip layout if necessary.
   b. Decide on number of views, arrangement and kind of views needed.
   c. Fix the scale and position of the component parts of the drawings with a minimum amount of calculation or estimation.
2. *Locate center lines or base lines in all views* from which the details will be measured. It is important to skeleton *all views* of a drawing before trying to complete individual parts. See Fig. 9-6b.
3. *Block in the basic skeleton outlines* very lightly, starting with the contour view. After locating center lines, circles and arcs can be drawn as finished outlines without blocking in. Notice the lines which have been omitted in the skeleton on Fig. 9-6c.
4. *Sharpen visible outlines and add hidden outlines and specifications* as necessary for a finished drawing. See Fig. 9-6d.
5. *Hinge lines* are omitted on simple multiview drawings.

9.7 Reading a Drawing

The mental process of arriving at a complete understanding of every detail of the object by composite study of the several views is known as reading the drawing. *Reading is essentially the reverse of the process used to make the drawing.*

ORTHOGRAPHIC PROJECTION—SOLIDS

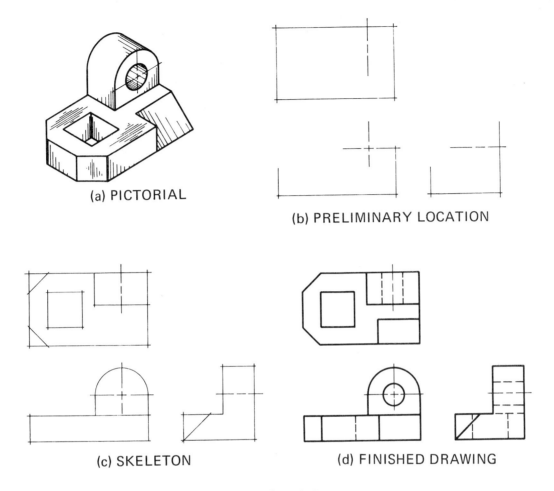

Fig. 9-6. *Order of Drawing.*

In drawing the descriptive views the writer has a mental image of the object, the details of which he describes. In reading the drawing, the interpreter must put together the details of the component parts of the object to form his mental image.

Each line on the drawing represents, a) a plane or curved area as an edge, b) the intersection of two areas, or c) the boundary or limit of an area. The reader should first obtain an overall idea of the basic shape of the object by relating the given views. He then breaks down the details by reading the contours as he refers to the views for a clear understanding of the relative size and shape of component parts.

A Picture Sketch. To assist in reading a drawing, a pictorial sketch is often employed. The development of a picture follows the steps of reading and assists in clarifying the details. Fig. 9-7 shows the steps for making the picture from the orthographic drawing using the basic steps given in Art. 3-5. Pictorial drawing is more completely described in Unit 10.

Models. Modeling clay or soap can be used to help the student visualize a drawing. The modeling procedure is very similar to that described for making a picture sketch.

Completion Drawings. The ability to read a drawing may be improved by completing partially drawn orthographic views. Good reading practice is obtained by the addition of extra views or the missing lines of incomplete views. More than one solution is often possible in a missing view problem if critical contour has not been completely described in the given projections.

9.8 Auxiliary Views

Some objects may have an intricate contour which cannot be completely described on one of

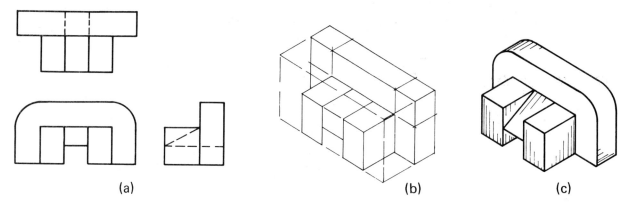

Fig. 9-7. *Reading a Drawing.*

the principal planes of projection. Therefore, depending on the detail, an auxiliary elevation, an inclined auxiliary view, or an oblique view taken from an auxiliary may be necessary. The student will find that the desired view may be easily obtained if he applies the theory of projection to show the *true size of a plane*, Art. 8.8.

Auxiliary Elevation Views. If a plane area, which is to be shown in true size to explain its detail, appears as an *edge* in the top view, the projection of the area upon a plane of projection parallel to the edge view will show the plane in true size. Fig. 9-8 shows a geometrical solid which needs an auxiliary elevation view to give the true area of part of the cutaway surface. Fig. 9-9 shows an object which needs an auxiliary elevation for complete description.

Auxiliary Inclined Views. If an area appears as an edge in a front or side elevation, a true-shape inclined view may be obtained upon a plane of projection that is parallel to the edge. See Fig. 9-10 and Fig. 9-11.

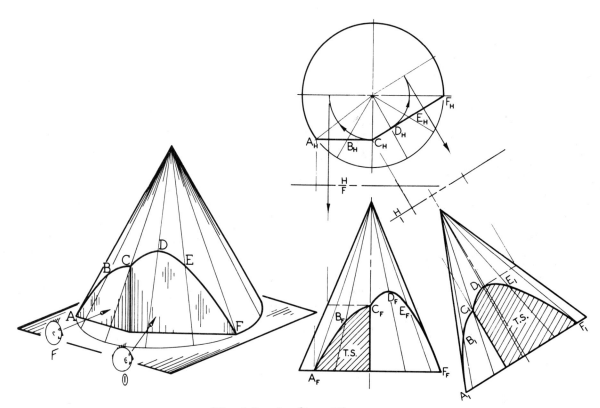

Fig. 9-8. *Auxiliary Elevation.*

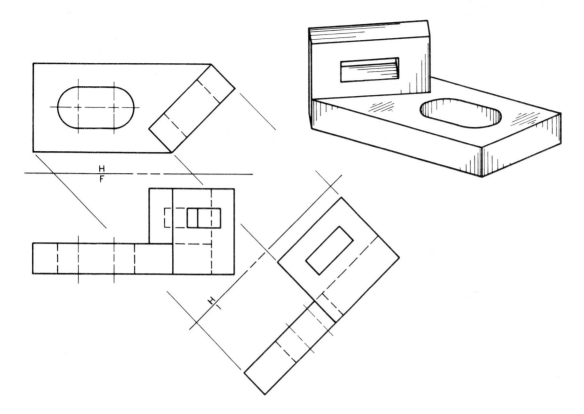

Fig. 9-9. *Auxiliary Elevation.*

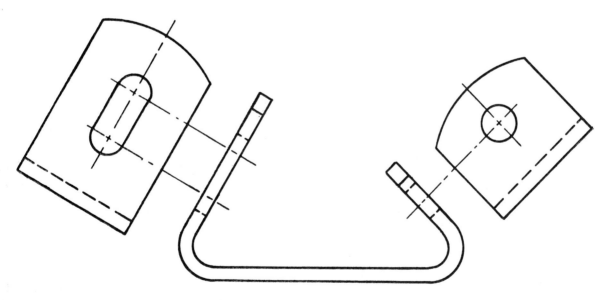

Fig. 9-10. *Partial Auxiliary Inclined Views.*

Oblique Views. If a contour face does not appear as an edge in one of the principal views, it will be necessary to obtain its edge view before its true size can be shown. An auxiliary elevation (partial) has been drawn in Fig. 9-11 to show the inclined face as an edge. From the edge view the true size and shape of the area may be found. Note that hinge lines and view numbers have been omitted.

Partial Views. For the sake of clarity, ease of drawing, and time consumed, views are sometimes left incomplete if the particular detail omitted has been described in one of the other views. In Fig. 9-11, a portion of the object was omitted in each of the views.

9.9 Conclusions

Shape description orthographic views are the type used for the production and fabrication of all kinds of parts, materials, and structures. A facility for reading and writing such drawings is of prime importance to the engineer. Such ability is accomplished and perfected by solving a number of problems of progressive difficulty. This will be the procedure as the many standards, methods, and specialties of the graphical language are developed.

Pictorial drawing affords an excellent aid for reading and visualizing the relationship of adjacent surfaces and particular shapes of an object. It will be found helpful to make pictorial sketches while developing the orthographic projections of solid objects.

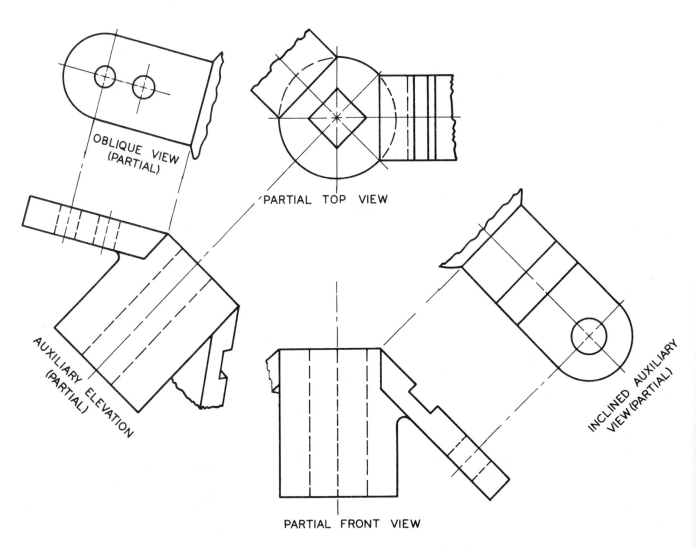

Fig. 9-11. *Partial Views.* (ANSI Y14.3)

PROBLEMS

The following problems on solids may be drawn to scale or sketched as reasonable facsimiles. Observe all principles of Orthographic relationship and alignment.

1. Show the necessary Orthographic Views of one or more of the following familiar objects. Use reasonable proportions and supply a title for each drawing. Include a thumbnail sketch.
   a. Hexagonal-based Pyramid
   b. Square-based Wastebasket
   c. Circular Ashtray
   d. Plastic Case for Table Radio
   e. Frame for a Deck Calendar
   f. Stop Sign mounted on Square Post
   g. 90° Shelf Bracket
   h. Door Knocker
   i. Rural Mail Box
   j. Bowling Pin
   k. Coffee Cup
   l. Hacksaw
   m. Semicircular Protractor
   n. Three-step Entranceway with Hand Rail
   o. Watering Can with Spout
   p. Roller Skate

2. Sketch the orthographic views necessary to completely describe the shape of each object, Figs. 9-12 and 9-13. Hidden outline should be included as well as all necessary center lines.

3. Completion problems. In Fig. 9-14, complete views, add lines, add views as indicated in (a), (b), (c), (d), (e).

4. a. Given the two related views, sketch an additional orthographic view necessary to completely describe the shape of each solid object, Figs. 9-15, 9-16, 9-17. Some problems may have more than one solution.
   b. Make a pictorial sketch of each object.

5. a. Given the front view only of each object, draw the horizontal and profile projections, Fig. 9-18. Many answers are possible for each.
   b. Given each drawing in Fig. 9-18 as a horizontal view, draw the front and profile projection of each object.
   c. Sketch a pictorial of each object drawn in (a) and (b).

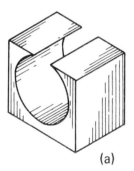 (a) 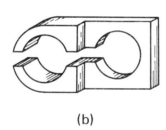 (b) 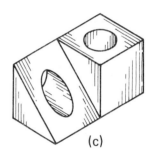 (c)

Fig. 9-12.

 (a) 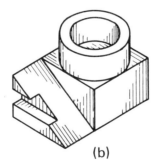 (b) 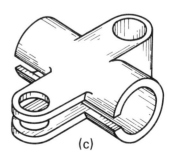 (c)

Fig. 9-13.

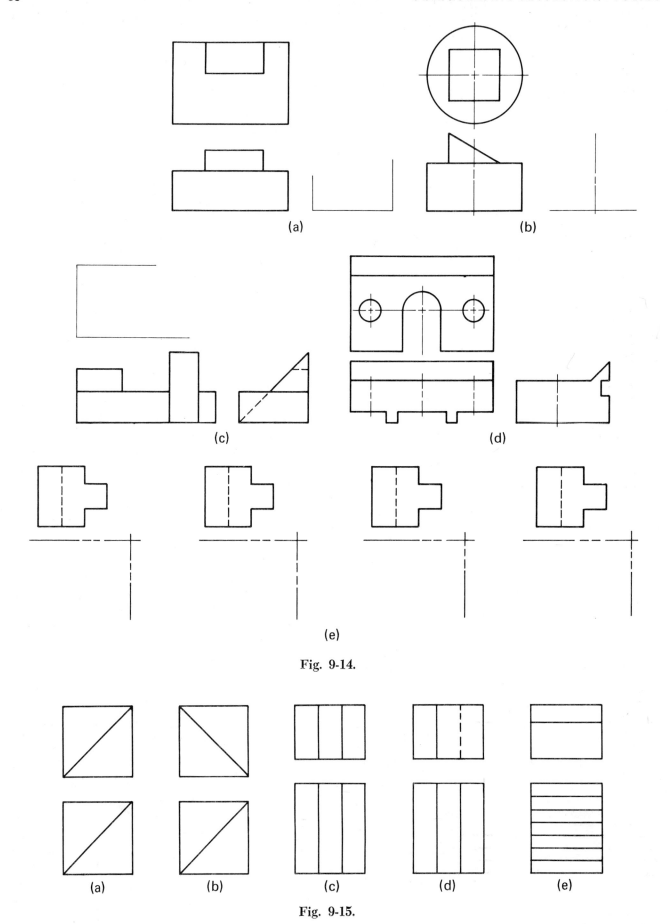

Fig. 9-14.

Fig. 9-15.

ORTHOGRAPHIC PROJECTION—SOLIDS

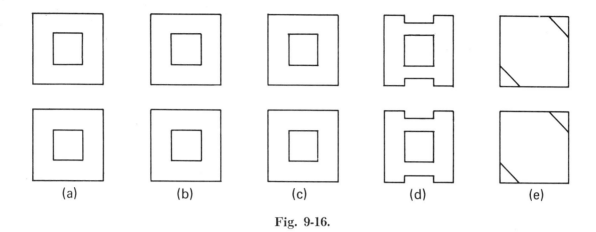

Fig. 9-16.

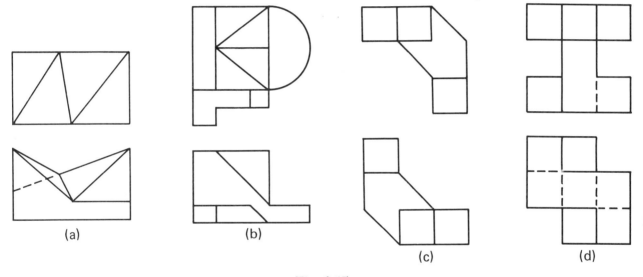

Fig. 9-17.

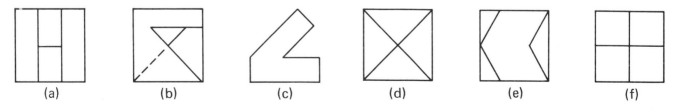

Fig. 9-18.

unit 10

Pictorial Systems

10.1 Pictorial Drawing Systems

Pictorial drawings, the oldest form of written communication known to man, are used to convey ideas; to explain complicated engineering drawings to those who do not have the ability or patience to read multiview drawings; to assist the designer with his visualization problems; to illustrate catalogs and service manuals; to explain machine assembly; and to assist a student in developing his power of visualization. The system of pictorial to be used is determined by its intended purpose. The following pages describe the systems in three parts: Part I, *Axonometric*, Part II, *Oblique*, and Part III, *Perspective*.

Part I
AXONOMETRIC

10.2 Axonometric Projection

Definition: An orthographic view in which all three faces of a rectangular object are neither parallel or perpendicular to the image plane is called an Axonometric Projection.

10.3 Types of Axonometric Projection

Isometric Projection. If a rectangular object is positioned with the three principal faces and the principal axes (X, Y, Z) at *equal angles* with the plane of projection, the orthographic image will be an isometric projection.

Dimetric Projection. If only two of the three principal faces and two principal axes of an object make equal angles with the plane of projection, the resulting image is a dimetric projection.

Trimetric Projection. When an object is positioned so that all three principal faces and principal axes form unequal angles with the plane of projection, the orthographic image will be a trimetric projection.

Fig. 10-1 illustrates an isometric projection. The construction should be easily followed from the

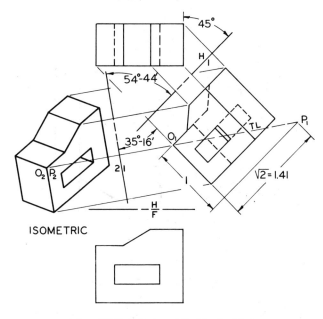

Fig. 10-1. *Isometric Projection.*

labeled drawings. It is obvious that the amount of details involved for an axonometric projection of a complicated object would become very tedious. However, an *approximate projection* can be made directly by measurements taken from the orthographic shape description. The image resulting from this operation is called an *axonometric drawing*.

10.4 Isometric Drawing

A full scale illustration to distinguish an isometric projection from a isometric drawing of a one inch cube is shown in Fig. 10-2. The obvious difference between the two is in the size of the drawing caused by the foreshortening of axis lengths in isometric projection.

The fact that *Isometric* (equal measure) axes make equal angles (120°) with each other and that measurements are laid off at the same scale on each axis makes Isometric Drawing relatively realistic and therefore popular to use. The only dimensions that are true length on the isometric drawing are the values of length, width, and height along or parallel to the respective axes. *No angle within the object can be constructed at true size in an Isometric Drawing.*

10.5 Positions of the Axes

Normal position of axes, the one most commonly used in isometric drawing, is shown in Fig. 10-3a. This position is appropriate for an object that has considerable detail in two or three of the principal planes and when a view of the top area is desired.

The *Reversed* axes position of isometric drawing shows details on the bottom of the object. In some instances the object can be turned over and drawn with normal axes and still show all necessary details. If, however, an object such as the shelf shown in Fig. 10-3b has an unnatural appearance when turned over, yet details on its under side need illustrating, the reversed axes position should be used.

The *Horizontal* position of axes produces a less distorted pictorial of an object having one principal dimension considerably greater than the other two, Fig. 10-3c.

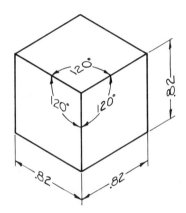

ISOMETRIC PROJECTION

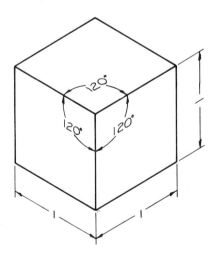

ISOMETRIC DRAWING

Fig. 10-2. *Comparison of Isometric Projection and Drawing.*

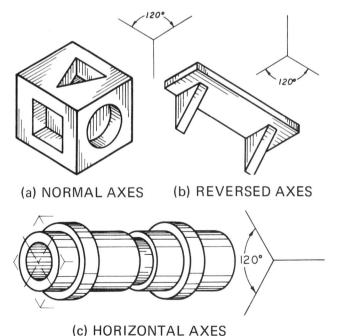

(a) NORMAL AXES   (b) REVERSED AXES

(c) HORIZONTAL AXES

Fig. 10-3. *Position of Isometric Axes.*

10.6 Basic Isometric Construction

To obtain the basic skeleton of an isometric drawing, the object may be enclosed in a box whose dimensions are equal to the overall length, width, and height of the object. Some objects lend themselves to a combination of two or more box outlines when the shape is such that the object can be divided into more than one component part. The skeleton construction for the object in Fig. 10-4 is composed of two boxes. Before constructing the axes of an isometric skeleton box, careful *planning* is necessary. The following *steps of procedure* illustrate the considerations which are involved.

1. Select the position of axes: normal, reversed, or horizontal.
2. Decide which corner of the object should be turned to the front. This in general should be the corner with the *least height*. It is sometimes desirable to make several "thumbnail" sketches to select the best position. See Fig. 10-4b.
3. Fix the scale or select a size of sheet to best show the object in isometric drawing.
4. Draw lightly the skeleton box into which the object will fit. It requires good judgment and some experience to center the object on the sheet. Sometimes the skeleton box may consist of several boxes wherein the details of the object are broken down into separate parts.
5. Plot the details of the object, bearing in mind that dimensions of length, width, and height *only* can be laid on or parallel to the three axes. The visible outline should be sharpened leaving, very lightly, the construction used to obtain any of the details. Invisible outline is generally omitted from the pictorial unless absolutely necessary to convey some particular information. The next three articles describe method for plotting angles and curves.

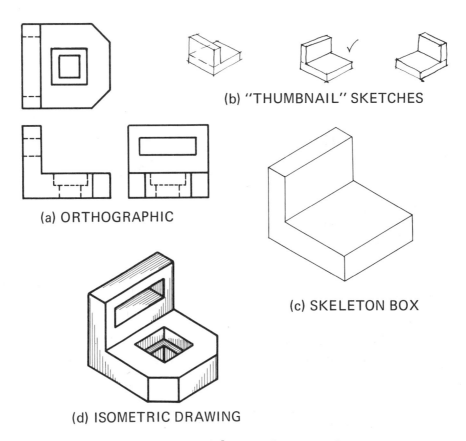

Fig. 10-4. *Steps in Making an Isometric Drawing.*

10.7 Non-isometric Lines

It is stressed again that oblique lines will not show in true length in isometric but must be located by plotting the coordinate values of their end points. Fig. 10-5 shows orthographic and isometric drawings where offsets must be used.

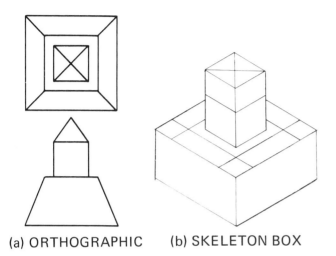

(a) ORTHOGRAPHIC (b) SKELETON BOX

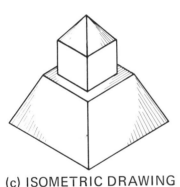

(c) ISOMETRIC DRAWING

Fig. 10-5. *Isometric Offset Locations.*

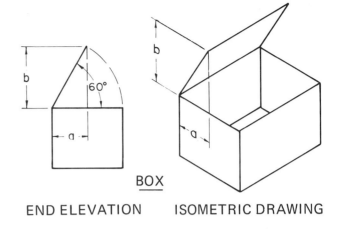

END ELEVATION   ISOMETRIC DRAWING

Fig. 10-6. *Plotting an Angle in Isometric.*

The orthographic projection of an angle must be used to determine its coordinate values. These coordinates can then be located on the isometric drawing. In Fig. 10-6, the box cover is located in isometric by plotting the width and height coordinates "a" and "b" found on the end elevation.

10.8 Isometric Curves

Circles or arcs on any of the three principal planes of an object become ellipses when plotted in the isometric planes. They can be plotted by means of their coordinate values by locating enough points so that a smooth curve can be drawn. Fig. 10-7 shows a curve plotted by means of coordinates.

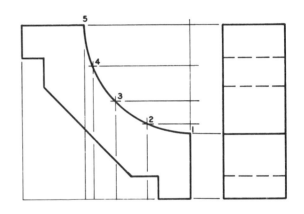

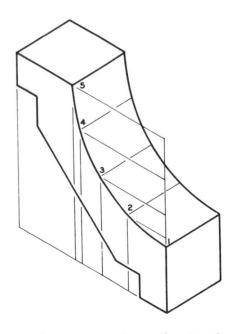

Fig. 10-7. *Isometric Curves by Coordinates.*

Ellipse Guides. An ellipse guide is a time saver when drawing circles in isometric, and is of particular value for those of small diameter. Ellipse guides are available in a variety of sizes and under a number of trade names. They usually consist of graduated sizes of ellipses cut in sheets of plastic. The isometric center lines for each ellipse are extended on the template to align them on the drawing. On some guides the sides of the template are cut at 30° angles so that isometric alignment can be made along a T-square. Fig. 10-8 shows isometric ellipse guides.

lines a circle in orthographic projection is distorted into an equilateral parallelogram in the corresponding isometric plane with the angles respectively equal to 60° and 120°. Fig. 10-9 shows the construction of the four-center ellipse in each of the isometric planes.

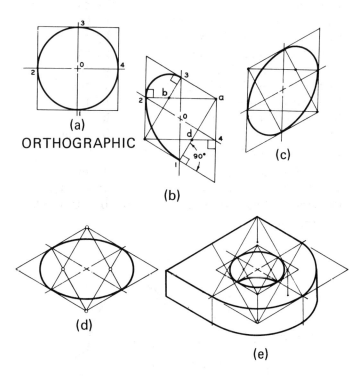

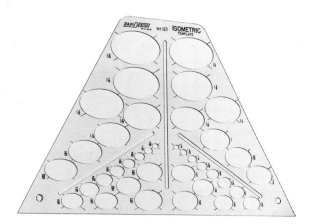

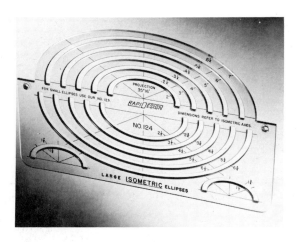

Fig. 10-8. *Isometric Templates.*
(Courtesy Rapidesign, Inc.)

Fig. 10-9. *Four-Center Isometric Ellipse.*

Steps for constructing the four-center ellipse:
1. Locate and draw the *center lines* in the desired isometric plane. These center lines are always parallel to the isometric axes.
2. Locate *with care* the four points where the curve crosses the center lines. These are points 1, 2, 3, 4 in Fig. 10-9b, found by laying off the radius of the circle at 0-1, 0-2, 0-3, 0-4.
3. Draw a parallelogram through the four points with the sides respectively parallel to the center lines.
4. The *center for any quadrant* of the curve is located on the *perpendicular bisector* of the sides of the parallelogram. In this manner the center for the arc between points 1 and 2 is found at point "a" where perpendicular bisectors of the sides (points 1 and 2) of the parallelogram intersect each other. The same procedure is used between 2 and 3, 3 and 4, and 4 and 1 to locate centers b, c, and d.

The Four-Center Ellipse. Approximate isometric ellipses can be drawn with a compass using a four-center construction. The square which out-

PICTORIAL SYSTEMS

Fig. 10-9 shows the minimum construction for four-center ellipses in each of the principal planes.

10.9 Isometric Dimensioning and Shading

While an orthographic drawing is better for size specifications, there are times when it is desired to call out sizes on a pictorial. The extension and dimension lines should all be isometric lines; that is, they should be parallel to the isometric axes. Since numerals should lie in the particular isometric plane which is formed by the intersection of the dimension and extension lines, draw light guide lines for the numeral height parallel to the dimension lines, and slope lines parallel to the extension lines. See Fig. 10-10 for an example of isometric dimensioning.

For realism or sales appeal, a small amount of simple line shading can be added to any pictorial drawing to "bring out" the object for better depth perception.

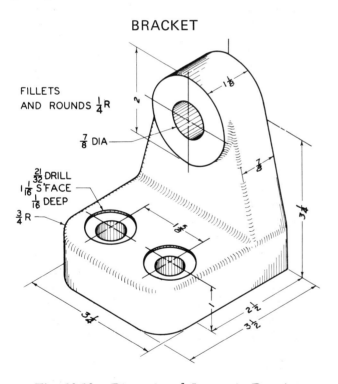

Fig. 10-10. *Dimensioned Isometric Drawing.*

10.10 Advantages and Disadvantages

The fact must be recognized that, even though isometric drawing is only an approximate picture, it does provide a fairly accurate delineation with minimum constructive effort.

Isometric drawing, when compared with the other types of pictorial representation to be studied in Parts II and III of this unit, has certain advantages and limitations. Each of the three principal planes of an isometric drawing is equally distorted. This gives the advantage that a hole or other detail will show the same proportional size or shape in whichever plane it appears. Isometric is appropriate for illustrating an object which has contour details in at least two of the principal planes.

One inherent disadvantage of isometric drawing is evident on illustrations of symmetrical objects or details. If an object has equal X and Y dimensions, the back corner will plot exactly above the front corner, as shown on the isometric cube in Fig. 10-2. Likewise, as shown in Fig. 10-4, an outside corner cut by a vertical plane at a 45° angle will plot as a single straight line.

Part II
OBLIQUE

10.11 Oblique Projection

Definition: *If parallel projectors from the outline of an object intersect an image plane at an acute angle (not 90°), the resulting outline is an Oblique Projection of the object.*

Oblique projection is used to best advantage for objects that have a series of circles, curves, or irregular outlines in the *same or parallel planes* and relatively short dimensions perpendicular to these planes.

The object to be shown in Oblique Projection is placed, for practical purposes, so that its front principal plane contains the greatest curve detail or intricate contour. Thus the projectors from the area selected as the front of the object, or planes parallel to the front, will form an image of identical size and shape upon the image plane.

10.12 Positions for Oblique Axes

In Oblique the length and height (X and Z axes) are always at 90° to each other and therefore any area in this or parallel planes is in true or scale size. The width or receding axis Y is at a convenient angle θ to the horizontal axis. See

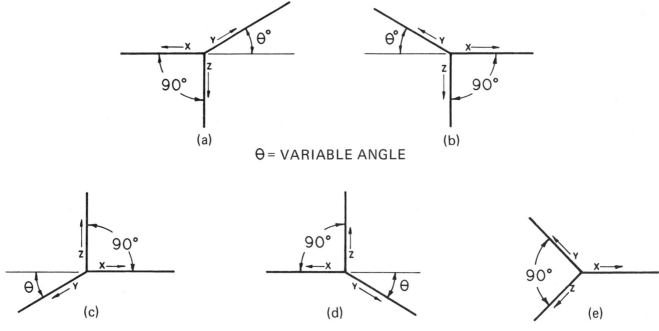

Fig. 10-11. *Positions of Axes for Oblique.*

Fig. 10-11 for several oblique axes positions. Oblique axes, like isometric, can be classified as Normal, Reversed, or Horizontal.

10.13 Types of Oblique

Oblique projections may be classified as Cavalier, Cabinet, or General, depending on the different scales used for measurements on the receding axes.

1. *Cavalier Projection.* When the measurements used on the receding axes are at the *same scale* as used for the length and height, the image is called a Cavalier Projection.

2. *Cabinet Projection.* When the measurements on the receding axis are at *half* scale to those used on the length and height, the image is a Cabinet Projection. The name Cabinet originated from furniture illustration where an attempt was made to give more emphasis to the front and less to the width of the object.

3. *General Oblique* results from measurements on the receding axis at any other than full or half the scale used on the other two axes. For practical purposes it is generally between half and full such as five-eighths or three-quarters. Refer to the full size oblique drawings of a cube with a horizontal hole, in Fig. 10-12. Which pictorial most nearly resembles a cube?

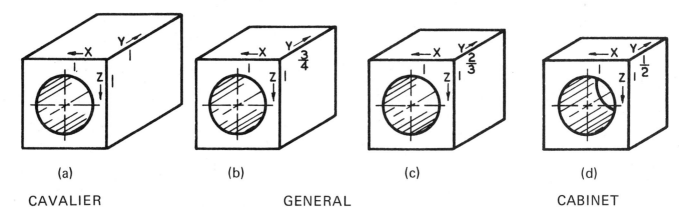

Fig. 10-12. *Comparison of Cavalier, General, and Cabinet.*

PICTORIAL SYSTEMS

10.14 Orienting the Object

An advantage of the oblique drawing lies in the ease of construction when the details of the object are in the front or parallel to the front principal plane. It may be necessary to turn the object so that this condition exists.

Two important rules should be observed when orienting the object for its best oblique viewing position:

1. Place the face or parallel faces having intricate contour, especially circles, toward the front.
2. The receding axis should be relatively short with little detail in the receding planes.

Rule one should take precedence when considering a particular object, otherwise another type of pictorial should be used. If the receding axis is relatively long compared to the X and Z dimensions of the object, Cabinet drawing will improve its appearance. In the oblique drawing shown in Fig. 10-13, the object has been turned from its given orthographic position to correspond to the above two rules.

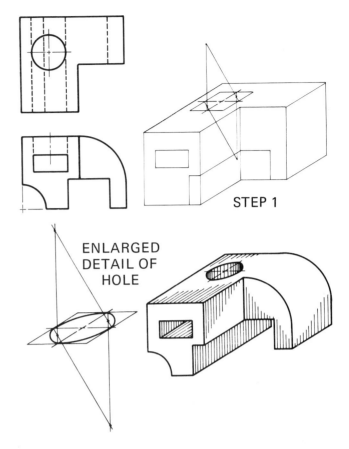

Fig. 10-14. *Cavalier Drawing.*

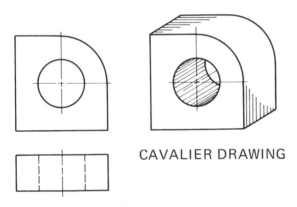

Fig. 10-13. *Object Turned to Fulfill Rules for Oblique.*

10.15 Oblique Curves

The four center ellipse (Art. 10.8) can be used in a receding plane in *cavalier* projection. However, if an ellipse is necessary in a receding plane for cabinet or general oblique, it must be plotted by offsets because the two dimensions of the enclosing parallelograms are not to the same scale; the four center method will fail because the sides of the parallelogram are unequal. Fig. 10-14 shows the cavalier projection of an object with some contour in the top plane in which the four center ellipse has been used. Fig. 10-15 shows

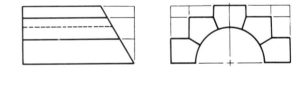

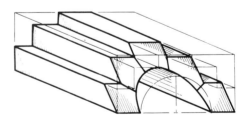

Fig. 10-15. *Locating Points by Coordinates.*

an object with a sloping plane in which the curve is plotted by coordinates or offsets and drawn with an irregular curve. An ellipse guide is useful for drawing a curve after its center lines have been located.

10.16 Advantages and Disadvantages

The most important advantage in using one of the forms of oblique projection lies in the *ease and speed of drawing*. It can be said that oblique offers a desirable method of representing an object with curves or intricate contour in one or parallel planes and with a relatively short dimension for the receding axis. If the object does not fit these two criteria, the distortion of oblique projection makes one of the other pictorial methods more desirable.

The shaded and dimensioned cabinet drawing, shown in Fig. 10-16, is constructed more easily and rapidly than any of the other forms of pictorial drawing.

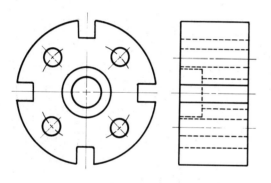

KEYED FLANGE

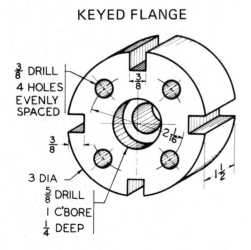

Fig. 10-16. *Dimensioned Cabinet Drawing.*

Part III

PERSPECTIVE

10.17 Principles of Perspective

Definition: *A perspective drawing is the outline formed by a series of lines of sight intersecting a picture plane as they converge from the object toward the sight point or eye of the observer.* The pictorial formed is similar to that seen by the eye or recorded by a camera.

Perspective projection differs from orthographic or oblique projection because the projectors or lines of sight are not parallel to each other but radiate from the sight point. Fig. 10-17 illustrates the space relationship of the object, picture plane, ground plane, and sight point showing the perspective outline formed where the lines of sight pierce the picture plane.

The student can visualize a perspective outline by looking at a distant object through a window. The perspective image would appear on the pane of glass where the lines of sight or rays from the eye to the object pierce the window.

Perspective drawing made by the visual ray method requires the top and front views showing the object, picture plane and sight point. The front view shows the true shape of the picture plane and therefore the true perspective view of the given object. The perspective picture is constructed by locating the points where the lines of sight pierce the picture plane.

10.18 Definitions of Terminology

The following definitions should be carefully studied in order to become familiar with the nomenclature used in perspective.

Sight Point is the point from which the visual rays radiate to the outlines of the object and corresponds to the observer's eye.

Ground Plane is a horizontal datum plane to which the object and sight point are related.

Picture Plane is the image plane upon which the perspective picture is outlined.

Ground Line is the line of intersection of the picture plane with the ground plane.

Axis of Vision is a line through the sight point perpendicular to the picture plane.

PICTORIAL SYSTEMS

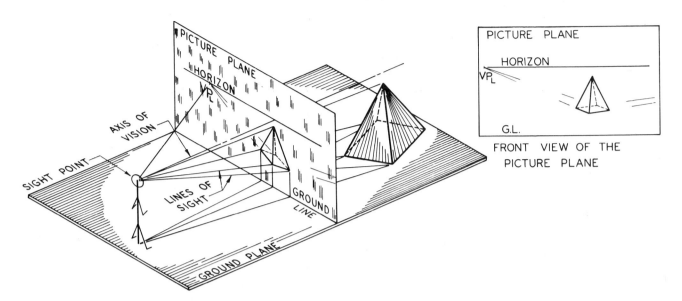

Fig. 10-17. *Space Relationships for Perspective Drawing.*

Horizon is the line formed by a level plane through the sight point intersecting the picture plane.

Vanishing Point (VP) is the point at the limit of vision where parallel lines appear to intersect. It is the point on a perspective picture where any family or system of parallel lines converge. *The vanishing point for a system of parallel lines is located where a ray from the sight point parallel to the given family of lines pierces the picture plane.* In most problems it is only necessary to locate the vanishing points of families of horizontal lines.

Initial Point (IP) is the point where an object line in space, if extended, would meet the picture plane.

Indefinite Perspective is the perspective image of a line of infinite length. It is the straight line joining the given line's initial point (IP) and its vanishing point (VP).

Definite Perspective of a line is that portion of the indefinite perspective which represents the image of a line of fixed length.

10.19 Parallel or One-Point Perspective

PRINCIPLE. *Lines parallel to the picture plane have parallel images in perspective.*

If one of the three principal planes of an object is parallel to the picture plane, the perspective image is known as parallel or one-point perspective. The X and Z coordinate lines for this case are parallel to the picture plane and therefore will be respectively parallel in perspective. The Y coordinate lines are perpendicular to the picture plane and their vanishing point will be located where the axis of vision (from S perpendicular to the picture plane) pierces the picture plane, Fig. 10-18.

Parallel perspective is comparable to oblique projection because one principal plane of the object is parallel to the picture plane. Also, like oblique projection, it shows to best advantage when intricate contour of the object is in the front or parallel planes. For example, circles parallel to the picture plane will show as circles in perspective. However, distortion due to relatively long receding dimensions appears to be less in perspective projection due to foreshortening.

The location of the sight point and the object with respect to the picture plane determines the proportions of the resulting picture. This is comparable to the choice of direction of the X, Y, and Z axes in axonometric or oblique. For a natural picture, the sight point should be located far enough from the object in the H projection so that a 30° cone, whose vertex is at the sight point S_H and whose axis lies along the axis of vision, will encompass the plan view of the object. This cone is known as the *Cone of Vision.*

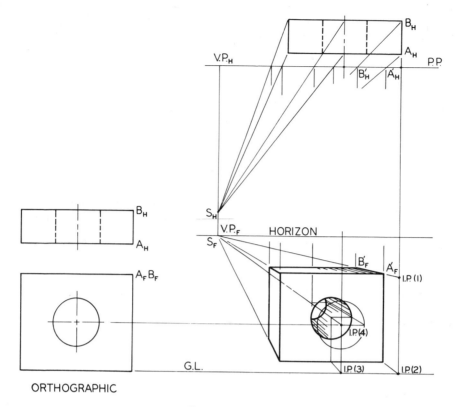

Fig. 10-18. *Parallel or One-Point Perspective.*

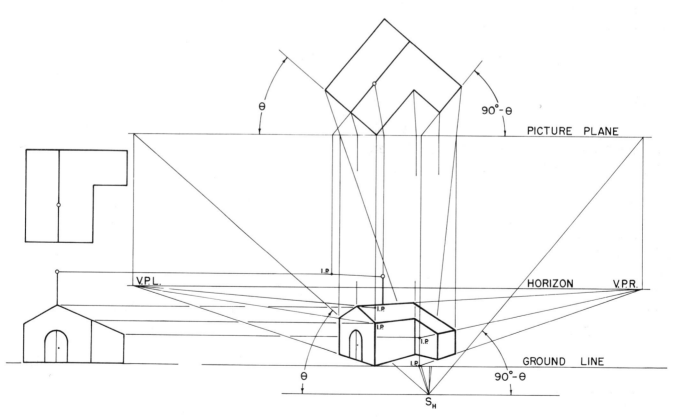

Fig. 10-19. *Angular or Two-Point Perspective.*

Fig. 10-18 has been drawn in an extreme position in order to separate the construction for easier interpretation. Follow the construction for this one-point perspective drawing.

In order to conserve paper, or to produce a larger perspective picture on a sheet of given size, the front view is often superimposed on the top view in the area between S_H and the edge view of the picture plane (PP).

10.20 Angular or Two-Point Perspective

When two of the principal planes of an object are neither parallel nor perpendicular to the picture plane, horizontal lines in the front and side planes will have different vanishing points. They are located by following exactly the definition of a *Vanishing Point*. For rectangular objects, the sum of the two angles formed by the front and side planes of the object with the picture plane is equal to 90°.

Objects with oblique faces (not in principal planes) are drawn in perspective by locating the corners of the oblique faces independently. This is comparable to the location of oblique lines in axonometric or oblique drawing.

The construction details for a two-point perspective are shown in Fig. 10-19.

10.21 Perspective Sketching

Freehand perspective sketches, involving all the details of layout construction, can be made. However, one principal objective of sketching is the rapid production of a clear illustration which eliminates the time consuming construction of details necessary for mechanical perspective.

A parallel perspective sketch can be constructed rapidly with reasonable accuracy and good proportion if the illustrator uses good judgment in the selection of horizon line, vanishing points, and ground line locations.

Angular perspective sketching can be rapidly accomplished by carefully choosing the vanishing points and ground line, and locating the definite perspective of the object by estimating the foreshortening effect of perspective. In this manner, the desired proportions of the finished sketch can be reasonably predetermined. Fig. 10-20, an example of an angular perspective sketch, illustrates the minimum construction necessary.

10.22 Advantages and Disadvantages

The obvious advantage of a properly constructed perspective drawing is its "true to life" appearance. An exact photographic image is made by judicious choice of picture plane, horizon, and sight point; while distortion will result from a poor selection of their relative positions in perspective representation. Perspective might well be chosen when a photograph is unavailable and it becomes necessary to draw and render the best possible likeness of a product for a catalog illustration. For similar reasons an architectural study of a new building design is usually made in perspective.

The often difficult and time-consuming construction makes persepective drawing more expensive than other types of pictorial which in many cases may be as satisfactory and can be drawn much quicker.

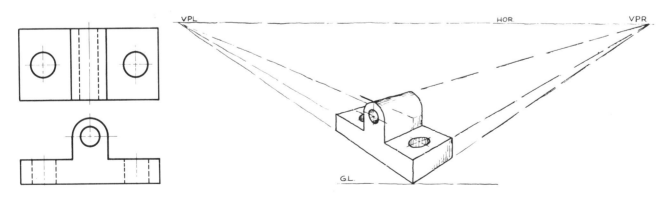

Fig. 10-20. *Perspective Sketch.*

PROBLEMS — ISOMETRIC

The following problems on axonometric drawing, instrument construction, or freehand sketching, provide a variety of conditions for pictorial drawing. In many cases the choice of the type of axonometric, position and orientation of the object, is left to the judgment of the student.

1. Draw or sketch the isometric projection of the objects shown in Fig. 10-21. The choice of the position of the views on the paper is important if overlapping is to be avoided.
2. Make an isometric drawing or freehand pictorial sketch of the following figures: 9-14, 9-15, 9-16, 9-17, 9-18, 11-2, 11-13, 11-24, 12-5, 12-21, 12-23. Maintain the relative proportions of the various features and select an appropriate scale for those that are drawn with instruments.
3. Make an axonometric drawing or sketch of the objects shown in Fig. 10-22. Draw to as large a scale as practicable.
4. Make a freehand axonometric sketch of the following familiar objects:
   a. Railroad Tank Car
   b. Room Interior
   c. Coffee Table
   d. Desk
   e. Chair
   f. Building
   g. Stairs and Railing
   h. Drawing Table
   i. Composition of Geometrical Solids
   j. An Arrangement of a Scale, Drafting Tape, and Triangle
   k. Antenna attached to a Roof

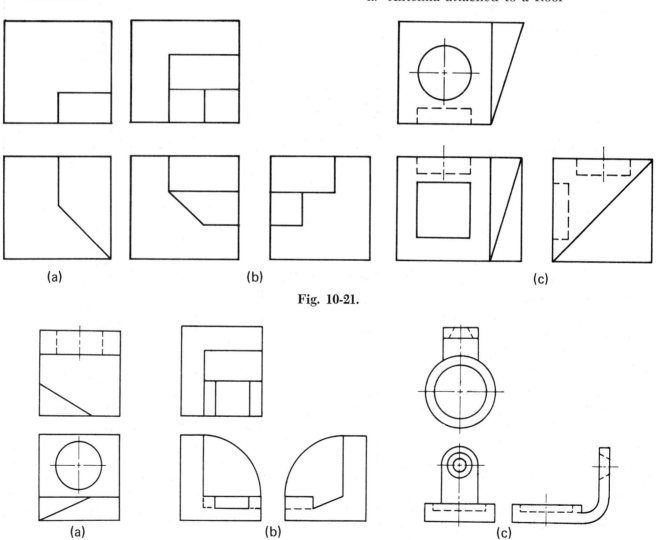

Fig. 10-21.

Fig. 10-22.

PROBLEMS — OBLIQUE

Decisions with respect to the position of the object and the type of oblique drawing or sketch are left to the student in many of the following oblique pictorial problems.

1. Make appropriate oblique drawings or sketches of the objects shown in Fig. 10-23.

   Use as large a scale as practicable. Choose the type of oblique, the receding axis angle, and its orientation to show the object to best advantage.

2. Apply the instructions of problem 1 to the following Figs.: 9-4, 9-1, 11-5, 11-12, 11-14, 11-17, 11-20, 11-23, 11-24, 11-25, 12-5, 12-25.

3. Make a freehand oblique sketch of suitable items with which you are familiar. The orientation of the object for a pleasing view is important. The following list offers some suggestions of objects appropriate to oblique drawing.
   a. Instruments in your drawing set
   b. Wrist Watch
   c. Wall Clock
   d. Tie Clasp
   e. Bicycle
   f. Chest of Drawers
   g. T.V. Set
   h. Fraternity Pin
   i. Tools, such as: Wrench, Hacksaw, Clamp, Hammer, Saw, Screwdriver
   j. Composition of Appropriate Objects

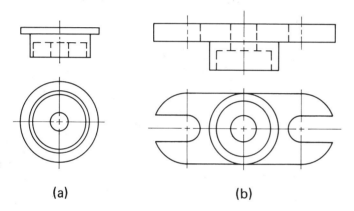
(a)　　　(b)

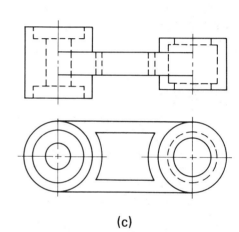
(c)

Fig. 10-23.

PROBLEMS — PERSPECTIVE

The choice of the relative locations of picture plane, sight point, ground line and the relative position, orientation and proportion of the selected objects to these locations is an important part of each of the following problems.

1. With an appropriately assumed layout, construct a parallel perspective of the objects given in Fig. 10-23.
2. Construct or sketch a parallel perspective of the objects shown in Figs. 9-4, 11-1, 11-5, 11-12, 11-14, 11-17, 11-20, 11-23, 11-24(a), 11-25, 12-5, 12-25.
3. Construct or sketch in parallel perspective, the following or other familiar objects or structures:
   a. Diving Board
   b. Fluorescent Desk Lamp
   c. Piece of Furniture
   d. A Doorway
   e. A Hallway
   f. Entrance Steps with Railing
   g. A Theater Entrance
   h. Window—showing construction
4. Construct an angular perspective of the objects shown in Fig. 10-24. Use careful consideration of the size of the layout, and angles used.
5. Make an instrument drawing in angular perspective or an approximate angular perspective sketch of the objects shown in Figs. 9-14, 9-15, 9-16, 9-17, 9-18, 11-2, 11-13, 11-24, 12-5, 12-21, 12-23, 12-25.

6. Construct or sketch the following familiar objects in angular perspective.
   a. A Building (garage, church)
   b. Stairway with Handrail
   c. Composition of Geometrical Solids
   d. Telephone
   e. Desk Calendar
   f. A Chair

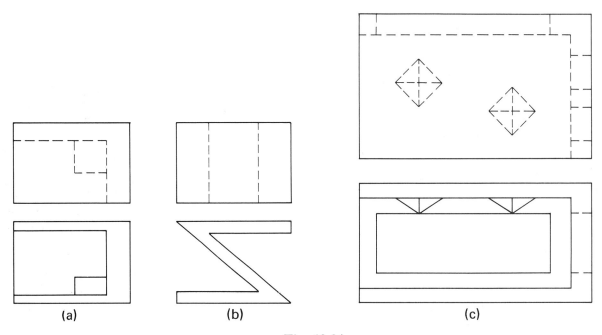

Fig. 10-24.

unit 11

Sections and Conventional Practices

11.1 Sectional Views

Definition: *If a portion of an object is cut away by an imaginary cutting plane, the view taken normal to the cut surface of the remaining part is a Sectional View.*

A Sectional View of an object is used to clarify any projection which otherwise would be confused by hidden or invisible outline. In other words, an object having peculiar or complicated interior features is usually sectioned. Sectioning is not only applicable to individual parts but is also used on assembly drawings (Unit 26) in order to better define interior features. Crosshatching, more often called section lining, is used on sectional views to accent those areas which have been exposed by the removal of a portion of the object.

The position and extent of the section cutting plane determines the resulting type of sectional view such as a) full section; b) half section; c) offset section; d) broken out section; e) revolved section; f) removed section; g) phantom section.

11.2 The Cutting Plane

The cutting plane is an imaginary plane passing through the object. In the construction of a sectional view, that portion of the object on the *near side* of the cutting plane is removed in order to reveal the section. See Figs. 11-1a and 11-1b. The cutting plane is usually indicated on a view adjacent to the section in order to show the location and extent of the cut. Arrowheads at the ends of the cutting plane line, as shown in Figs. 11-1c and 11-1d, indicate the viewing direction for the section. Note that the cutting plane line is heavy and composed of two short dashes between long dashes. The standard of a continuous series of long dashes is sometimes used.

11.3 Full Section

When the cutting plane passes "fully" through the object, the view normal to the cut surface is known as a full section. A pictorial of an object with cutting plane is shown in Fig. 11-1a. When the near half of an object is removed to show a sectional view, the area of the cut material with all *visible* background lines are drawn as shown in Fig. 11-1c. Hidden outlines behind the cutting plane are *omitted* from the sectional view unless they are essential for the description of the object.

When the object has one major center line and the position of the cut is obvious, the indication of the cutting plane on the drawing is sometimes omitted. This is particularly true for full and half sections.

11.4 Half Section

If one half of an orthographic view is sectioned while the other half shows the interior details, the result is a half section. Half sections are applicable to *symmetrical objects*. In Fig. 11-1d the position of the cutting plane is shown for the construction of a half section view. Notice that the exterior details are separated from the sectional portion

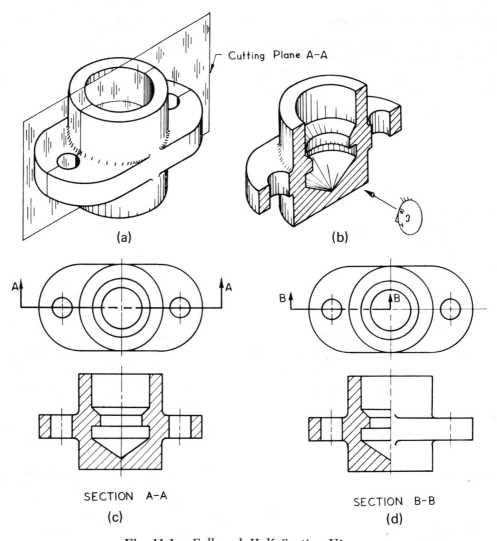

Fig. 11-1. *Full and Half Section Views.*

by a center line rather than a solid line. Hidden details are omitted from the exterior half unless they are essential for clarity. One of the most common uses of half sections is on assembly drawings as shown in Fig. 11-10.

11.5 Offset Sections

If the object is cut by a staggered plane, as shown in Fig. 11-2, the sectional view is called an offset section. The student should take particular note of two things:

1. The cutting plane and view direction arrows are essential.

2. No line of demarcation in the crosshatching is shown on the sectional view where the cutting plane changes direction.

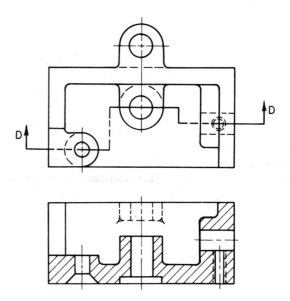

Fig. 11-2. *Offset Section. (ANSI Y14.3)*

SECTIONS AND CONVENTIONAL PRACTICES

11.6 Broken Out or Partial Sections

A broken out or partial section may be used when only a small interior portion of an object needs exposure for clarification of details. The location of the cutting plane is not shown, but the extent of the break is shown by an irregular boundary line around the sectional view of the object. This type of section is illustrated by Fig. 11-3.

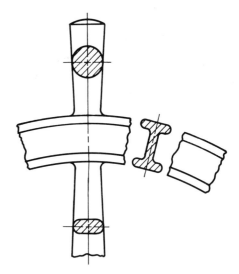

Fig. 11-4. *Revolved Sections. (ANSI Y14.3)*

the direction of vision must be indicated and the resulting section identified by proper labels. Advantages of this type of section are that the original view of the object is uncluttered and the removed section may be drawn to a larger scale in order to show the detail. Figs. 11-5 and 11-6 illustrate applications of removed sections. A detail section B-B is shown in Fig. 11-22.

11.9 Phantom Lines and Sections

Sometimes phantom outlines of related parts are used to clarify the operating principles of a particular mechanism. Phantom lines are also useful in depicting alternate or consecutive positions of a mechanism or for representation of repeated details such as threads, springs, and gear teeth.

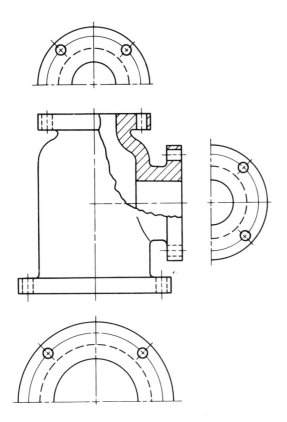

Fig. 11-3. *Broken Out Section and Half Views. (ANSI Y14.3)*

11.7 Revolved Sections

Revolved sections are used to show the cross section of a special portion of the object such as a rib, spoke, rim, handle, etc. The cut area is revolved 90° to bring the resulting sectional view into the plane of the paper. This type of section is illustrated in Fig. 11-4.

11.8 Removed or Detail Sections

A removed or detail section is similar to a revolved section except that it is not superimposed on the view of the object. The cutting plane and

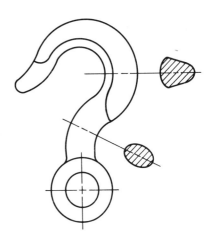

Fig. 11-5. *Removed Sections. (ANSI Y14.3)*

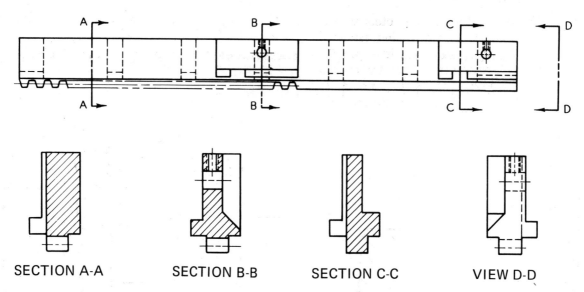

Fig. 11-6. *Enlarged and Removed Sections and View. (ANSI Y14.3)*

Fig. 11-7 shows the use of phantom outlines in representing alternate position and repeated parts.

When it is desired to accent interior detail without destroying the continuity of the exterior view, a phantom section may be shown by dashed section lining. Such treatment gives the effect of a phantom or hidden appearance. See Fig. 26-7.

11.10 Section Lining

Metallic pieces are section lined by means of uniformly spaced, fine, full lines at an angle, usually 45°. The spacing depends somewhat upon the size of the area to be sectioned; the larger the area, the wider the spacing, with a usual spacing of approximately 1/8 inch or wider. Some

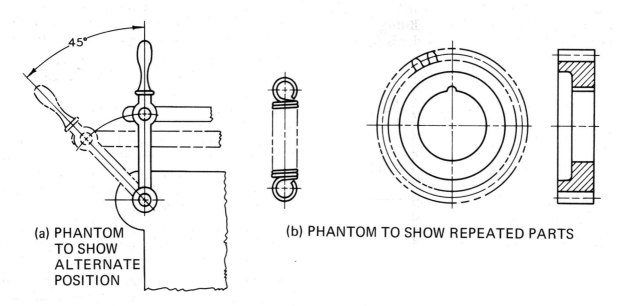

Fig. 11-7. *Phantom Lines.*

companies omit section lining completely when it is obvious or use partial section lining especially on large areas, Fig. 11-8. After a little practice, section lines can be spaced evenly by eye, although instruments such as a section liner or the Ames lettering device can be used. Be sure that

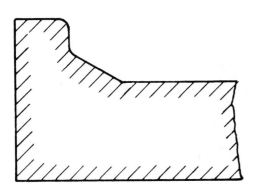

Fig. 11-8. *Section Lining in Large Areas.*

the spacing is uniform and that the section lines are light and all of equal weight so that the sectional view will be clean in appearance.

The angle of section lining remains constant for all cut areas of any one single object. Refer to Figs. 11-12, 11-14, 11-15.

Adjacent Parts. On assembly drawings adjacent parts should be section lined in different directions as shown in Figs. 11-9, 11-10, and 11-15.

Standard Symbols. On sectioned assembly drawings, it may be desirable to distinguish between different materials. The standards for symbolic section lining are shown in Fig. 11-11. There is a trend to use fewer symbols, relying on accurate word descriptions in the Materials List or Bill of Materials.

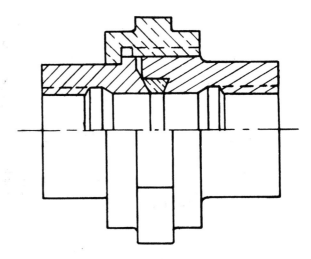

Fig. 11-10. *Half Section in Assembly. (ANSI Y14.3)*

11.11 Conventional Practices

Definition: *A conventional practice is an accepted standard which has been adopted for clarity although it violates true orthographic projection.* Some conventional practices apply to sections and others refer to exterior views.

Ribs. If the cutting plane is parallel to and slices lengthwise through a rib, web or thin portion of an object, the sectional view of this part is not section lined to avoid the misleading impression that the part is massive. Fig. 11-12a shows a full section taken through the supporting ribs.

Alternate section lining may be used in cases where the actual presence of a thin elements is not sufficiently clear without section lining, or whereby clarity may be improved. For example, in Fig. 11-12b the presence of the rib is not immediately clear in the sectional view, while in Fig. 11-12c the alternate lining is used to distinguish the rib. Note the use of hidden outline around the rib. Fig. 11-22 also shows where alternate section lining is used.

When the cutting plane is perpendicular to, or cuts across ribs, webs, etc., the sectional view is section lined, as shown in Fig. 11-13.

Spokes. When a section plane passes lengthwise through a spoke of a wheel, the spoke should not be section lined to avoid the effect of a disc

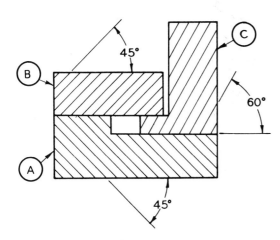

Fig. 11-9. *Section Lining on Adjacent Parts.*

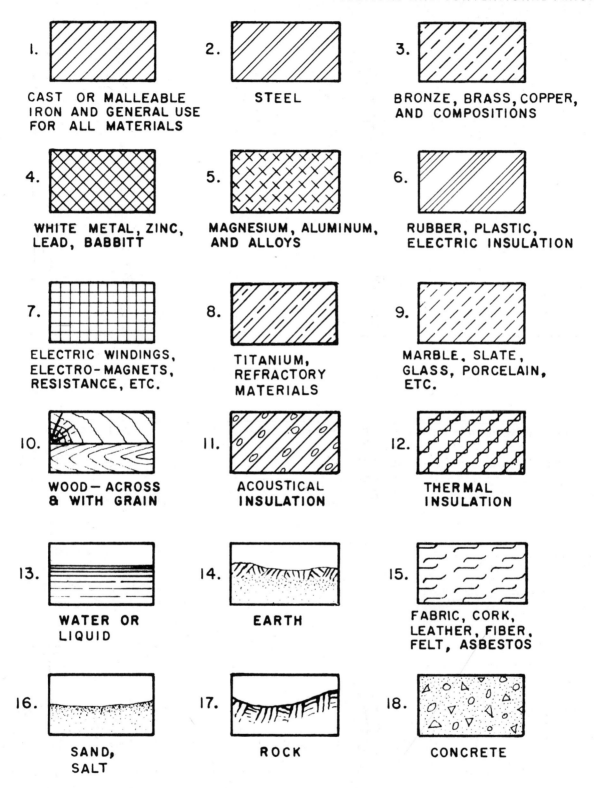

Fig. 11-11. *Material Symbols for Sectioning*

SECTIONS AND CONVENTIONAL PRACTICES 111

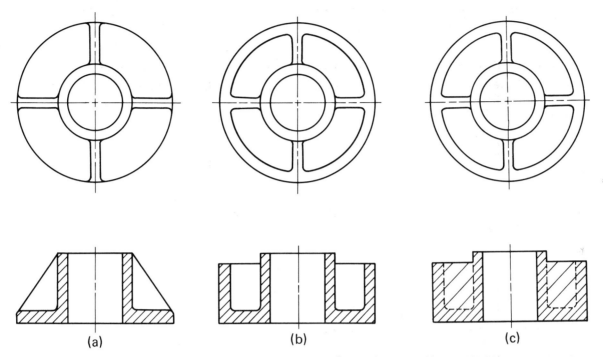

Fig. 11-12. *Conventional Practice—Ribs in Section. (ANSI Y14.3)*

wheel. The approved treatment is shown in Fig. 11-14. When the cutting plane contains the center lines of such elements as shafts, bolts, nuts, rods, rivets, keys, pins, screws, ball or roller bearings, or similar shapes, these elements should *not* be sectioned. See Fig. 11-15. If the plane cuts *across the axes* of ribs, spokes, or thin material, they should be sectioned in the usual manner as shown in Fig. 11-13.

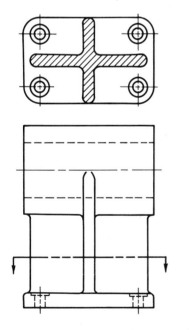

Fig. 11-13. *Cross Section of Ribs. (ANSI Y14.3)*

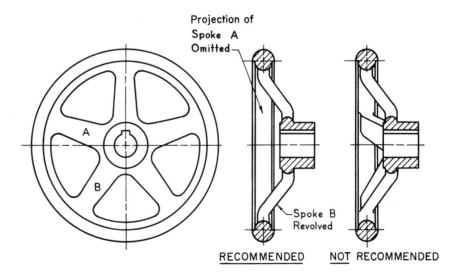

Fig. 11-14. *Conventional Practice—Spokes in Section. (ANSI Y14.3)*

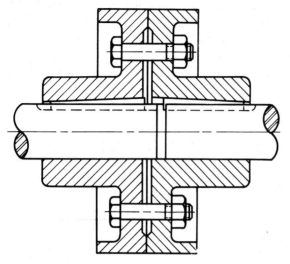

Fig. 11-15. *Shafts, Keys, Bolts, and Nuts in Section View. (ANSI Y14.3)*

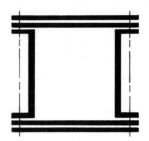

Fig. 11-16. *Thin Plates in Section.*

Thin Plates. Solid shading may be used on the cut area, Fig. 11-16, of small scale sectional views of thin plates such as sheet metal, packing, gaskets, shims, and structural shapes.

Foreshortened Projections and Rotated Features. Many objects contain features such as ribs, arms, and holes which become foreshortened or superimposed in the true projection of the object. In order to avoid confusing overlapping or foreshortening, the elements are rotated into the plane of projection. Figs. 11-14 and 11-17 illustrate this conventional projection of objects in which a feature forms an oblique angle to the plane of projection. Fig. 11-3 shows holes which have been rotated so that the adjacent projection indicates their true distance from the center of the flange. These conventional practices are used to aid clarity and readability.

Conventional Breaks. In order to shorten a view of an elongated object, conventional breaks are used. Samples are shown in Fig. 11-18.

Half or Partial View Drawings. Where space is limited it is permissible to represent symmetrical objects by half views. If the adjacent view is an exterior one, the *near half* of the symmetrical view is drawn. If the adjacent view is a full or half section, the *far half* of the symmetrical view should be used. This is illustrated in Fig. 11-3.

Intersections in Section. If a section cuts through the intersection of a hole in a curved surface and the exact curve of intersection is small or of no consequence, the curve or figure of intersection may be simplified as shown in Fig. 11-19a and c. Larger figures of intersection may be projected as shown at Fig. 11-19b or d.

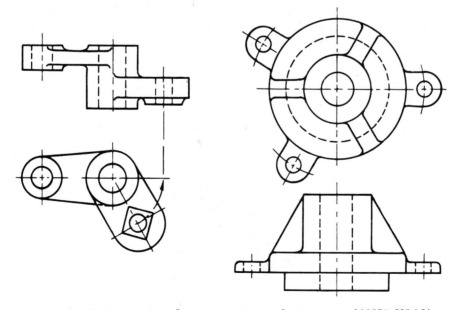

Fig. 11-17. *Conventional Practice—Rotated Features. (ANSI Y14.3)*

SECTIONS AND CONVENTIONAL PRACTICES

Rounded and Filleted Intersections. The intersection of two unfinished surfaces generally should be shown rounded or filleted. This intersection, which actually is not a sharp corner, may be indicated by a conventional line, the location of which should be at the intersection of the principal surfaces, disregarding the fillet or round. Figs. 11-20 and 11-21 illustrate such contours.

Fig. 11-22 illustrates a combination of several conventional practices.

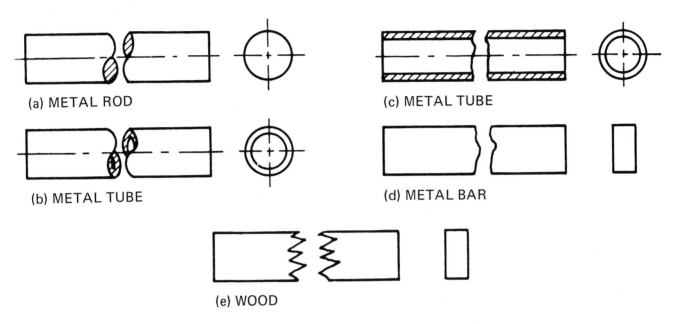

Fig. 11-18. *Conventional Breaks.* (ANSI Y14.3)

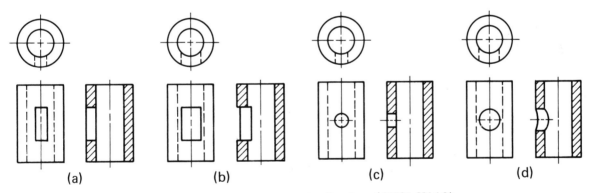

Fig. 11-19. *Intersections in Section.* (ANSI Y14.3)

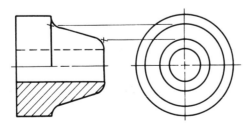

Fig. 11-20. *Representation of Fillets and Rounds.* (ANSI Y14.3)

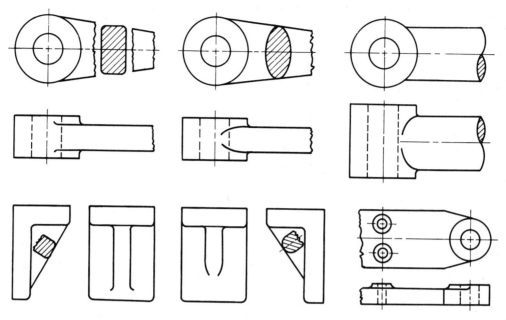

Fig. 11-21. *Runouts and Filleted Intersections. (ANSI Y14.3)*

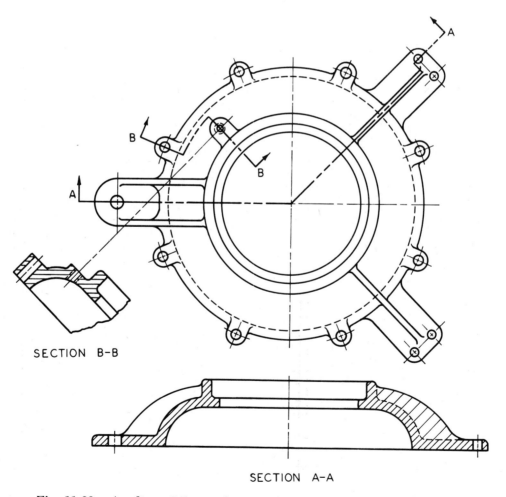

SECTION B-B

SECTION A-A

Fig. 11-22. *Auxiliary Offset, Ribs, and Rotated Sections. (ANSI Y14.3)*

SECTIONS AND CONVENTIONAL PRACTICES

PROBLEMS

The following problems concern the drawing or sketching of appropriate sectional views, some of which involve conventional practices. Three types of problems are included: (1) to show a sectional view which will satisfy the conditions of a given one view drawing; (2) given two exterior views, to construct an appropriate sectional view; (3) to show the necessary views, to include sections, of an object or assembly shown in pictorial form. In each case indicate the position of the cutting plane.

1. Show a sectional view consistent with the given view of each object in Fig. 11-23.

2. Show an appropriate additional view which is sectioned, or change one of the given views to a section for each of the objects in Figs. 11-24 and 11-25.

3. Draw or sketch the necessary views, one of which is sectioned, for each object or assembly in Fig. 11-26.

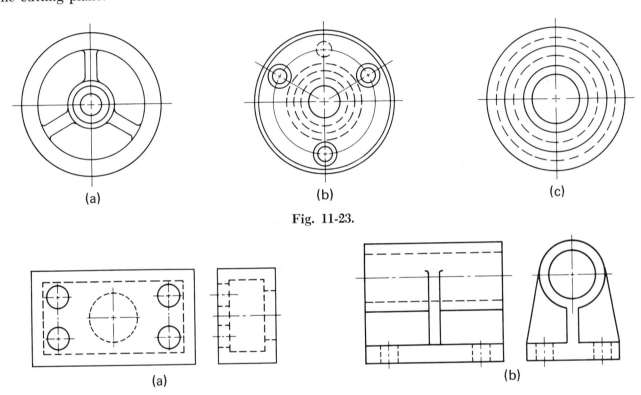

Fig. 11-23.

Fig. 11-24.

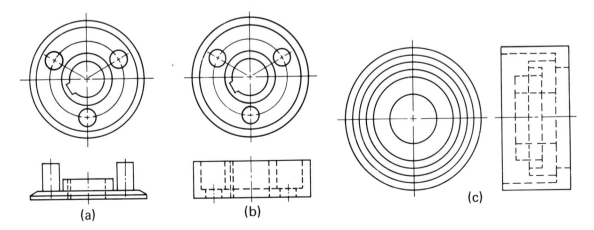

Fig. 11-25.

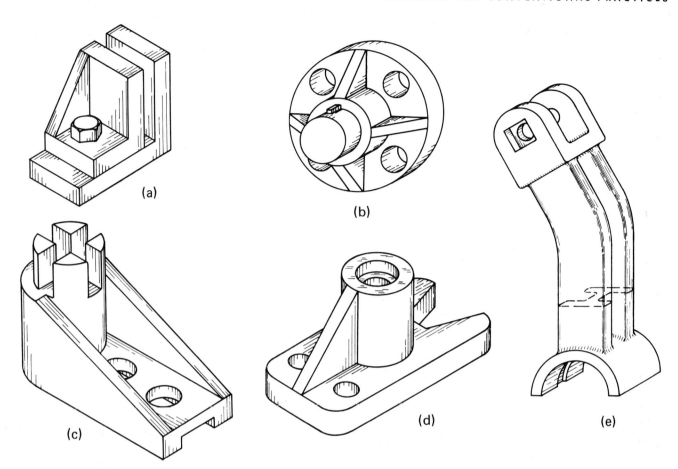

Fig 11-26.

unit 12

Dimensioning-Basic Concepts

12.1 Purpose of Dimensioning

The purpose of a treatise on the basic concepts of dimensioning is to direct student thinking toward fundamental size specifications immediately after completing a study of the standards of shape delineation. Creative Design Projects are logical exercises to be integrated throughout a graphics course of study rather than to delay until a study of all facets of representation have been completed. The finalization of a project should include the specifications for manufacture or construction as was indicated in Unit 2. More complete dimensioning for production will be possible after studying *Fasteners* in Units 19 and 20, and *Manufacturing Processes* and *Production Dimensioning* in Units 21 and 22. Standards of drawing layout and dimensioning in specialized fields will be considered in Units 23 through 26.

An important consideration in dimensioning is that the values used should be those needed for the production of the object. These dimensions sometimes differ from those used by the draftsman. For example, the draftsman uses a radius to draw a circle which describes a drilled hole. The machinist, however, must know the diameter of this same hole before he can select the proper drill.

Four important *criteria* guide the application of dimensions to a drawing:

1. *Accuracy.* The dimension values must be correct.
2. *Clearness.* Each dimension must be placed in its most appropriate position.
3. *Completeness.* There can be no omissions of specifications.
4. *Readability.* Lettering, numerals, and dimension lines must be neat, uniform in size, and very distinct.

If a simple procedure for dimensioning is followed, observing the standards outlined in this unit, the resulting finished drawing will contain accurate, clear, complete, and readable specifications.

12.2 Standards of Dimensioning

Standards are those practices which have been recognized by industry as being desirable and adopted for company manuals and text books to promote uniformity.

The three best recognized drafting standards are the *American National Standards Institute* (ANSI), formerly the American Standards Association; the *Aerospace-Automotive Drawing Standards* (SAE); and *Military Standards* (MIL-STD), of the Department of Defense. The particular standards that are concerned with dimensioning and from which both written material and illustrations are used copiously in the following articles include: ANSI Y14.5, 1966; SAE, section A; and MIL-STD 100A.

The standard for the proportion of arrow heads was given on Fig. 4-4. As well, the illustrations in this unit should be studied for dimension line weight and spacing, and the relative size and length of arrow heads.

12.3 Definition of Terms

Dimensions are numerical values expressed in appropriate units of measure and indicated on a drawing along with lines, symbols, and notes to define a geometrical characteristic of an object.

Datums are points, lines, planes, cylinders, etc. assumed to be exact for purposes of computation. The location of the geometric relationship of features may be established from the datums.

Features are specific characteristics or component portions of a part and may include one or more surfaces such as holes, screw threads, profiles, faces, or rabbets.

Undimensioned drawings are prepared on stable material with only necessary control dimensions. These include loft, printed wiring, templates, master layouts, tooling layout, etc.

12.4 Fundamental Rules (abridged from ANSI Y14.5—1966)

The following are basic for beginning dimensioning. Many are repeated and supplemented with other standard practices in Art. 12.16.

1. Dimensions for *size, form,* and *location* of features must be complete so that no scaling of drawings is required, and so that the intended sizes and shapes can be determined without assuming any distances.
2. Each dimension must be expressed clearly so that it will be interpreted in only one way.
3. Dimensions and related data must be given only once.
4. Dimensions must be shown between points, lines, or surfaces which have a necessary and specific relation to each other or which control the location of other components or mating parts.
5. Where practicable, the finished part should be defined without specific manufacturing methods. Thus only the diameter of a hole is given without indication as to whether it may be drilled, reamed, punched, or made by any other operation.
6. Dimensions must be selected to give required information directly. Dimensions should preferably be shown in *true length views* and refer to visible outlines rather than to hidden lines.
7. *Thou shalt not crowd dimensions.*

12.5 Units of Measurement

The conventional United States linear unit used on engineering drawings is the *inch,* although increased pressure is being placed on industry in this country to adopt the metric system whose linear unit is the millimeter. Some drawings used for overseas work have dual dimensions, the millimeters being enclosed in parentheses and identified by symbol, for example, (12 MM), the inches being placed in their normal manner as .473.

Angular units are most commonly expressed in degrees, minutes, and seconds. A symbol should follow the numeral; for degrees used alone, 32°, or the abbreviation DEG; for minute and seconds, the standard symbol is used, for example, 2°–57′–7″.

Normally no unit symbol is indicated with the numeral dimensions, although a note on the drawing may indicate units if there is any doubt, for example: ALL DIMENSIONS IN INCHES. Exceptions are found on some surveying and structural drawings where units may be in feet or other units. If feet and inches are in combination, the symbol (′) is used for feet but the inch symbol is omitted, for example, 42′–8. *Always separate feet and inches with a dash.*

12.6 Subdivisions of Units

Using the inch as the standard linear unit, dimensions may be expressed in decimals or common fraction.

Decimal dimensions are now used by a large number of industries especially in such fields as automotive, electronic, and aerospace. The SAE standard recommends the use of decimals and the latest ANSI manual contains nothing but illustrations using decimal dimensions as will be used for most illustrations in this text. Decimal dimensions are specified to two, three, or more places. The *preferred* basis is a *two place decimal dimension* employing an increment of .02 in., such as .04, .24, 5.20, etc. When such dimensions are divided by two the quotient is still a two-place decimal. All dimensions on the drawing will have no fewer than two decimal places. When drawings are con-

DIMENSIONING—BASIC CONCEPTS

verted from fractional to decimal dimensions, three- or four-place decimals may be necessary to bring agreement to the fractions and permit continued use of existing gages and tools.

The *three place decimal system* is, where practicable, expressed as multiples of .010, such as .100, and 1.060. In converting fractions such as 3/32 the value would be .094. All dimensions in this system will have no fewer than three decimal places.

Fractional dimensions are still used in many places, especially by the older companies who have been in existence many decades. Civil engineering work and structural design still use fractions as their standard. The reader will notice several cases of dimensioning with fractions especially in Unit 23. When fractions are used it should be recalled that the overall fraction height is double the height of the whole number of feet or inches and that the fraction bar *must not* be omitted and *should not* be slanted except when typewritten.

12.7 Application of Dimensions

Some Standards (see Art. 12.16 for more)

1. *Dimension lines* with their arrowheads and numerals show measured distances on an object. They are drawn parallel to the direction of measurement, broken for the numerals, aligned and grouped for uniform appearance. They should *not be crowded;* the space between parallel lines being not less than .24 inch, and between the dimension line and the part, about .40 inch. Refer to Fig. 12-1 for good spacing of dimensions.

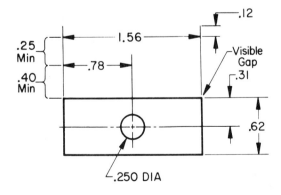

Fig. 12-1. *Standards for Dimension Lines. (ANSI Y14.5)*

2. *Extension or Witness Lines* are used to indicate the extension of a surface or point to a location outside the part outline. Where possible, avoid crossing dimension lines, object lines, or other extension lines. Where crossing is unavoidable they should be continuous unbroken lines.
3. *Leaders* are generally inclined straight lines to direct a note to the intended place on the drawing. A short horizontal portion of the leader extends to mid-height of the first or last letter or digit of the note. When pointing to a circle or arc, they should be radial. See sample, Fig. 12-2.

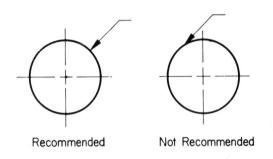

Fig. 12-2. *Radial Leaders. (ANSI Y14.5)*

4. *Direction of Reading* dimensions on a drawing is either *unidirectional or aligned.* Unidirectional is preferred, that is, so all numerals read from the bottom of the drawing. See illustrations, Fig. 12-3. Aligned expressions are placed parallel to their dimension lines, the numerals

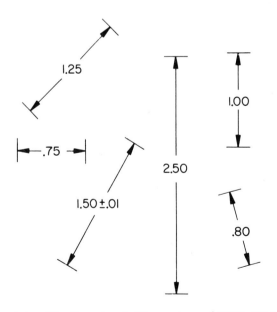

Fig. 12-3. *Unidirectional Dimensions. (ANSI Y14.5)*

FULL SIZE	or	1.00 = 1.00	or	1 = 1	or	1/1
HALF SIZE	or	.50 = 1.00	or	$\frac{1}{2}=1$	or	1/2
QUARTER SIZE	or	.25 = 1.00	or	$\frac{1}{4}=1$	or	1/4
EIGHTH SIZE	or	.125 = 1.00	or	$\frac{1}{8}=1$	or	1/8
TWICE SIZE	or	2.00 = 1.00	or	2 = 1	or	2/1
TEN TIMES SIZE	or	10.00 = 1.00	or	10 = 1	or	10/1

Fig. 12-4. *Specification of Scales.* (ANSI Y14.5)

reading from the *bottom* and the *right side* of the drawing. See Figs. 23-17 and 23-19.

5. *Avoid Dimensions within the Outline* of a view whenever practicable. Clarity and avoidance of crowding will dictate when dimensions within the outline are necessary. Otherwise place dimensions outside and preferably between views.

6. *The Scale* of a drawing should be stated on the drawing and, of course, the sizes carefully laid out to the stated scale. If more than one detail is drawn on a sheet and different scales are used, each should be stated. Fig. 12-4 shows the most used scales for production drawings as well as four different forms for their specification. *Sketches* are not drawn to a scale, only in *proportion*. Sometimes a note— APPROX. HALF SIZE, or similar statement may be used.

12.8 Size and Location Dimensions

Any object, no matter how complicated its structure, consists of an assemblage of simple geometrical shapes, such as prisms, cylinders, cones, pyramids, and spheres. These geometric shapes, either positive or negative, can be readily dimensioned to show individual sizes and locations. A shaft is a positive cylinder while a hole is a negative cylinder.

Contour dimensions show the *size* of the geometric shapes and are therefore known as *Size Dimensions*.

The dimensions that indicate the positions of the geometric shapes with respect to a datum or

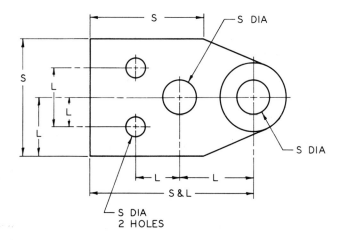

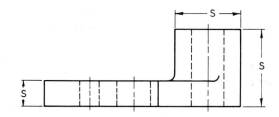

Fig. 12-5. *Size and Location Dimensions.*

to each other are known as *Location Dimensions*. Care must be used in the choice and position of location dimensions in order to assure ease and accuracy of manufacture. The correct placement of the dimensions on an object depends upon its ultimate use and its position with respect to the other parts of the structure.

The two types of dimensions, *Size and Location*, cannot always be sharply defined since in many cases size dimensions are also location dimensions. Fig. 12-5 illustrates size and location dimensions.

DIMENSIONING—BASIC CONCEPTS

12.9 Dimensioning Geometrical Components

The Prism. Fig. 12-6 shows two acceptable methods for placing size dimensions on a prism. Note that two dimensions are referred to the principal view and one to an adjacent view without repetition of the dimension.

The Cylinder. The standards of dimensioning require that two dimensions be placed on the rectangular view of a positive cylinder as shown in Fig. 12-7. On a negative cylinder such as a hole, the diameter and depth is usually given by a local note referred to the circle view, Fig. 12-5.

The Cone. A cone may be dimensioned by one of two standard procedures. In Fig. 12-8a, the height and base diameter are indicated on the triangular view of the cone. The other acceptable method, illustrated in Fig. 12-8b, combines the base diameter and the included angle at the vertex. The cone can be fully described by a dimensioned one-view orthographic drawing. For the frustum of the cone in Fig. 12-8c, the dimensions applied to the bases are followed by DIA to designate its circular shape.

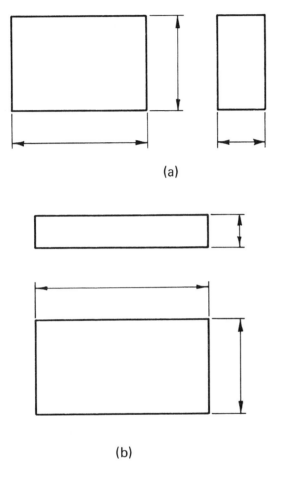

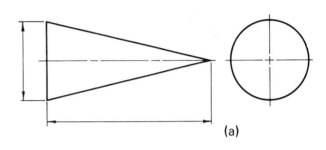

(a)

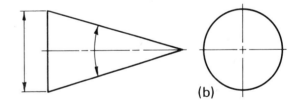

(b)

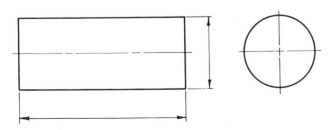

Fig. 12-6. *Prism Dimensions.*

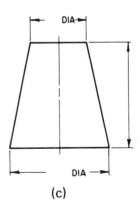

(c)

Fig. 12-7. *Cylinder Dimensions.*

Fig. 12-8. *Cone Dimensions.*

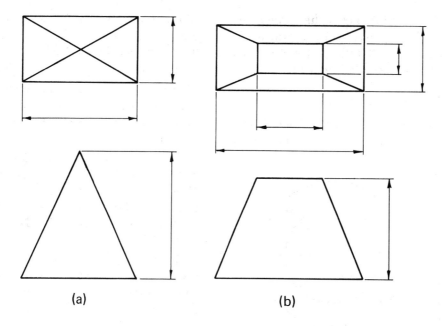

Fig. 12-9. *Pyramid Dimensions.*

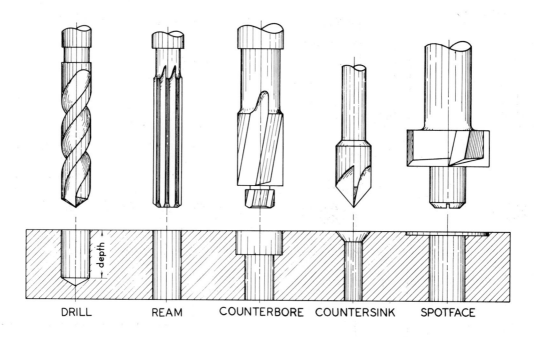

Fig. 12-10. *Standard Machine Tool Operations.*

DIMENSIONING—BASIC CONCEPTS

The Pyramid. Fig. 12-9 shows the dimensions of a right rectangular pyramid and a frustum of a pyramid. Oblique cones or oblique pyramids are dimensioned in a comparable manner.

12.10 Holes, Cylinders and Arcs

An illustration of several Machine Tool Operations for making holes is shown in Fig. 12-10.

Many other pertinent operations including machining, casting, forging, stamping, grinding, etc., are described in Unit 21.

Diameters of positive cylinders are preferably placed on the rectangular view. Large diameters may be located as spokes on the circular view. Leaders are used especially for the hole sizes. See Fig. 12-11.

Circular arcs are dimensioned by giving their radii. Several samples are shown in Fig. 12-12. When a drawing has a number of radii of the same dimension, it is preferable to use a note such as: ALL RADII .12 UNLESS OTHERWISE SPECIFIED.

Rounded Ends. Overall dimensions *should* be used with parts having rounded ends. For *fully rounded ends* the radius is indicated as R but not dimensioned. See Fig. 12-13. For a partially

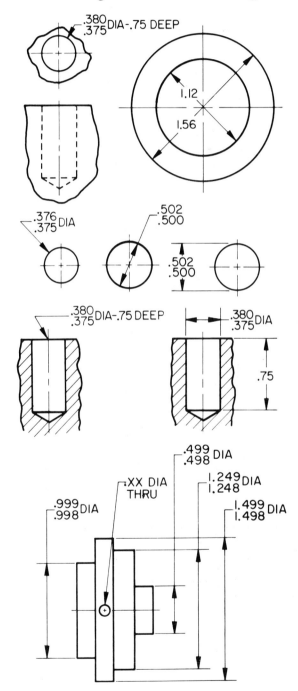

Fig. 12-11. *Diameter of Cylinders and Holes. (ANSI Y14.5)*

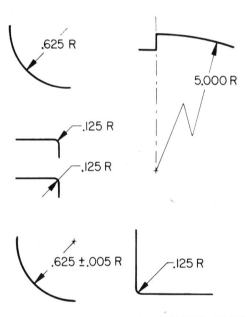

Fig. 12-12. *Radii of Arcs. (ANSI Y14.5)*

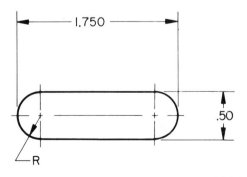

Fig. 12-13. *Rounded Ends. (ANSI Y14.5)*

rounded end the radius is dimensioned, Fig. 12-14. Compound, circular arc curves should be dimensioned by giving the radii and center locations, Fig. 12-15. Symmetrically curved outlines may be dimensioned on only one side of the axis of symmetry, Fig. 12-16.

Slotted holes of regular shape are dimensioned for size by overall length and width, and for location by a dimension to its longitudinal center and to either rounded end. See Fig. 12-17. Exception to this standard is made for slots to be machined between two drilled holes; in which case the dimension from center line to center line of the round ends is given and one center line located.

Counterbored, countersunk, and spotfaced holes are dimensioned as shown in Fig. 12-18. Refer to Fig. 12-10 for method of making these holes.

12.11 Other Features

Chamfers are dimensioned by giving an angle and a length. Fig. 12-19 illustrates several chamfers.

Keyseats are specified by width and depth. The depth is dimensioned from the opposite side of the shaft or hole. See Fig. 12-20.

Knurls should be specified in terms of type, pitch, and diameter before and after knurling. The

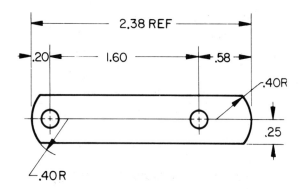

Fig. 12-14. *Partially Rounded Ends. (ANSI Y14.5)*

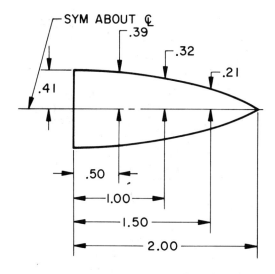

Fig. 12-16. *Symmetrical Curves. (ANSI Y14.5)*

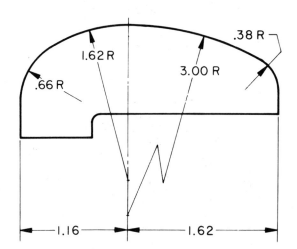

Fig. 12-15. *Circular Arc Curves. (ANSI Y14.5)*

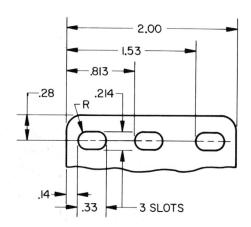

Fig. 12-17. *Slotted Holes. (ANSI Y14.5)*

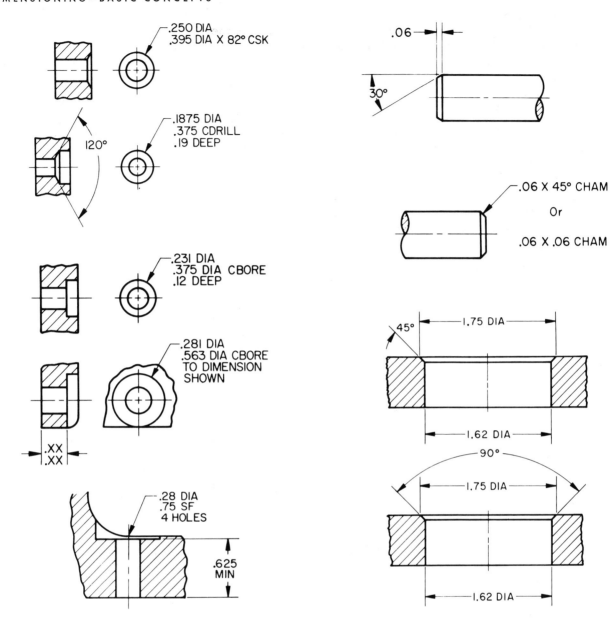

Fig. 12-18. *Countersunk, Counterbored, and Spotfaced Holes. (ANSI Y14.5)*

Fig. 12-19. *Chamfers. (ANSI Y14.5)*

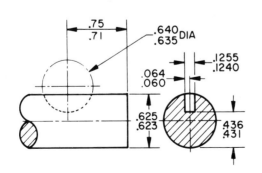

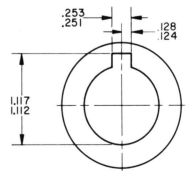

Fig. 12-20. *Keyseats. (ANSI Y14.5)*

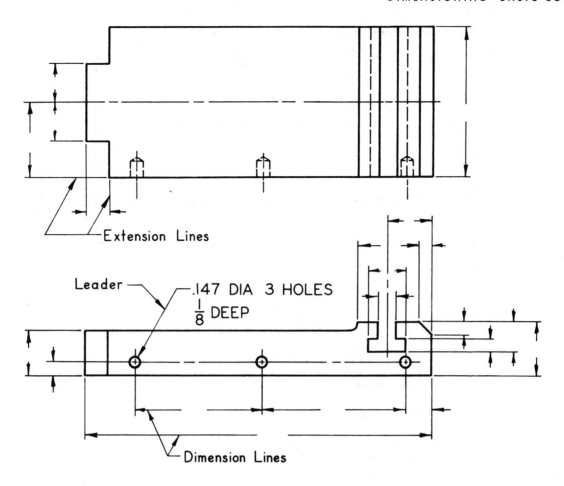

Fig. 12-21. *Placing Dimensions by Contour Rule.*

axial length of knurling may be required where only a portion of a total length is to be knurled.

Screw Threads are specified as shown in Unit 19.

12.12 Contour Rule for Dimensioning

The contour view is the one which will show a *characteristic shape* more clearly than any of the other views. Two of the three necessary size dimensions and the location of the detail should be placed on this contour view. From this consideration the Contour Rule may be stated as follows: *Dimensions should be related to the view that shows the most characteristic shape.*

Refer to Fig. 12-21 and note that the size and location of the Tee Slot is dimensioned on the front view. The projection on the left end is located and specified on the top view which shows its shape. The three small holes are fixed as to location and size on the front projection.

12.13 Overall Dimensions

Generally, the three overall dimensions of length, width, and height of any object are shown on the drawing. The overall dimensions are always placed outside of any series of small size or location dimensions referred to the same view.

Overall Dimensions are usually given with one intermediate dimension omitted. Where the intermediate dimensions are all important, the overall dimension should be marked REF. Refer to Fig. 12-14.

12.14 Dimensions from Datum vs. Chain Dimensions

Where positions are specified by dimensions from a datum line, different features of the part are located with respect to this datum and not with respect to one another. All parallel, independent dimensions are referred to a center line,

DIMENSIONING—BASIC CONCEPTS

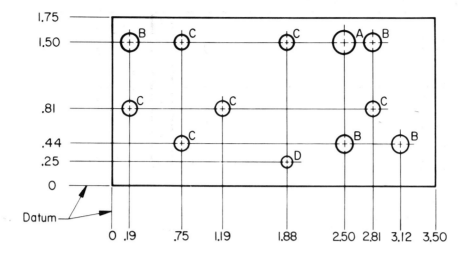

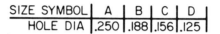

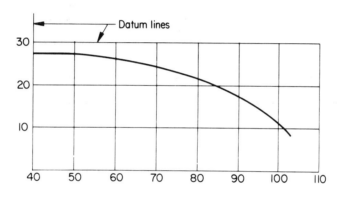

Fig. 12-22. *Dimensions from Datum. (ANSI Y14.5)*

finished surface, or grid datum. The latter is used for programming information for a computer to produce a graphical plot of the required drawing. Also a grid datum is used for large layouts, such as automobile bodies and airplane fuselages. See Fig. 12-22 for datum dimensions.

There are instances where *chain* or continuous dimensions are appropriate. This situation occurs when different features of a part are dependent and therefore located with respect to each other; or where an accumulation of individual variations will not effect the total size. Note that, as Art. 12.13 indicated, one intermediate dimension is omitted where a chain is enclosed by an overall dimension, or the least critical is marked REF. Fig. 12-14 illustrates chain dimensions.

12.15 Dimensioning Procedure

The following steps of procedure are recommended:

1. Use the minimum number of orthographic views necessary for complete shape delineation of the object. Choose a scale for the drawing and a reasonable spacing of views on the sheet to provide ample room for dimensions.
2. For each view, list the features whose contour shape is best delineated by that view.
3. Show the size dimensions (usually two) of each contour shape and the coordinate locations with respect to the most important datum or reference center line.
4. Add overall sizes, notes, and title.

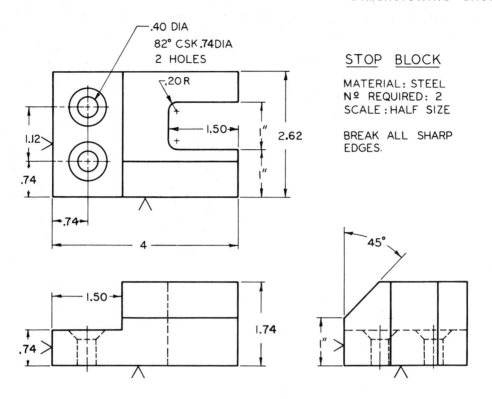

Fig. 12-23. *Dimensions for Production.*

Referring to Fig. 12-23 these steps are illustrated.

Step 1. Three views of the Stop Block are drawn, each being necessary to show critical contour.

Step 2. a. "H" projection shows:
  Size and Location of two countersunk holes.
  Size and Location of the slot in right end.
  b. "F" projection shows:
  Size of the L-shaped cut.
  Relative heights.
  c. "P" projection shows:
  Shape and Location of beveled corner.

Step 3. a. Dimensions placed on H:
  Holes—three dimensions of location and the size specification.
  Slot—three size dimensions and one location.
  b. Dimensions placed on F:
  Size of the left L shape.
  c. Dimensions placed on P:
  Height to the bevel.
  Angle of the bevel.

Step 4. Three overall sizes added.
  Title added, including name of object, material, number required, scale, and general notes.

Fig. 12-24 shows a reproduction of an industrial drawing with complete specifications.

12.16 Good Dimensioning Practices Review

1. Dimensions should be placed outside of the views. Clearness, ease of reading, and shorter extension lines sometimes make it practical to place some dimensions within the outline.
2. Dimensions should generally be placed between views rather than on the outside of adjacent views. However, due to the position of the contour shape or to other reasons this standard may be violated for clearness.
3. A particular dimension should apply to one view only. Do not extend lines between adjacent views.
4. Sufficient dimensions should be given so that the piece can be made without any calculations such as addition or subtraction.
5. Dimensions should not be duplicated.

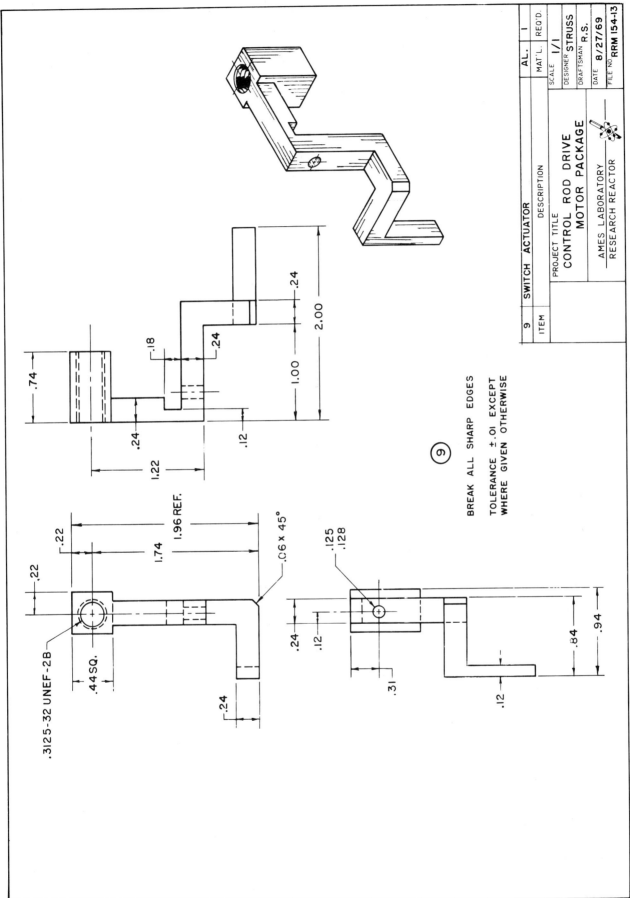

Fig. 12-24. *Industrial Detail Drawing.*
(Courtesy Ames Laboratory, Atomic Energy Commission)

6. Dimensions should be specified only on that view which shows the true length of the dimension.
7. Dimensions should be given from a datum line, center line or mating surface whenever possible.
8. Dimensions should never be crowded. A space of .40 inch should be left between an outline and a parallel dimension and about .24 inch between dimension lines.
9. Numerals are inserted about midway between the arrowheads except they may need to be staggered on adjacent dimension lines.
10. When a series of dimensions relate to one view, the smaller dimensions should be nearer to the view, longer ones further away. Overall dimensions should be outside of all others.
11. Center lines are extended as witness lines, but never used as dimension lines. Fig. 12-1 shows standard size and spacing of numerals, dimension lines, extension lines, and leaders.
12. If possible, extension, or dimension lines should not cross each other.
13. Dimensions should not refer to hidden lines if the specification can be given otherwise.
14. Most center lines are located by coordinate dimensions.
15. Chain dimensioning is poor practice except in special cases. If necessary to use chain dimensions, leave one "link" out of the chain and include an overall dimension. If no dimension is omitted from the chain, the overall or an intermediate dimension of less importance should be marked REF.
16. The point of all arrowheads should *touch* the line to which the dimension refers. Leaders used for circle diameters should proceed *radially* from the circumference and terminate in a horizontal bar at the mid-height of the note. The arrow end of the leader touches the circumference, *not* the center of the circle.
17. Avoid the use of "spoke dimensions" for circles except for large diameters. It is better to use a leader from the circular view or best to relate the dimension to the rectangular view of the cylindrical shape.
18. Dimension a circle by giving the diameter; arcs by giving the radius. When not obvious, DIA follows a diameter dimension; the letter R follows a radius.
19. Blank out section lines for the dimension line and numerals if a dimension must be placed within a sectioned area.
20. All numerals and lettering should be of uniform height, in most cases .10 inch. Fractions should be .20 inch high with the bar in line with the dimension line, not at an angle.
21. Rounding off decimal equivalents to the required number of decimal places follows standard mathematical rules.
22. Numerals for angles in degrees are placed to read from the bottom of the sheet. Other dimensions also are generally unidirectional.
23. Finish marks should show on all edge views of machined or smoothed surfaces.
24. If a dimension given on a drawing is not to the scale of the drawing because of a change or an error in the layout, rather than remaking the drawing a wavy line can be placed under the dimension thus: 4.75.
25. The four essentials of dimensioning are *Accuracy*, *Clearness*, *Completeness*, and *Readability*.

PROBLEMS

The following dimensioning problems include given layouts as well as references to previous units. The necessary orthographic views may be sketched or drawn with instruments. Observe orthographic principles and avoid crowding of dimensions. Each competed detail drawing must contain, in addition to all size specifications, a title indicating the name of the object, what it is made of, how many are required and any other pertinent general notes.

1. Sketch or draw the necessary detail views of the objects described in the following figures and problems.
    a. All Problems listed at the end of Unit 3.
    b. Figs. 9-2, 9-3, 9-4, 9-6, 9-7, 9-9, 9-12, 9-13, 9-14, 9-15, 9-16, 9-17, 9-18.
    c. Figs. 11-1, 11-2, 11-3, 11-4, 11-5, 11-6, 11-12, 11-13, 11-17, 11-19, 11-20, 11-21, 11-22, 11-23, 11-24, 11-25, 11-26.

DIMENSIONING—BASIC CONCEPTS

2. Dimension the objects shown in Fig. 12-25. Using an appropriate scale, measure the drawings for sizes.
3. Given partial views showing three possible **assembly positions** of a PUSH ROD and LEVER, Fig. 12-26. A LINK must be designed so that it can be attached to the Push Rod in its given neutral position, to actuate the Lever from any of the three initial positions, (1), (2), or (3) indicated. By assuming reasonable sizes in proportion to those given, sketch and completely detail the Link.
4. Make completely dimensioned detail drawings of the objects in Fig. 12-27. Supply reasonable sizes for good proportion and design. See Fig. 12-23 for a detail drawing complete for manufacture.

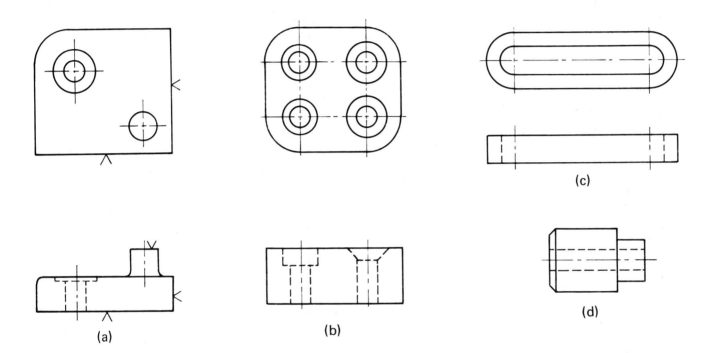

Fig. 12-25.

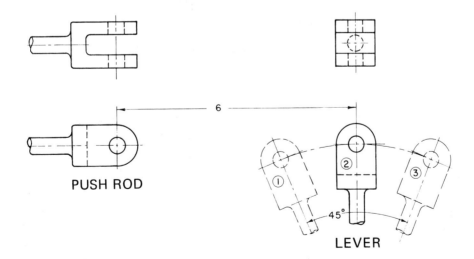

Fig. 12-26.

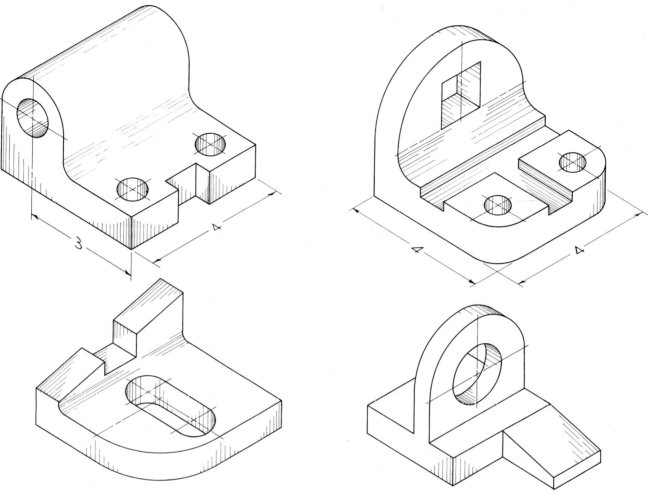

Fig. 12-27.

unit 13

Space Geometry – Points and Lines

13.1 Introduction

Design projects are composed of many different types of problems whose solutions comprise the solution of the project itself. Many of the problems encountered are three dimensional or spatial, and in general lend themselves to a graphical solution. These spatial problems involve relationships existing between points, lines, planes, and solids, either singly or in combination. The many different combinations make the study of these problems important to the understanding of space relationships. Clearance and connector problems are among a few of the applications of point and line fundamentals.

13.2 Perpendicularity of Two Lines

PRINCIPLE. *If two lines are perpendicular to each other, their projections on any plane of projection will appear perpendicular if one or both of the lines appear in true length. See Fig. 13-1.*

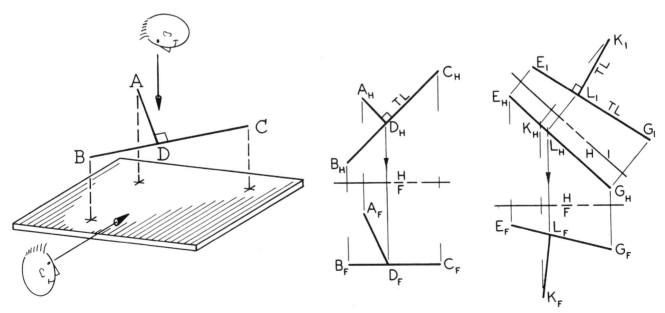

Fig. 13-1. *Perpendicular Lines.*

13.3 Clearance Distance Between a Point and a Line

The clearance distance between a point and a line is the perpendicular distance from the point to the line. Either the Line Method or the Plane Method of solution can be applied to this type of problem.

Problem 1. To find the clearance distance between point P and the line RS. See Fig. 13-2.

SPACE ANALYSIS—Line Method. *The true clearance distance between a point and a line is found in a view showing the given point and the given line as a point.*

Procedure—Line Method:

1. Find true length of line RS with point P shown.
2. In this true length view of RS, locate the perpendicular PX from the point to the line.
   a. See Art. 13.2.
   b. Remember that PX does not appear in true length in this particular case.
3. Obtain the point view of the line RS. This point view shows the desired clearance distance PX in true length.
4. Project point X into the H and F views.
5. Draw the H and F views of the connector PX.

Problem 2. To find the perpendicular distance between point Q and the line TV. See Fig. 13-3.

SPACE ANALYSIS—Plane Method. *The true perpendicular distance between a point and a*

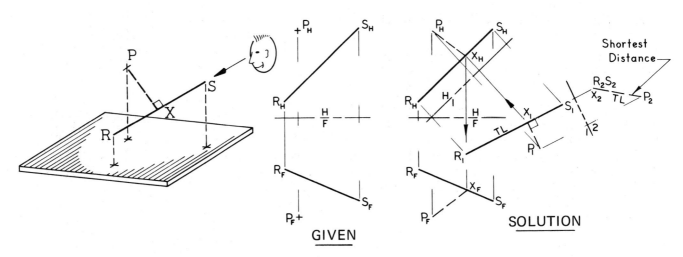

Fig. 13-2. *Distance from a Point to a Line—Line Method.*

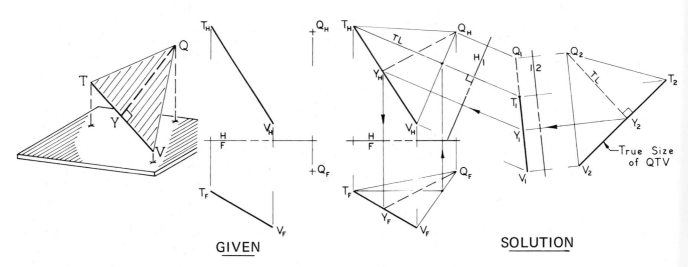

Fig. 13-3. *Distance from a Point to a Line—Plane Method.*

line is found in a view showing the true size of the plane formed by the given point and the given line.

Procedure—Plane Method:
1. Form the triangular plane QTV.
2. Draw the true size view of plane QTV.
3. Draw the perpendicular QY to line TV.
4. Project point Y into the H and F views.

13.4 Connectors

Many types of connector pipes are used to transport liquid, gaseous or bulk materials in industrial installations. The most frequently used are the perpendicular, horizontal, and grade connectors. The next few articles will consider the geometry involved for their solution. Careful examination of Fig. 7-1 will show some practical applications.

13.5 Clearance Between Two Nonparallel, Nonintersecting Lines

The clearance distance is the common perpendicular to each of the given lines. Visualize this space problem by use of two pencils.

Two methods of solution, the Line Method and the Plane Method are applicable to the perpendicular connector.

Problem. To find the length and location of the shortest connector XY between the nonparallel, nonintersecting lines MN and RS. See Fig. 13-4.

SPACE ANALYSIS—Line Method. *The true clearance is found in a view which shows the point view of one of the two given lines.*

The shortest connector will be the perpendicular from the point view to the other line. Visualize this by viewing two pencils held in space so that one of the pencils appears as a point.

Procedure:
1. Construct a view showing one of the given lines in true length. In this case, true length of RS was determined.
2. Obtain the point view of line RS.
3. Construct the perpendicular connector XY. This perpendicular is in true length and can therefore be measured to determine the clearance.
4. Project the connector XY into the other orthographic projections of lines RS and MN. Note that XY is perpendicular to the true length view of line RS.

13.6 Horizontal and Grade Connector Solution

In order to solve problems involving Horizontal and Grade Connectors it becomes necessary to construct an elevation view in which the two given nonparallel, nonintersecting lines appear to be parallel. This can be done by forming a plane

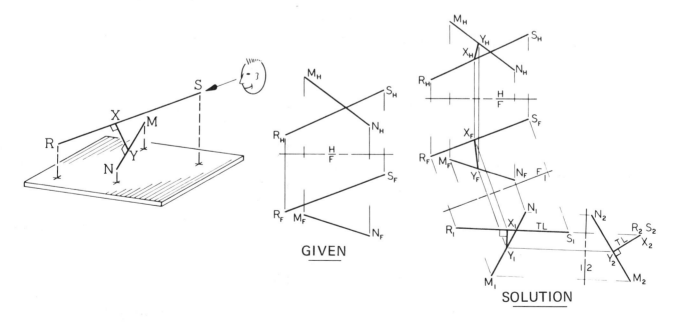

Fig. 13-4. *Shortest Distance between Two Nonparallel, Nonintersecting Lines—Line Method.*

containing one of the lines and parallel to the other line.

The perpendicular, horizontal, and grade connectors will all appear in true length in the elevation view where the given lines appear to be parallel.

13.7 Plane Containing One Line and Parallel to Another Given Line

PRINCIPLE. *A plane is parallel to a line in space if the plane contains a line parallel to the given line.*

Problem. To construct a plane containing line AB and parallel to the line CD. See Fig. 13-5.

SPACE ANALYSIS. *The desired parallel plane will be formed by AB and a line parallel to CD intersecting any point on the line AB.*

Procedure:
1. Through point A of line AB construct the H and F projections of a line parallel to line CD.
   a. A plane has been constructed since, by definition, two intersecting lines determine a plane.
   b. It can be shown graphically that the plane is parallel to CD by construction of an edge view, as in View 1 of Fig. 13-5.

13.8 Clearance Between Two Nonparallel, Nonintersecting Lines—Plane Method

There are many practical applications such as connecting pipes using standard 90° fittings, braces, cables, and so on, for this type of connector problem.

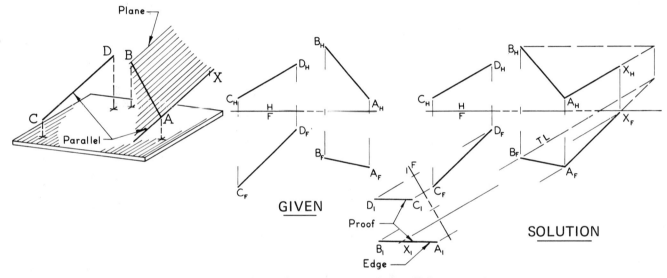

Fig. 13-5. *Plane Through One Line and Parallel to Another Line.*

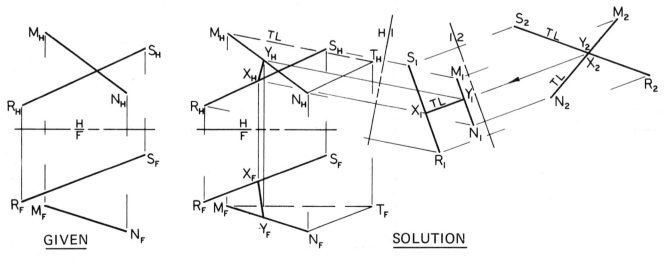

Fig. 13-6. *Shortest Distance between Two Nonparallel, Nonintersecting Lines—Plane Method.*

SPACE GEOMETRY—POINTS AND LINES

Problem. To find the shortest connector XY, and the clearance distance between the nonparallel, nonintersecting lines MN and RS. See Fig. 13-6.

SPACE ANALYSIS—Plane Method. *The true length of the common perpendicular to two nonparallel, nonintersecting lines is found in an elevation view where the two lines appear to be parallel. Its location is determined in an adjacent view showing the given lines in true length.*

Procedure:
1. Construct the plane MNT through MN parallel to RS. See Art. 13.7.
2. Obtain the edge view of plane MNT.
   a. Note that RS is parallel to MNT in this view.
   b. The perpendicular distance between the edge view of MNT and the projection of RS is the true clearance, but its location is undetermined.
3. Construct a view showing lines RS and MN in true length. The exact location of the shortest connector XY is at the apparent point of intersection of the true length views of RS and MN.
4. Project XY to the plan and elevation views.

13.9 Shortest Horizontal Connector Between Two Nonparallel, Nonintersecting Lines

Problem. To find the shortest horizontal connector VW and its exact location between the two nonparallel, nonintersecting lines AB and CD. See Fig. 13-7.

SPACE ANALYSIS. *The true length of the shortest horizontal connector is found in an elevation view where the nonparallel, nonintersecting lines appear to be parallel. Its exact location is found in a point view of the horizontal connector adjacent to the elevation view.*

Procedure:
1. Pass a plane ABT through line AB parallel to line CD.
2. Construct the edge view of plane ABT in an elevation view, and project line CD into this view.
   a. The lines AB and CD should appear parallel.
   b. The direction and true length of the horizontal conector is known in this elevation view, but its location is undetermined.
3. Project the given lines AB and CD onto a plane (plane 2) perpendicular to the true length projection of the horizontal connector. The horizontal connector VW is located at the apparent point of intersection of AB and CD.
4. Project the horizontal connector into the other views of lines AB and CD. Note that the projections of VW on planes 1 and H are parallel and that VW is a horizontal line in the F projection. Why?

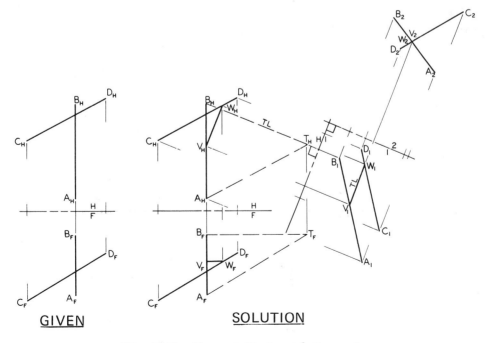

Fig. 13-7. *Shortest Horizontal Connector.*

13.10 Shortest Horizontal Connector—Short Cut Method

Problem. To find the shortest horizontal connector VW and its exact location betwen the nonparallel, nonintersecting lines AB and CD. See Fig. 13-8 (Similar to Fig. 13-7).

SPACE ANALYSIS. *A point view of the shortest horizontal connector is found in an auxiliary elevation view whose hinge line is parallel to a horizontal line lying in a plane containing one of the given lines and parallel to the remaining line.*

Procedure:
1. Pass a plane ABT through AB and parallel to CD.
2. Construct the auxiliary elevation view of lines AB and CD on a plane *parallel to a horizontal line lying on plane ABT*. This view shows the shortest horizontal connector as a point at the apparent intersection of AB and CD. Why?
3. Project the connector VW into the horizontal and front views of lines AB and CD.

13.11 Shortest Grade Connector Between Two Two Nonparallel, Nonintersecting Lines

The solution of this type of problem is similar to that for the shortest horizontal connector in Art. 13.9, and shortest connector in 13.8, except that the direction and amount of grade must be established in the elevation view where the given lines appear parallel.

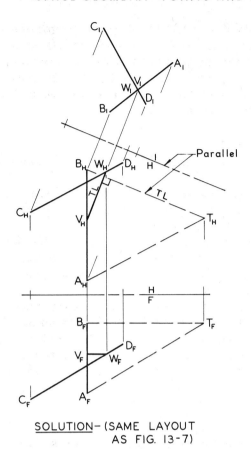

Fig. 13-8. *Shortest Horizontal Connector—Short Cut Method.*

Problem. To locate the shortest 80% grade connector between two nonparallel, nonintersecting lines KL and MN. See Fig. 13-9.

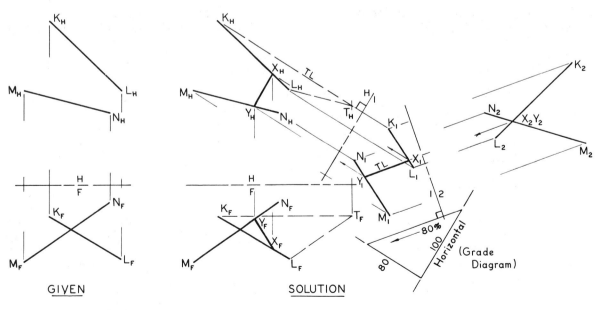

Fig. 13-9. *Shortest + 80% Grade Connector.*

SPACE ANALYSIS. *The true length of a grade connector is found in an auxiliary elevation view where the given nonparallel, nonintersecting lines appear parallel. Its exact location is found in a point view of the grade connector adjacent to the auxiliary elevation view.*

Procedure:
1. Construct plane KLT containing KL and parallel to MN.
2. Construct auxiliary elevation where KL and MN appear parallel. The desired grade line is true length in this view.
3. In this elevation view, establish the direction of the shortest 80% grade connector.
   Construct the grade triangle, Run = 100, Rise = 80.
4. Construct a view of the given lines on a plane perpendicular to the direction of the desired 80% grade line. The point view of the grade connector XY is at the apparent point of intersection of lines KL and MN.
5. Project the grade connector into the Auxiliary Elevation, Horizontal, and Front views of lines KL and MN.

13.12 Vertical Connectors Between Nonparallel, Nonintersecting Lines

Problem. Determine the location and length of a vertical connector between the nonparallel, nonintersecting lines shown in Fig. 13-6.

Vertical connectors appear frequently in industrial applications. The solution of these problems becomes very simple if the definition of a vertical line is kept in mind.

What changes in the lines shown in Figs. 13-5 and 13-9 are necessary before a vertical connection can be made between them?

When is it impossible to construct a vertical connector between two nonparallel, nonintersecting lines?

The above problem and questions are left for the student.

PROBLEMS

In the following space problems locate and find the true length of the shortest connectors between specified points and lines. If the layout drawings given in Figs. 13-10 through 13-15 are doubled, the solution will fit easily on an 8 1/2 x 11 sheet of paper.

1. See Fig. 13-10.
   a. Find the shortest connecting line between point A and the line CD.
   b. Find the shortest connector between lines AB and CD.
2. Using the views of lines AB and CD in Fig. 13-10, determine the shortest connector by means of rotation.
3. Determine the H and F views of the shortest horizontal connector for the lines AB and CD of Fig. 13-10.
4. See Fig. 13-11. JK and LM are the center lines of two oblique pipe lines of equal di-

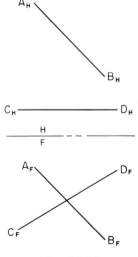

Fig. 13-10.

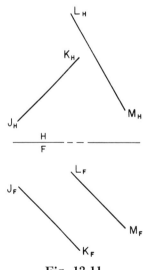

Fig. 13-11.

ameter. Find the maximum possible outside diameters of the pipes.

5. Find the H and F views of the shortest horizontal connector for the pipe lines JK and LM shown in Fig. 13-11.

6. See Fig. 13-12. BC and EF are possible routes for lubrication lines leading to the oil seals on a set of steam turbine reduction gears. If A and D are existing connections on the oil storage tank, which route should be employed for the most economical and efficient gravity flow from the tank to the lines. Why? Show the H and F projections of the line connecting A and BC, and the line connecting D and EF.

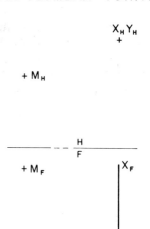

Fig. 13-13.

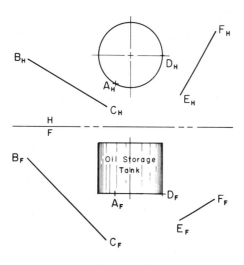

Fig. 13-12.

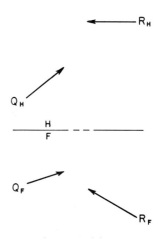

Fig. 13-14.

7. See Fig. 13-13. A line through M has a slope angle of –30° and is 1 1/2 in. from line XY at the nearest point. Show the H and F projections of the line through M and the shortest horizontal connector between it and line XY.

8. See Fig. 13-14. If two airplanes, R and Q, are traveling on the separate paths indicated, what is the closest the planes can possibly approach each other? Show the H and F projections of the planes when they are closest together.

9. See Fig. 13-15. Given the H and F projections of pipes ST and UV. Show the following in both given views: a) shortest horizontal connector; b) shortest vertical connector; and c) shortest 60% grade connector.

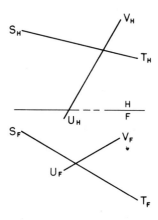

Fig. 13-15.

SPACE GEOMETRY—POINTS AND LINES

10. See Fig. 13-16. Given the H and F projections of points A and C, draw lines AB and CD to the following specifications. AB is horizontal and bears N 50° E. CD has a slope angle of +50° and bears due East. Show properly labeled H and F projections of the following connectors: a) Vertical; b) Horizontal; c) Perpendicular; d) 150% Grade.

11. Points A and C are at the same elevation. C is 1 in. due North of A. Line AB bears N 60° W and has a slope angle of –30°. Line CD bears S 60° W and has a slope angle of +20°. Show the H and F projections of a) shortest connector; b) shortest vertical connector; c) any specified grade connector.

12. Sketch the H and F views of two non-parallel, nonintersecting lines whose

    a. Shortest connector is also the shortest horizontal connector.

    b. Vertical connector is also the shortest connector.

    c. Shortest profile connector is the shortest horizontal and also the shortest connector.

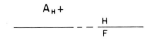

Fig. 13-16.

unit 14

Space Geometry — Lines and Planes

14.1 Introduction

Design problems are usually separated into subproblems and the final solution of the original problem is reached by solution of these components. During the breakdown process some of the components may involve the solution of space relationships involving lines and planes. Problems concerning the location and anchoring of guy wires, intersection of planes and angles formed between them, location of holes to admit cables, pipes or wiring, mining problems, aircraft design and many others may be encountered during the breakdown process. Many of these problems lend themselves to more than one method of solution and no attempt will be made to discuss all possible methods. Reference should be made to the many problems of points, lines and planes found on the Lunar Module shown in Fig. 2-1.

14.2 Lines Piercing a Plane—Edge View Method

Problem. To find the piercing point of a line AB and a plane CDE. See Fig. 14-1.

SPACE ANALYSIS. *The intersection of a line and a plane is found in a view showing the plane as an edge. The exact location of the point of intersection is found in a view adjacent to this edge view.*

Procedure:
1. Construct an edge view of plane CDE.
2. Project line AB into this edge view of CDE.
3. Locate the piercing point X at the intersection of the line AB and the edge view of the plane CDE.
4. Determine the piercing point by projecting X into the H and F views of the plane CDE.

14.3 Visibility of a Line and an Opaque Plane

Problem. To determine the proper visibility of a line AB and an opaque plane CDE. See Fig. 14-2.

SPACE ANALYSIS. *The adjacent projection of any apparent intersection of two nonintersecting lines will show that one of the lines is closer to the observer and is therefore a visible line.*

Discussion.

1. *Visibility in Horizontal View.*

    As the projection $A_H B_H$ of the line crosses the projection of the plane, there is an *apparent* point of intersection Q_H between the edge of $C_H E_H$ and the line $A_H B_H$. To determine the visibility in the H view, construct a projector from point Q_H to the front view of the given line and plane. If this projector crosses the edge $C_F E_F$ before it crosses $A_F B_F$, edge CE lies above the line and is visible in the Horizontal View. The line AB, therefore, passes below the plane and is hidden until it emerges at the piercing point X.

2. *Visibility in the Front View.*

    The visibility of the line and the plane can be determined in the *Front* view by the same

SPACE GEOMETRY—LINES AND PLANES

procedure used in part (1) except to interchange the projections, i.e., change the H subscripts to F and F subscripts to H.

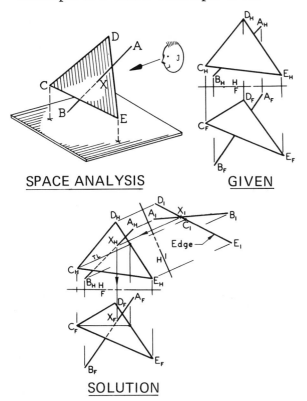

Fig. 14-1. *Line Piercing a Plane—Edge View Method.*

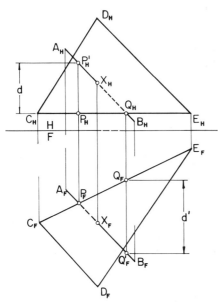

Fig. 14-2. *Visibility of Lines.*

14.4 Line of Intersection of Two Planes—Edge View Method

Intersections of surfaces are very common in everyday life. All branches of engineering are concerned with intersections in one form or another from printed circuits to construction.

Two common methods, *edge view* or the *auxiliary cutting plane* can be used for the solution of intersection problems.

Problem. To locate the line of intersection of planes MNOP and QRST and to show proper visibility of the planes. See Fig. 14-3.

SPACE ANALYSIS. *If two intersecting planes are viewed so that the edge view of one cuts across the other plane, this line of cut is the line of intersection of the two planes.* Visualize this by viewing two intersecting triangles, oblique to each other, held so that one of the triangles appears as an edge.

Procedure:

1. Construct an edge view of plane QRST, projecting plane MNOP into this same view.
2. Locate the projection of the line of intersection X_1Y_1 where the edge view of plane QRST cuts across plane MNOP in View 1.
3. Project the intersection line XY into the given H and F views.
4. Considering the planes to be opaque, determine the correct visibility of the planes in the H and F views, using the method explained in Art. 14.3.

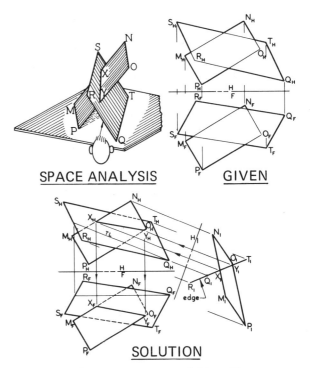

Fig. 14-3. *Intersection of Two Planes—Edge View Method.*

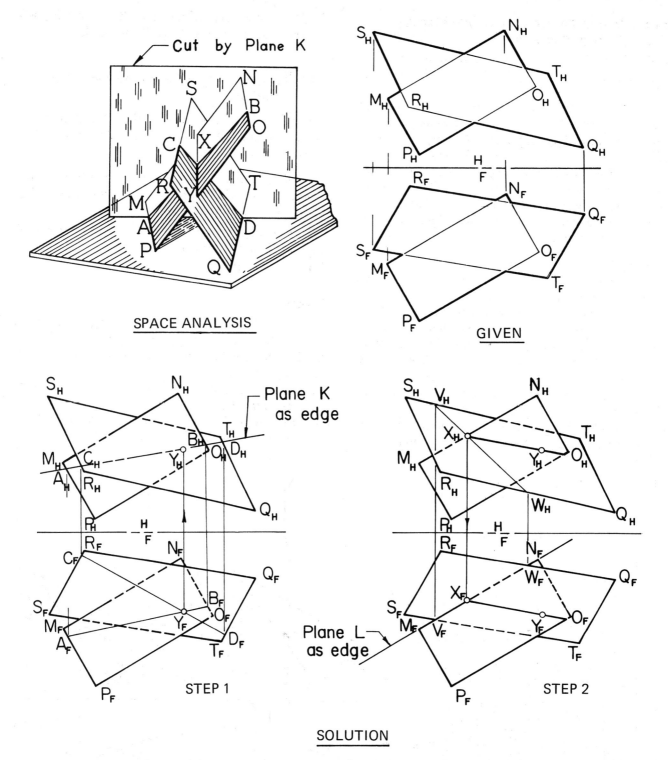

Fig. 14-4. *Intersection of Two Planes—Cutting Plane Method.*

14.5 Auxiliary Cutting Plane Method

An auxiliary cutting plane, appearing as an edge is introduced into one of the two given views of two intersecting planes, Fig. 14-4, step 1. This cutting plane forms two lines of intersection, AB and CD, on the two original planes. If these two lines of intersection are extended they will meet at point Y, common to the given planes and the cutting plane.

SPACE GEOMETRY—LINES AND PLANES

14.6 Line of Intersection of Two Planes—Cutting Plane Method

Problem. To locate the line of intersection of two opaque planes MNOP and QRST. See Fig. 14-4.

SPACE ANALYSIS. *If two given oblique planes are cut by a third auxiliary plane, the two resulting lines of intersection will have one point common to both given planes. A second auxiliary plane is used to find a second point. These two points will determine the indefinite line of intersection. For ease of solution the auxiliary planes should be selected to appear as edges in either the H or F views.*

Procedure:

1. Pass the cutting plane K appearing as an edge in the horizontal projection of the given planes.
2. Locate the lines of intersection AB on MNOP and CD on QRST formed by plane K.
3. Project AB and CD into the front view to locate the point Y common to all three planes.
4. Pass a second plane L, Fig. 14-4, step 2.
5. Locate the point X, also common to all three planes.
6. Connect points X and Y, extending this line to form the desired line of intersection.
7. Determine the correct visibility of the boundary lines of planes MNOP and QRST.

14.7 Line Piercing a Plane—Cutting Plane Method

Problem. To find where line AB pierces an opaque plane CDE, using only the given two views. See Fig. 14-5.

SPACE ANALYSIS. *An auxiliary cutting plane, containing the line, forms a line of cut intersecting the given line at the piercing point.*

This analysis is illustrated in the pictorial of Fig. 14-5. MN is the line of cut and point O is the piercing point.

Procedure:

1. Pass a cutting plane K appearing as an edge in either of the two given views.
2. Locate the line of intersection MN in both given views.
3. Locate the piercing point O common to lines AB and MN.
4. Determine the visibility of the line AB and the opaque plane CDE. See Art. 14.3.

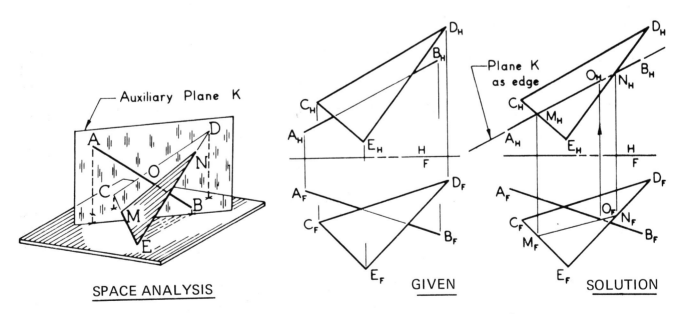

Fig. 14-5. *Line Piercing a Plane—Cutting Plane Method.*

14.8 Dihedral Formed by Two Planes—Edge View Method

Problem. To find the dihedral formed by planes GKL and MNO. See Fig. 14-6.

SPACE ANALYSIS. *The true angle between two planes is found in a view showing their line of intersection as a point and both planes as edges.*

Procedure:
1. Locate line of intersection XY. (Step 1)
2. Obtain point view of line of intersection. (Step 2)
3. Measure the angle β formed between the planes as edges.

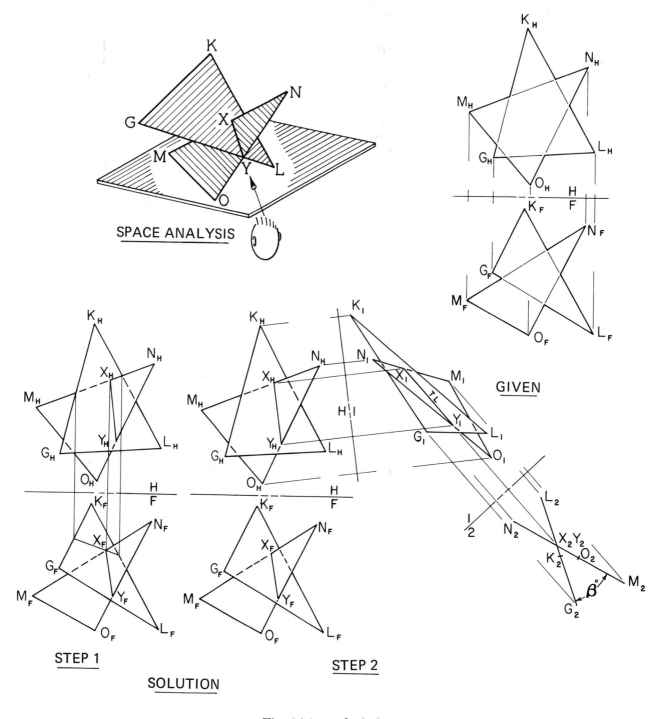

Fig. 14-6. *Dihedral.*

SPACE GEOMETRY—LINES AND PLANES

14.9 Perpendicularity of a Line to a Plane

PRINCIPLE. *A line is perpendicular to a plane if it is perpendicular to every line lying on the plane.*

Each projection of a perpendicular from a point to a plane is located in every view by drawing from the projection of the point perpendicular to a true length line on the plane in that view. See projections $P_F X_F$ and $P_H X_H$ in solution 1, Fig. 14-7.

14.10 Projection of a Point Onto an Oblique Plane

Problem. To project a point P onto the plane KLM. See Fig. 14-7.

SPACE ANALYSIS. *The projection of a point onto an oblique plane lies at the piercing point of the line from the point perpendicular to the plane.*

Procedure:

Solution 1

1. Find projections of true length lines lying on the horizontal and front views of the plane KLM. Art. 14.9.
2. From point P establish the direction of perpendicularity of a line by dropping perpendiculars *toward* the true length projections in the H and F views of the plane.
3. Using the auxiliary cutting plane method, Art. 14.5, find the piercing point P′ of the perpendicular line from point P to the plane KLM.

Solution 2

This problem can also be solved by use of an auxiliary view showing the plane KLM as an edge shown in Fig. 14-7, solution 2. Careful examination of the given figure will show the procedure used to locate the piercing point P′.

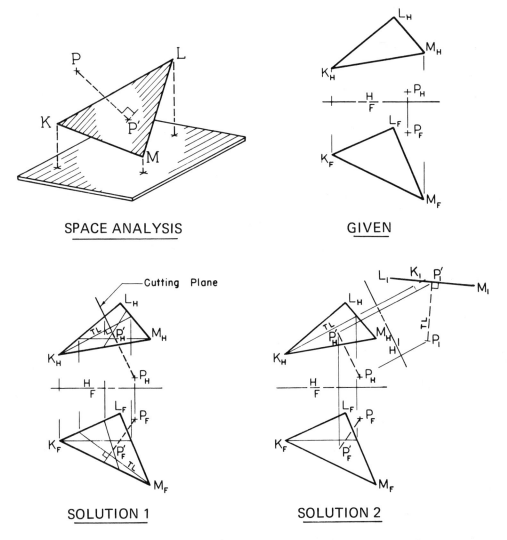

Fig. 14-7. *Perpendicular to a Plane from a Point in Space.*

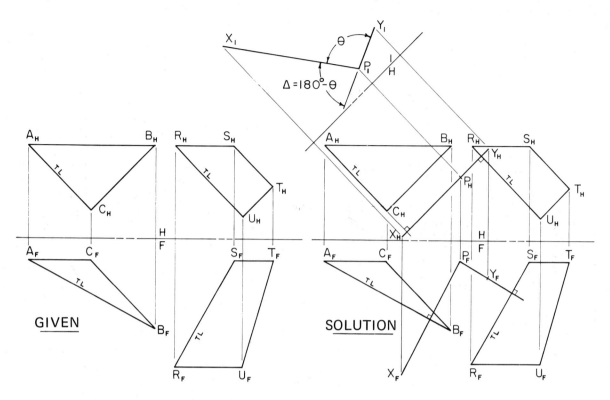

Fig. 14-8. *Dihedral–Method of Perpendiculars.*

14.11 Dihedral Between Two Oblique Planes—Method of Perpendiculars

Sometimes it is inconvenient to establish the line of intersection formed by two planes before determining the dihedral.

Problem. To find the dihedral between two given oblique planes ABC and RSTU. See Fig. 14-8.

SPACE ANALYSIS. *The dihedral is found in the true size view of a plane formed by dropping perpendiculars toward the given planes from a point in space. The desired angle is the supplement of the actual angle between the two perpendiculars.*

Procedure:

1. Select a convenient point P in space.
2. From point P establish the directions of perpendiculars PX and PY to the planes ABC and RSTU. Note that it is *not* necessary to determine the piercing points of the perpendiculars PX and PY.
3. Construct a true size view of the plane XPY formed by the perpendiculars.
4. Measure the angle θ between PX and PY.
5. Graphically subtract θ from 180° to determine the desired angle.

14.12 Angle Between a Line and a Plane—Line Method

Problem. To measure the angle between the line AB and the plane DEG. See Fig. 14-9.

SPACE ANALYSIS. *The angle between a line and a plane is found in a view showing the true size of a plane formed by the line and its projection upon the given plane.*

Procedure:

1. Drop perpendiculars from the line AB to the given plane DEG.
2. Determine the piercing points X and Y of the perpendiculars. Join X and Y to form the projection of AB on the plane.
3. Construct a true size view of the new plane ABYX.
4. Measure the angle θ between AB and XY.

SPACE GEOMETRY—LINES AND PLANES

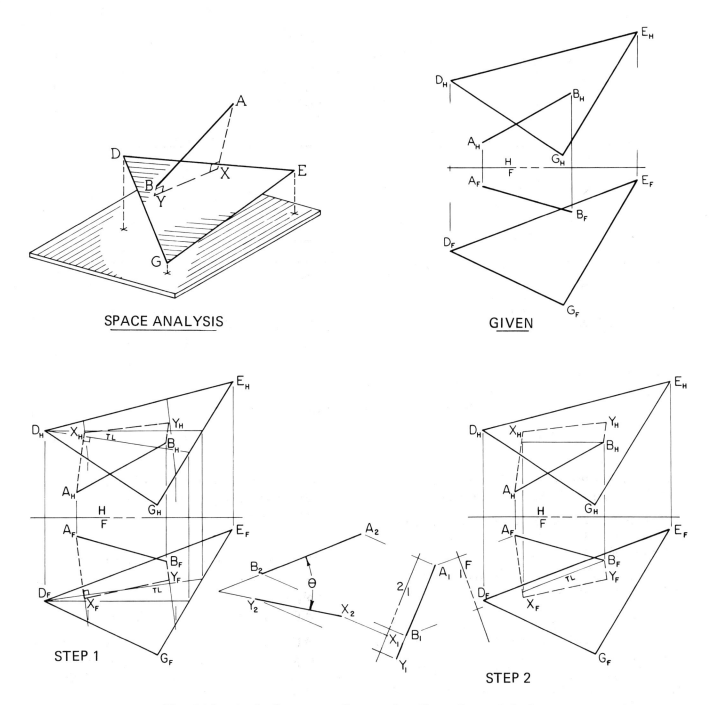

Fig. 14-9. *Angle Between a Line and a Plane—Line Method.*

14.13 Angle Between Line and Plane—Edge View Method

Problem. To measure the angle between the line JK and a plane MNO. See Fig. 14-10.

SPACE ANALYSIS. *The angle between a line and a plane is found in a view which shows the plane as an edge and the line in true length.*

Procedure:
1. Construct an edge view of the plane MNO. The line JK is not in true length in this projection.
2. Construct a true size view of the plane MNO.
3. Adjacent to the true size view, construct a view showing the line JK in true length.
4. Measure the angle θ between the edge view of MNO and the true length view of JK.

14.14 Location of Objects Upon a Plane

It often becomes necessary to locate a point, line or solid upon the surface of an oblique plane. This locating must be done in a view showing the true size of the given oblique plane in order that the true measurements can be determined.

Problem. A hole for a 24 inch circular, vertical stack centered on point P is to be cut in the roof AB. Find the H and F projections and a scale size template of the hole to be cut in the roof. See Fig. 14-11.

No discussion of procedure is made because it is felt that careful examination of the given data and the figure showing the solution will provide all necessary explanation.

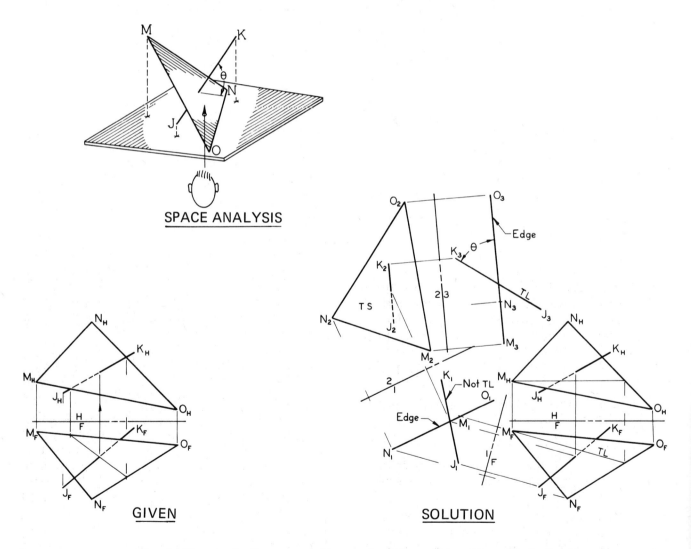

Fig. 14-10. *Angle Between a Line and a Plane—Edge View Method.*

SPACE GEOMETRY—LINES AND PLANES

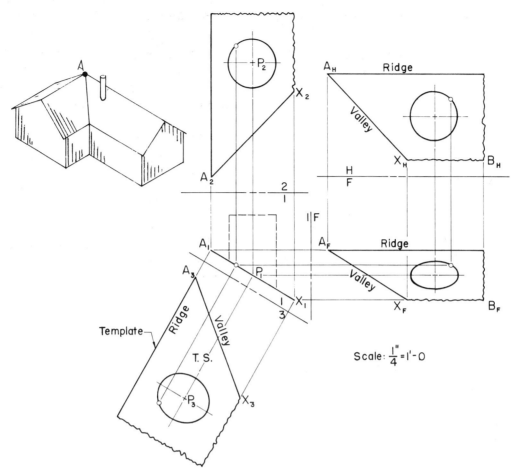

Fig. 14-11. *Locating a Curve on an Oblique Plane.*

PROBLEMS

Line Piercing a Plane

1. Fig. 14-12. Use the auxiliary-plane method to find where the line pierces the plane in each of the given cases. Use the edge-view method to check your results. Show proper visibility of line and plane in each instance. Sketch a pictorial view of each problem showing the space analysis for its solution.

2. Using rotation, determine the piercing point of the lines and planes as shown in Fig. 14-12.

Line of Intersection

3. Fig. 14-13. Find the line of intersection of planes PDQ and JKL by the edge-view method. Show proper visibility.

4. Fig. 14-14. Find the line of intersection between the two planes using the auxiliary-plane method. Check your results by the edge-view method.

5. Fig. 14-15. Plane CDE dips 60° SE and plane XYZ dips 45° SW. The strike line CD is 1/8 in. higher than strike line XY. Find the line of intersection and determine proper visibility in the given H projection.

Dihedral Angle

6. Fig. 14-16. Find the value of the dihedral angle.

7. Fig. 14-17. A pattern maker needs a template to check the angle formed by the planes K and L of the given wooden pattern. Determine the value of this dihedral angle.

8. Fig. 14-18. Given two views of a transition piece, determine the dihedral angles formed by the surfaces A and B, B and C, C and D, D and A.

Dihedral Angle and Line of Intersection

9. Fig. 14-19. Determine the dihedral angle and show the H and F projections at the line of

intersection between a) planes CDOP and RQT: b) planes CDOP and QRS; c) planes CDOP and SQT.
10. Fig. 14-20. Walls A and B are faces of a wing wall abutment for a new superhighway. The sloping face of wall A has a batter $\left(\dfrac{\text{rise}}{\text{run}}\right)$ of 3/4. Wall B has a slope ratio $\left(\dfrac{\text{rise}}{\text{run}}\right)$ of 1. Show the complete H and F projections of both walls and their line of intersection. Determine graphically the true angle between their sloping surfaces.
11. Determine the dihedral formed by the planes shown in Fig. 14-14. Use method of perpendiculars.
12. Determine the line of intersection and the dihedral of the planes shown in Fig. 14-21.

Projecting Points and Lines on a Plane

13. Fig. 14-12(a). Project point M onto plane ABC using the edge view method.
14. Fig. 14-12(b). Project points P and Q onto plane DEF using only the two given views.
15. Fig. 14-12(a). Project line YZ onto plane ABC using the edge-view method. Show proper visibility.
16. Fig. 14-12(b). Project line WX onto plane DEF using only the two given views. Show proper visibility.

Angle Between a Line and a Plane

17. Fig. 14-12(b) and (c). Find the true angle between the line and the plane in each of the given problems. What would be the angle between a vertical line and the given plane in each case?
18. The bottom portion of a right regular, triangular pyramid with a vertical axis, is cut off by a rectangular plane, strike N 30° E and dip 30° NW, which makes a horizontal line of intersection with one of the faces of the pyramid.
    a. Construct the H and F views showing the remaining upper portion of the pyramid sitting on the top surface of the rectangular plane. Show correct visibility.
    b. Show a true size view of the intersection formed by the pyramid and the rectangular plane.
    c. Determine the value of the dihedral angles formed by the faces of pyramid and the rectangular plane.
    d. Determine the true angles formed by the edge lines of the faces and the rectangular plane.

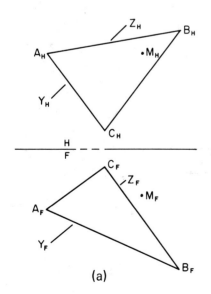 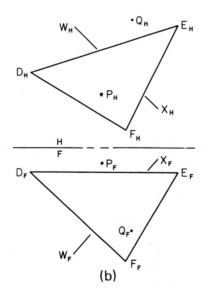 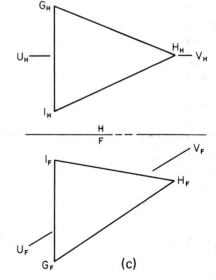

Fig. 14-12.

19. A rectangular mirror ABCD has a strike N 30° W and a dip of NE 60°. A horizontal ray of light, bearing due west strikes the center of the surface of the mirror and is reflected. What is the angle between the ray and the mirror? Show the projections of the reflected ray.
20. Motor boat windshield problem Fig. 14-22. Find the following angles:

a. between planes M and N.
b. between planes N and P.
c. between planes N and R.
d. between planes M and R.
e. between line AB and plane M.
f. between line DC and plane M.
g. between line EB and plane N.

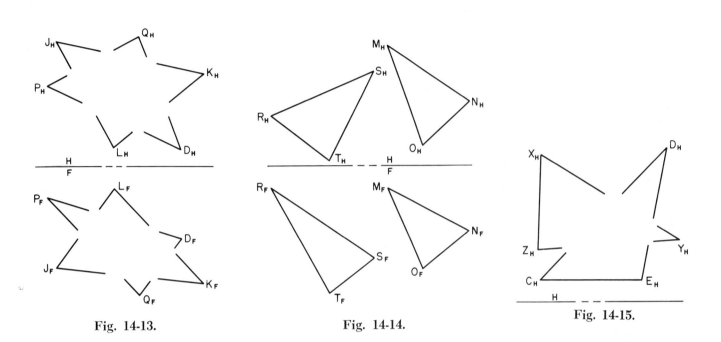

Fig. 14-13.　　　　Fig. 14-14.　　　　Fig. 14-15.

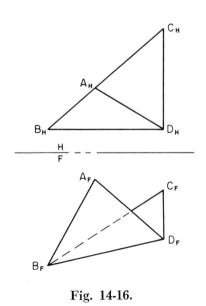
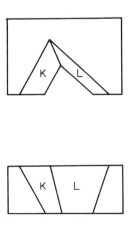
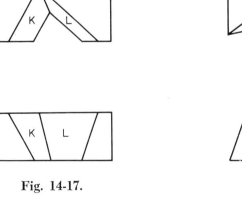
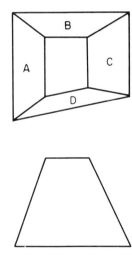

Fig. 14-16.　　　　Fig. 14-17.　　　　Fig. 14-18.

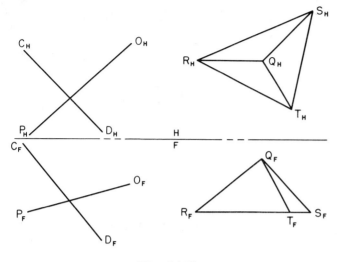

Fig. 14-19

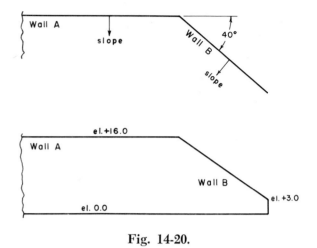

Fig. 14-20.

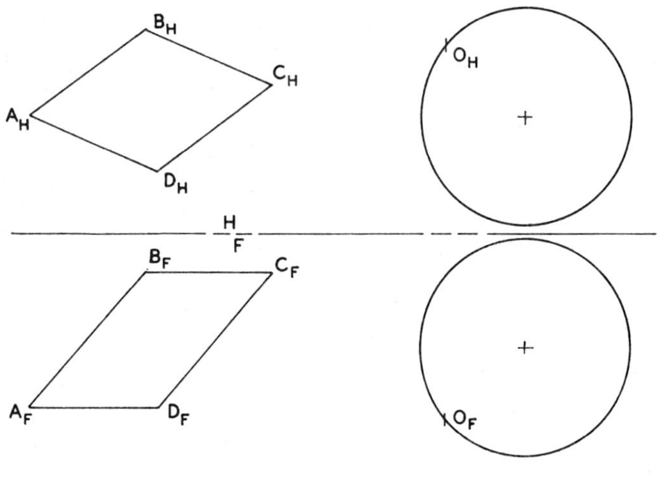

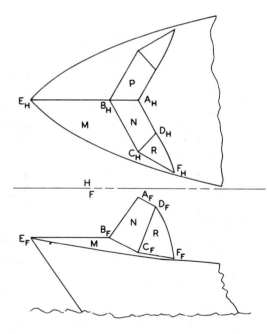

Fig. 14-21.

Fig. 14-22.

unit **15**

Vector Geometry

15.1 Graphical Solutions

Structural systems must be thoroughly analyzed for strengths and stresses before a design problem is complete. Forces of tension and compression can be represented vectorially during this analysis. Many quantities such as velocity, power, displacement, electrical properties and others may be represented by vectors.

The graphical method has considerable merit for solving many vector problems because it provides two important advantages when compared to mathematical solutions.

First, data can frequently be presented graphically with greater clarity than possible by any other method.

Second, the graphical method often reduces complex, time consuming computations whose equations require many stages of substitution before the unknowns can be calculated. In such cases the quicker graphical solution is often more efficient because it requires less time and the calculations can be easily checked.

The designer should strive to integrate all methods of solution available to him. Each method will serve as an effective check on solutions determined by other methods.

15.2 Definition of Terms

A knowledge and understanding of the terminology of graphical vectors is necessary before the techniques of solution can be applied.

1. *Scalar Quantity* is one having magnitude only and whose measure is completely described by a single number. Quantities such as volume, temperature and pressure are scalar.
2. *Vector Quantity* is one having both magnitude and direction which can be represented by a directed line segment and which can be added graphically. Velocity, displacement, and force are vector quantities.
3. *Vector* is any directed straight line segment whose length is proportional to the magnitude of the vector quantity.
4. *Force* is a vector quantity, considered as a push or pull, which changes or tends to change the state of rest or motion of a body.
5. *Vector Chain* is a series of vectors drawn head to tail in proper continuity of direction, forming a closed circuit when the vector system is in equilibrium.
6. *Equilibrium* is a resulting state of rest or of uniform motion when the sum of all vector quantities acting on a body equals zero.
7. *Resultant* is the simplest single vector quantity which will produce the same effect when substituted for the vector quantities acting on a body.
8. *Equilibrant* is the vector quantity which will produce equilibrium. The magnitude of the equilibrant is equal to that of the resultant, but acts in the opposite direction.
9. *Composition* is the process of reducing a vector system to a simpler system.

10. *Component* is any one of two or more vector quantities into which a given vector can be resolved.
11. *Resolution* is the process of replacing a vector quantity by its components.

15.3 Definitions of Vector Systems

1. *Collinear* vector quantities act along a common line of action. These quantities may be added or subtracted arithmetically to determine the resultant acting at the point of application.
2. *Concurrent* vector quantities act at a common point of application.
3. *Nonconcurrent* vector quantities act at different points of application.
4. *Coplanar* vector quantities lie in the same plane. This vector system can be described by a single-view drawing.
5. *Noncoplanar* vector quantities do not lie in the same plane. This vector system requires two or more orthographic views for a full description.

Part I

PLANE VECTORS

15.4 Coplanar Vector Solution

Parallelogram Law. The resultant of two concurrent, nonparallel vector quantities is the diagonal of a parallelogram formed by the two vectors, Fig. 15-1a, 15-1b.

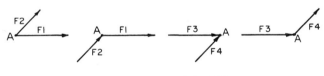

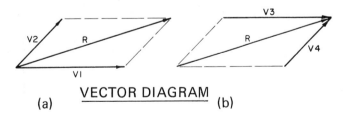

Fig. 15-1. *Composition—Parallelogram Method.*

Triangle Method. Examination of Fig. 15-1a and 15-1b shows that the diagonal representing the resultant divides the parallelogram into two triangles. Therefore, it is only necessary to construct either one of these triangles to determine the resultant.

If two vector quantities are represented by their vectors placed tip to tail, their resultant vector is the third side of the triangle. The direction of the resultant is from the tail of the first vector to the tip of the second, in other words, opposed in direction to the other two. Fig. 15-2.

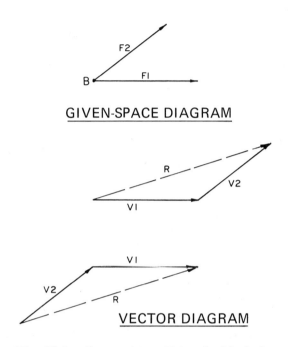

Fig. 15-2. *Composition—Triangle Method.*

Either the parallelogram of the triangle method can be used to determine the resultant of a system containing two or more coplanar, concurrent vector quantities. See Fig. 15-3a, 15-3b. The parallelogram method is very cumbersome if three or more vectors are involved.

Any number of forces may be outlined in a coplanar vector polygon but if the polygon is to be closed, it can contain no more than two *unknown* magnitudes, with known directions.

Polygon Method. A more convenient method for the composition of several concurrent, coplanar vector quantities is to construct a vector chain or polygon representing the given vectors and their lines of action, Fig. 15-3c. In order that the system shall be in equilibrium, close the vector circuit by a line drawn from the tail of the first vector V1 to the tip of the last vector V4. This line represents the magnitude and line of action of the resultant or the equilibrant. If the arrowhead its placed on this line so that its direction

VECTOR GEOMETRY

opposes the chain direction, the *resultant* has been determined. The line of action and magnitude of the *equilibrant* equals that of the resultant, but its direction is opposite.

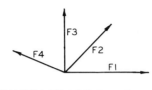

(a) GIVEN-SPACE DIAGRAM

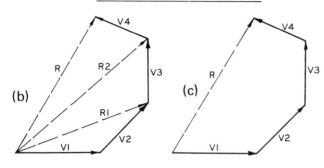

VECTOR DIAGRAM

Fig. 15-3. *Composition—Polygon Method.*

15.5 Resolution of a Force

A single force may be replaced by a system of forces whose vector sum equals the original force.

Problem. If a 100 pound weight is suspended by ropes, determine the pull in the two ropes MA and MB. See Fig. 15-4.

SPACE ANALYSIS. *The components of a given load in a coplanar force system are found by closing a vector diagram whose sides are parallel to the given members of the system.*

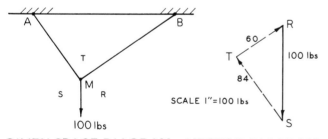

GIVEN-SPACE DIAGRAM   VECTOR DIAGRAM

Fig. 15-4. *Coplanar Vectors.*

Procedure:
1. Select and record a suitable vector scale.
2. Construct a vector RS scaled to represent 100 pounds and parallel to the downward line of action of the load.
3. Through S draw a line parallel to line MA.
4. Through R draw a line parallel to line MB until it intersects the line paralel to MA at point T.
5. Add arrowheads at T and R to complete the vector chain, R to S to T to R.
6. Determine magnitudes of the two unknowns ST and TR.

15.6 Resultant of a Concurrent, Coplanar System

Problem. Four concurrent, coplanar forces acting at point A, Fig. 15-5, have the following specifications:

F1, N60W, 30 lbs.; F2, N45E, 35 lbs.; F3, Due E, 20 lbs.; F4, Due S, 29 lbs. Determine the magnitude and bearing of the resultant R for this system.

SPACE ANALYSIS. *The resultant of a concurrent, coplanar system is found in a view showing the vector chain and a closing link whose direction opposes the direction of the vectors in the chain.*

Procedure:
1. Select a satisfactory vector scale.
2. Construct a closed vector diagram by the method shown in Fig. 15-3c.
3. Measure the true magnitude of the resultant R.
4. Determine the bearing of the resultant R.

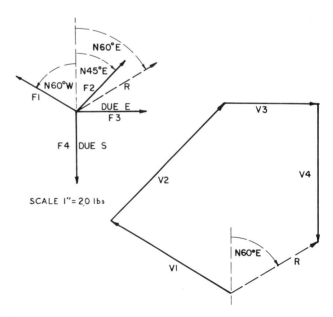

Fig. 15-5. *Concurrent Coplanar Vectors.*

Part II

NONCOPLANAR VECTOR SYSTEMS

15.7 Noncoplanar Vector Systems

The principles of orthographic projection are used to determine magnitudes, directions, and relationships of space vector quantities. For complete description and graphical solution of a three dimensional, or noncoplanar system of forces, it is necessary to construct two or more orthographic views of the space and vector diagrams.

15.8 Composition of Concurrent, Noncoplanar Vector Quantities

Problem. Determine the magnitude and direction of the resultant for the system of concurrent, noncoplanar forces shown in Fig. 15-6.

SPACE ANALYSIS. *The resultant and its direction is found by closure of a vector diagram representing the given forces and their directions.*

Procedure:
1. Determine the H and F views of the projected magnitude of the given forces. Fig. 15-6b.
2. Construct the H and F views of the vector diagram for the given system. Fig. 15-6c.
3. Close the vector diagram.
   a. This closing line establishes the line of action.
   b. The direction of the resultant opposes that of the vector diagram.
4. Determine the true magnitude of R in a true length view of the vector.
5. Draw the H and F views of the resultant and its line of action on the space diagram.

15.9 Non-vertically Loaded Tripod

Problem. Determine the type and magnitude of the stress in the members of a pin-connected tripod with a load of 500 pounds as shown in Fig. 15-7.

A practical application of this type of structure is illustrated at the top of the test stand in Fig. 15.8.

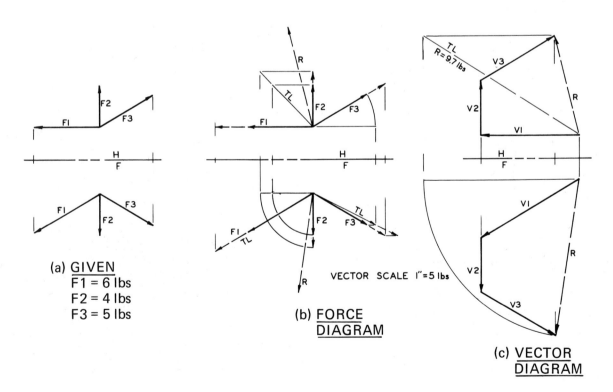

Fig. 15-6. *Concurrent Noncoplanar Vectors.*

VECTOR GEOMETRY

SPACE ANALYSIS. *The reaction in the members of a tripod (concurrent, noncoplanar) structure is found by resolving the given load into components parallel to each leg of the given structure.*

Procedure:

1. Determine the vector diagram of the tripod and the projections of magnitude of the 500 lb. load.
2. Select a plane containing two of the members. In this case, the plane AOC was chosen.
3. Through the free end of the *projected* true length load vector, draw a line parallel to the remaining member OB.
4. Determine the exact location of the piercing point P of the line parallel to OB and the selected plane AOC.
5. Through the piercing point P draw a line parallel to line OC until it intersects the line OA. These are the lines of action for OC and OA.
6. Proceeding around the vector diagram in the direction indicated by the load vector, place arrows on OB, OC, and OA to form the closed vector chain in both views.
7. Determine the true magnitudes represented by vectors OB, OC, and OA. See Fig. 15-9.
8. Determine the type of stress in member OA, OB, and OC. To find whether a member is in tension or compression, transfer the direction of the vectors, shown by the arrowheads on the vector diagram to the corresponding members of the tripod. If the indicated direction on the member is away from the joint, as on OA, the member is in tension. If the direction is toward the joint, as on OB and OC, the member is in compression.
9. Tabulate the results.

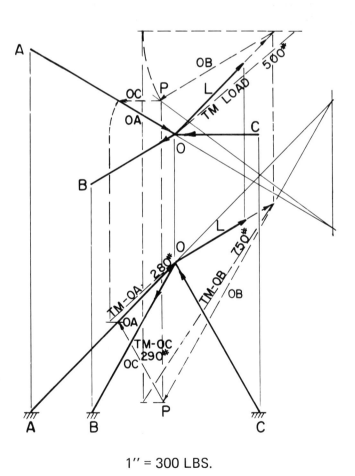

Fig. 15-7. *Tripod Structure—Nonvertical Load.*

Fig. 15-8. *Crane. (Courtesy N.A.S.A.)*

15.10 Vertically Loaded Tripod

Problem. Determine the type and magnitude of the stress in the members of a vertically loaded pin-connected tripod with a load of 500 pounds as shown in Fig. 15-9.

Procedure:

The solution of this problem is identical to the one in Art. 15.9. However, in this case the piercing point can be quickly found because one of the planes, BOC in Fig. 15-9 appears as an edge in the front projection.

Careful examination of the Dragline pictured in Fig. 15-10 will show many other problems which can be solved vectorially.

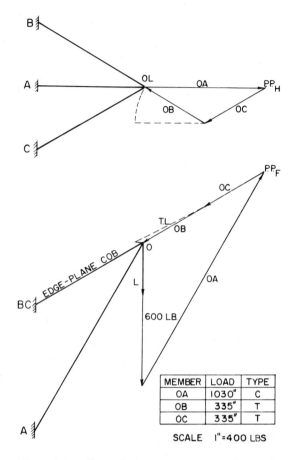

Fig. 15-9. *Tripod Structure—Vertical Load.*

Fig. 15-10. *Dragline. (Courtesy Bucyrus-Erie Company)*

VECTOR GEOMETRY

PROBLEMS

1. Fig. 15-11. Given the space diagrams for two coplanar force systems, both of which are in equilibrium. Determine the magnitude and kind of stress in each of the unknown forces in both sytsems.

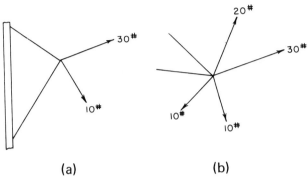

Fig. 15-11.

2. Fig. 15-12. A beam which is pivoted at A and resting on a roller at B is loaded as indicated. Determine graphically the reactions at A and B.

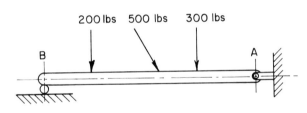

Fig. 15-12.

3. Fig. 15-13. Determine graphically the magnitude of the forces acting at F' and at the pin in the given coplanar system.

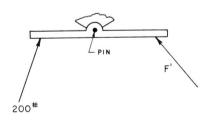

Fig. 15-13.

4. Fig. 15-14. A force of 50 lbs. applied at C will hold this coplanar system of ropes in equilibrium. Determine graphically the load at B. Choose a suitable scale.

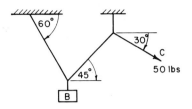

Fig. 15-14.

5. Fig. 15-15. Determine the readings of the two scales. The scales read zero before the three loads were applied.

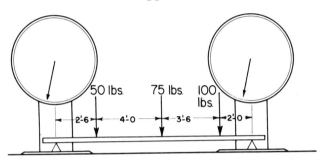

Fig. 15-15.

6. Fig. 15-16. A beam is loaded as indicated. Determine graphically the magnitudes of the reactions at A and B.

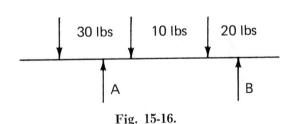

Fig. 15-16.

7. A ferry boat is to cross a one-mile wide river from the west bank to a point directly opposite on the east bank. The average velocity of the river which flows south is 4.5 knots. The boat speed in still water is 10 knots. The wind blowing from the northwest would impart a velocity of 1 knot to the ferry in still water. Determine the course steered by the boat and the time required to cross the river.

8. From a common point, a 15-pound force acts at an angle of 130 degrees with a second force. If the equilibrant of these forces equals 10 pounds, determine graphically the magnitude of the second force.

9. Fig. 15-17. Solve for magnitude, and type of stress (T or C) in the given noncoplanar force system.

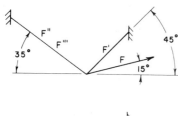

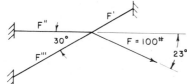

Fig. 15-17.

10. Fig. 15-18. Find the stress in the members of the given structure due to the imposed load.

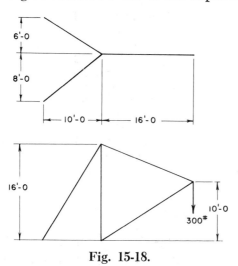

Fig. 15-18.

11. Fig. 15-19. Determine the value and type of stress in members A, B, and C of the given framework due to the applied force F.

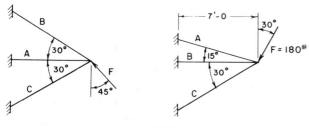

Fig. 15-19.

12. Fig. 15-20. Determine the nature and amount of stress in the boom, cable A, cable B and vertical supporting cable.

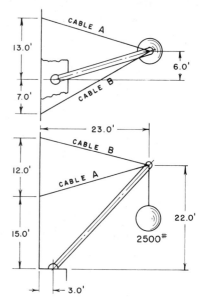

Fig. 15-20.

13. Fig. 15-21. The tow truck is raising a 2000 lb. load from the position indicated. Determine the nature and magnitude of the forces acting in the tripod members.

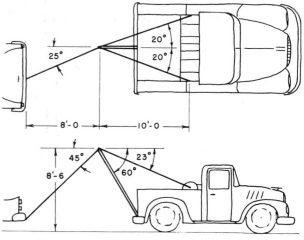

Fig. 15-21.

14. Fig. 15-22. A tripod supports a vertical load (W) of 185 pounds. Use the piercing point method to determine the magnitudes and types of loads in the members of the tripod.

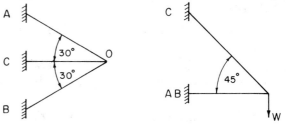

Fig. 15-22.

unit 16

Design Surfaces

16.1 Introduction

A study of surfaces is important to any engineer because all objects from the simplest toys to sophisticated rockets contain surfaces that vary from the simple to the complex. During the design of these objects it is often necessary to determine the contour of intersections and provide developments of the surfaces. These surfaces can be classified as plane, single curved, double curved, or warped.

The classification chart in Fig. 16-1 identifies many of the surfaces and solids common, either singly or as component parts, to many objects, mechanisms, and structures encountered in engineering practice.

The following definitions are concerned with the generation of the more important surfaces and solids. The delineation, intersection and development of some of these surfaces will be discussed in Units 17 and 18.

Part I
PLANE AND WARPED SURFACES

16.2 Definitions

The following definitions are illustrated in Fig. 16-2.

A *Surface* is considered as the area generated by the motion of a generatrix directed by either a law, line, or plane.

A *Generatrix* is a straight or curved line whose path of motion generates a surface.

An *Element* of a surface is any fixed location of the generatrix.

A *Directrix* is a straight or curved line which controls the motion of the generatrix.

A *Director* is a plane which controls the motion of the generatrix.

A *Ruled Surface* is any surface which is generated by a straight line. Ruled surfaces may be divided into two classes, single curved and warped.

A *Plane Surface* is formed by the motion of a straight line generatrix always in contact with two parallel straight lines, two intersecting straight lines or a line and a point not on the given line.

A *Warped Surface* is a ruled surface on which the consecutive elements are nonparallel and nonintersecting.

16.3 Surfaces—Plane and Warped

Many of the more common surfaces can be described in terms of generatrix and directrix. The table of Fig. 16-3 shows the method of generation in terms of generatrix and directrix together with the developability and a possible use of the more common plane and warped surfaces. Each of the listed surfaces can be classed as ruled surfaces because each one is generated by the motion of a straight line.

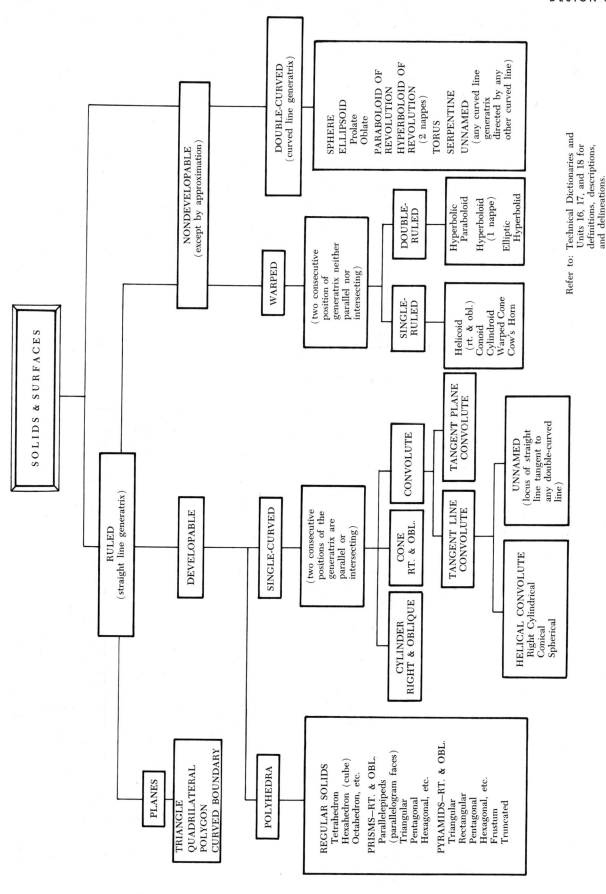

Fig. 16-1. *Classification of Solids and Surfaces.*

DESIGN SURFACES

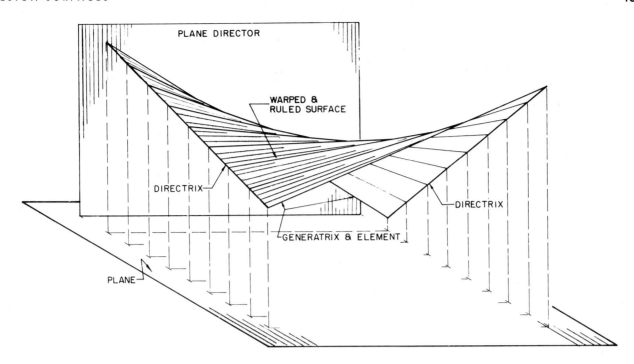

Fig. 16-2. *Composite Parts of a Surface. (Hyperbolic Paraboloid)*

SURFACE	TYPE	GENERATRIX	DIRECTRICES	PLANE DIRECTOR	TRULY DEVELOPABLE	POSSIBLE USE
Plane	Plane	St. Line	2 Str. Lines	None	Yes	Table
Prism	Planes	St. Line	2 Str. Lines	None	Yes	Box
Pyramid	Planes	St. Line	Line & a Point	None	Yes	Top of a spire
Hyperbolic Paraboloid	Warped	St. Line	2 Nonparallel, Nonintersecting St. Lines	Yes	No	Roof
Conoid	Warped	St. Line	One St. Line One Curved Line	Yes	No	Architecture
Cylindroid	Warped	St. Line	2 Curved, Nonparallel Lines	Yes	No	Fairing Surface
Hyperboloid 1 Nappe	Warped	St. Line	Nonparallel St. Line	None	No	Hypoid Gears
Oblique Helicoid	Warped	St. Line	Axis & Helix	None	No	Thread
Warped Cone	Warped	St. Line	St. Line & 2 Curved Line Perpendicular to Center of Circle	None	No	Math. Model
Cow's Horn	Warped	St. Line	2 Curved 1st Lines	None	No	Culvert
Right Helicoid	Warped	St. Line	Axis and Helix	None	No	Stairway

Fig. 16-3. *Plane and Warped Surfaces.*

The *true developability* of each surface is indicated. An object is not considered to be truly developable unless it can be rolled on a plane with a straight line element always touching the plane. The surfaces which are shown as being non-developable can be formed by methods of approximation. Developments are discussed in Unit 18.

The *Hyperboloid of One Nappe* can be generated by revolving a hyperbola about its conjugate axis. In the table this surface is generated by revolving a straight line at a fixed angle other than 90° to a base plane about a non-intersecting straight line perpendicular to the base plane. Fig. 16-4.

In the case of the *Oblique Helicoid*, the straight line generatrix intersects a helix and makes any angle except 90° with a straight line axis. If the generatrix intersects the helix and makes an angle of 90° with a straight line axis, the resulting surface is called the *Right Helicoid*. This surface has applications in circular stairs, screw threads, conveyor chutes and others. Practical applications of the Helicoids are shown in Figs. 16-5 and 16-6.

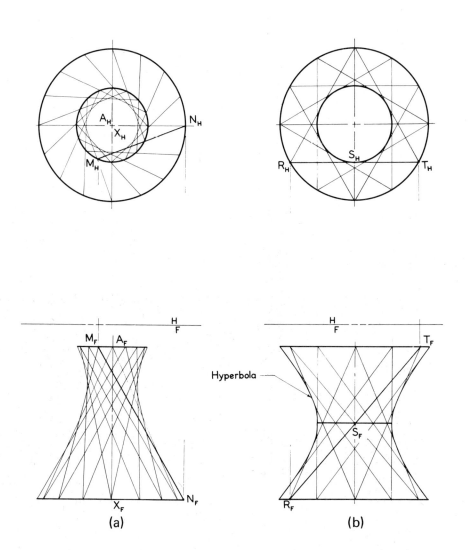

Fig. 16-4. *Hyperboloid, One Nappe.*

DESIGN SURFACES

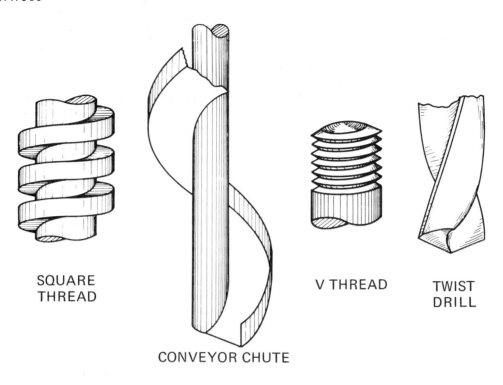

Fig. 16-5. *Application of Helicoids.*

Fig. 16-6. *Helicoid and Warped Cone. (Courtesy General Motors)*

Part II
SINGLE AND DOUBLE CURVED SURFACES

16.4 Single Curved Surfaces

A *Single Curved Surface* has either parallel or intersecting straight line elements and is truly developable. The cone, cylinder, and convolute are examples.

16.5 Cone

This single curved surface is generated by the motion of a straight line generatrix with one fixed point and any other point touching a curved line directrix.

A *Right Circular Cone* has an axis perpendicular to the circular base plane at its center. An *Oblique Cone* has an axis which makes an acute angle at the center of its base plane which may be circular or elliptical.

A *Cone of Rotation* of two nappes is the surface generated by rotating one of two intersecting lines about the other as an axis. The intersecting lines must form an acute angle.

16.6 Conic Sections

If a cutting plane is passed through a right cone the resulting line of intersection becomes one of the conic sections, namely *two intersecting straight lines, circle, ellipse, parabola, or hyperbola*. These sections can be described mathematically or they can be plotted graphically by locating the line of intersection between a cutting plane and the right cone. The sections of the cone, illustrated in Fig. 16-7, are formed as follows.

Straight Lines. If the cutting plane passes through the vertex of the cone, the line of intersection will be two intersecting straight lines. If the cutting plane is tangent to the cone, the line of tangency is a straight line.

Circles. If the cutting plane intersects the cone at right angles to the axis and does not pass through the vertex, the resulting line of intersection is a circle.

Ellipses. If the cutting plane is oblique to the axis and cuts all the elements of the cone, the line of intersection forms an ellipse.

Parabolas. If the cutting plane passes through the cone and is parallel to an element of the cone, the line of intersection is a parabola.

Hyperbola. If the cutting plane intersects the cone at an angle greater than the element angle of the cone, the line of intersection is a hyperbola.

16.7 Cylinder

A *cylinder* is generated by a straight line generatrix moving so that it is always parallel to its initial position while touching a curved line. A circular cylinder is generated if the generatrix re-

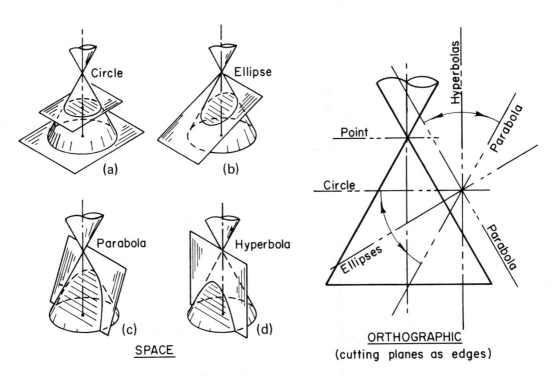

Fig. 16-7. *Conic Sections.*

DESIGN SURFACES

mains parallel to and rotates at a fixed distance about a straight line axis.

16.8 Convolute

Tangent Line. A *Convolute* may be generated by a straight line generatrix moving so that it is always tangent to a double curved line, usually the helix. Fig. 16-8.

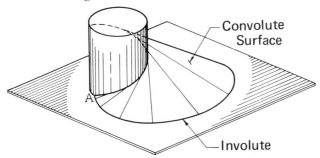

Fig. 16-8. *Tangent Line Convolute.*

Tangent Plane. If a plane is placed so that it is tangent to two dissimilar curved directrices lying in different planes, the straight lines joining the successive points of tangency become the elements of a convolute surface. Fig. 16-9.

16.9 Surface of Rotation

A surface of rotation is generated by rotating a straight or curved generatrix about an axis.

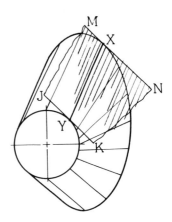

Fig. 16-9. *Tangent Plane Convolute.*

16.10 Double Curved Surfaces

A *double curved surface* is generated by the motion of a curved line and therefore cannot contain any straight line elements. In the majority of cases, the curved line generatrix is rotated about an axis lying in the plane of the curve, either internal or external, to form the surface.

The most common surfaces are spheres, ellipsoids, paraboloids, and the torus. The table, Fig. 16-10, describes these surfaces. Many are illustrated in Fig. 16-11. Further discussion of some of these surfaces can be found in Units 17 and 18.

SURFACE	TYPE	GENERATRIX	DIRECTRICES	TRULY DEVELOPABLE	POSSIBLE USE
Cylinder	S.C.	St. Line	Axis	Yes	Pipe
Cone	S.C.	St. Line	Fixed Point Curve	Yes	Spire
Tan. Line Convolute	S.C.	St. Line	Helix	Yes	Conveyor
Tan. Plane Convolute	S.C.	Plane	Two Curved Lines	Yes	Fairing Surface
Sphere	D.C.	Circle	Diam. of Circle	No	Ball
Prolate Ellipsoid	D.C.	Ellipse	Major Diameter	No	Spotlight
Oblate Ellipsoid	D.C.	Ellipse	Minor Diameter	No	Door Knob
Hyperboloid of Revolution 2 Nappes	D.C.	Hyperbola	Transverse Axis	No	Little Practical Value
Torus	D.C.	Circle	External Line	No	Tire
Paraboloid	D.C.	Parabola	Axis of Symmetry	No	Reflector

Fig. 16-10. *Single and Double Curved Surfaces.*

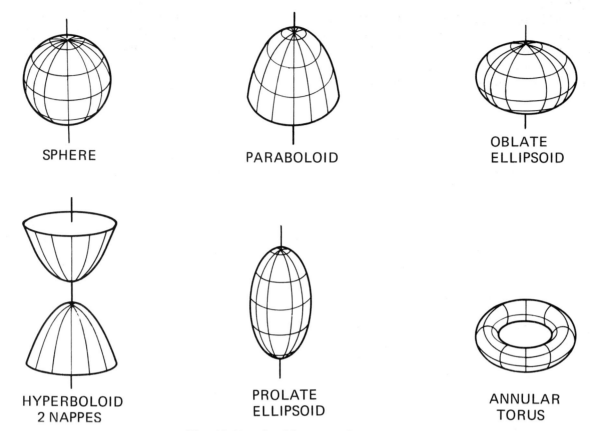

Fig. 16-11. *Double Curved Surfaces.*

PROBLEMS

Warped Surfaces

1. Sketch pictorials of the plane and warped surfaces defined in Fig. 16-3. State possible uses not already listed.
2. Sketch pictorials of the single and double curved surfaces defined in Fig. 16-10. State possible uses not already listed.
3. Design a robot figure composed of three or more of the surfaces defined. Label each surface used.
4. Design an ornamental table lamp composed of three or more of the surfaces defined. Label each surface used.
5. Construct the Plan and Elevation Views of 1 1/2 turns of a right helicoidal conveyor chute designed to carry packages through a vertical distance of 15 feet in 2 1/2 turns of the chute. The diameter of the core cylinder is 18 inches, the chute is 24 inches wide, and the guard rail is 12 inches high. Packages are delivered to the front. Show correct visibility, omitting hidden lines. Scale 1/2" = 1'-0.
6. Fig. 16-12. Show the H and F projections of a hyperbolic paraboloid with nine elements, using the given directrices and a horizontal plane director. Cite a possible use.

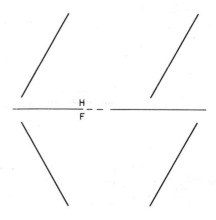

Fig. 16-12.

DESIGN SURFACES

7. Fig. 16-13. Show the H and F projections of a conoid with seven well placed elements, using the given directrices and a frontal plane director. Consider the conoid to be solid and show proper visibility. Cite a possible use.

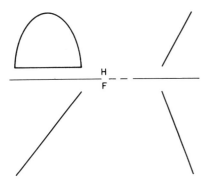

Fig. 16-13.

8. Fig. 16-14. Show the H and F projections of a cylindroid with seven elements, using the given directrices and a horizontal plane director. Cite a possible use.

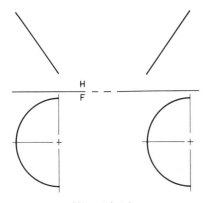

Fig. 16-14.

9. Fig. 16-15. Prepare a detail drawing of one of the hyperboloids of revolution for use on a machine employed to straighten rods and tubes. Be certain to include all necessary features of the hyperboloids of revolution used on this machine. Indicate the "circle of the gorge." Assume sizes and appropriate scale.

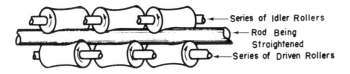

Fig. 16-15.

10. Fig. 16-16. An arch in the form of a conoid is generated by a horizontal straight line touching a semicircular directrix, SMU, and a vertical straight line directrix, AB.

　a. Plot the line of intersection and show the true shape of the cut formed by the vertical cutting plane (C.P.).

　b. A line, WH (true length 3 1/2 in., bearing S 30° W, slope angle −15°), pierces the conoid. Determine the points of intersection and show proper visibility of the line and the surface.

　c. The conoid surface is cut by a cylindrical pipe shown in the given elevation view. Plot the plan view of the resulting line of intersection.

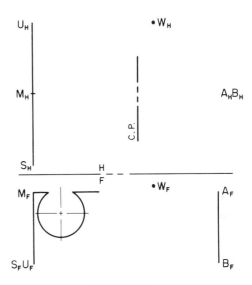

Fig. 16-16.

11. Draw three-fourths of a turn of a right-hand oblique helicoid. Use a cylinder with vertical axis and any convenient diameter. Make the lead equal twice the diameter. Construct the helicoid so that its elements make a 30° angle with the axis of the helix and are limited by a level plane.

12. Draw one turn of a left-hand, right helicoid around a cylinder with vertical axis and any convenient diameter. Make the lead equal twice the diameter. Limit the helicoid by a concentric cylinder whose diameter is twice that of the original cylinder.

Single Curved Surfaces

13. Draw the frontal and horizontal views of a tangent line convolute for each of the following cases. Use a vertical cylinder with any convenient diameter. Develop the surface generated.

    a. Using a left-hand cylindrical helix with a lead equal to twice the diameter of the cylinder, draw three-fourths of one turn of the convolute limited by a level plane.

    b. Using a right-hand cylindrical helix with a lead equal to twice the diameter of the cylinder, draw one full turn of the convolute limited by a concentric cylinder whose diameter is twice that of the original cylinder.

    c. Using a right-hand cylindrical helix with a lead equal to three times the diameter of the cylinder, draw one full turn of the convolute limited by a concentric cylinder whose diameter is twice that of the original cylinder.

14. Design a convolute conveyor component that might function in a crushed rock spreading mechanism which is to be attached to a dump truck for evenly distributing crushed rock on a road bed (principle similar to that of the auger conveyor on a coal stoker). Develop the convolute surface.

Double Curved Surfaces

15. Fig. 16-17.

    a. Which point, A or B, is closer to the surface of the sphere? How much closer? Project points A and B onto the surface of the sphere. Show the H and F views of the projected points.

    b. Find the horizontal projection of point P which lies on the surface and is visible in the frontal projection.

    c. Show the piercing points X and Y of line RS on the surface of the sphere. Show proper visibility of a line from R to X and Y to S.

16. Fig. 16-18. Points A, B and C are on the surface of the sphere with only A hidden in the horizontal projection.

    a. Show the H and F projections of the line of intersection formed on the surface of the sphere by a plane containing the given points A, B and C.

    b. Make an approximate development of the sphere and show the location of points A, B and C on the development.

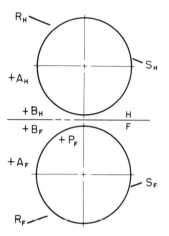

Fig. 16-17.

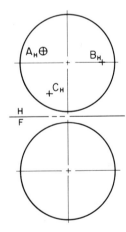

Fig. 16-18.

17. A level line bearing N 70° E passes directly over and 2 in. from the center of a 2 in. sphere. What is the angle between two planes containing this line and tangent to the sphere? Show the H and F projections of the tangent planes, with correct visibility.

DESIGN SURFACES

18. a. Delineate the H and F projections of an annular torus by revolving a 1 in. diameter circle about a vertical axis located 1 1/4 in. from the center of the circle.
    b. Show the piercing points of a line XY which bears due south, has a −30° slope angle, and intersects both the axis of rotation and the circular center line of the annular torus described in part (a). Indicate proper visibility.

19. a. Delineate the H and F projections of a prolate ellipsoid whose major axis is horizontal, 2 in. long, and bears due north. Its minor axis generates a circle of 1 in. diameter.
    b. Using the layout of part 19(a), show the piercing points of a line PQ which bears due north, intersects the midpoint of the axis of rotation, and makes an angle of 60° with the front plane of projection.

unit 17

Surface Intersections

17.1 Intersecting Surfaces

The determination of the line of intersection of the several types of solids or surfaces is of practical value. Inspection of any mechanism, machine, or structure reveals lines of intersection between the surfaces of the objects. Usually, these simple or complicated joints must be accurately located before the object or its pattern can be produced.

This unit will apply the principles of intersections to basic geometrical shapes since complex objects usually consist of combinations of these basic shapes. An understanding of the basic principles will provide a means of solution for practically any type of problem because practical problems involve variations of the basic principles of intersection.

17.2 Principles of Intersection

Two planes will form a straight line of intersection as shown in Figs. 17-1 and 17-2. The intersection between a plane and a curved surface or between two curved surfaces will result in a curved line as shown in Figs. 17-3, 4 and 5.

SPACE ANALYSIS. *Pass planes that cut element lines on each of the given surfaces. The intersection of two element lines cut by a plane is a point lying on the desired line of intersection.*

In order to facilitate the plotting of intersections apply the cutting planes in such a way that they cut the simplest possible elements on each of the given surfaces. Dependent upon the type of intersecting surfaces involved, the simplest elements may be straight lines, curved lines or combinations thereof. Also, it should be noted that it is wise to select a cutting plane which appears as an edge in a given view of the surfaces.

17.3 Intersection of Two Prisms

Problem. Determine the line of intersection of a vertical rectangular prism and a horizontal triangular prism shown in Fig. 17-1.

Procedure:
1. *Number the points of intersection.* The line of intersection of two solids, whose surfaces are made up of a series of straight or curved elements, is always continuous and forms a closed circuit. Plotting points and determining their visibility is simplified by numbering each point in consecutive order around the path of intersection. Place the number on the contour view of either or both solids which shows the elements as points. In Fig. 17-1, the top view of the vertical prism shows its contour. Study the pictorial view illustrating the numbering sequence for points around the triangular prism. Observe the order and position of this sequence of numbers on the three orthographic views.

SURFACE INTERSECTIONS

2. *Pass cutting planes.* A plane will appear as an edge where an element of either solid appears as a point.
3. *Locate points of intersection.* The coplanar elements can be seen as intersecting lines in the front view. Locate the corresponding elements and points in numerical sequence, 1, 2, 3, etc.
4. *Draw line of intersection.* Connect points in their numerical order and observe the standards of visibility.
5. *Sharpen edge elements.* The edge elements of each solid terminate where they meet the other solid. Note that no element line shows between point 3 and point 5. The visibility of the edge elements should always be considered.

17.4 Intersection of Two Prisms (One with Inclined Axis)

Problem. Determine the line of intersection between a vertical hexagonal prism and an inclined triangular prism shown in Fig. 17-2.

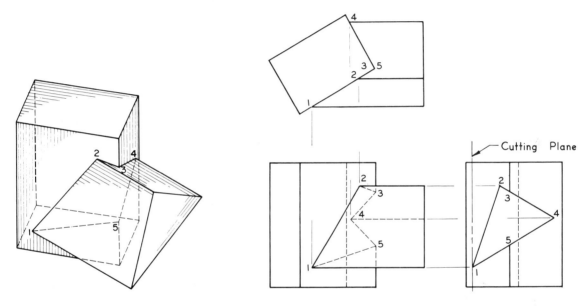

Fig. 17-1. *Intersection of Two Prisms.*

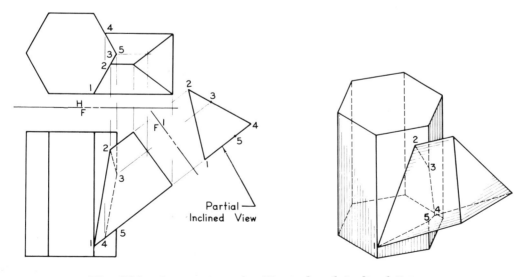

Fig. 17-2. *Intersection of a Vertical and Inclined Prism.*

Procedure:

Apply the steps of procedure listed in Art. 17.3.

Note that in this problem it is necessary to use an auxiliary inclined view in order to show the point views of the critical elements of the triangular prism. In the previous problem an auxiliary view was not necessary since the critical elements appeared as points in the given Horizontal and Profile projections of the rectangular and triangular prisms, Fig. 17-1.

17.5 Intersection of a Cylinder and a Prism

Problem. Determine the lines of intersection formed by a cylinder passing through a triangular prism. Fig. 17-3.

Procedure:

Apply the steps of procedure listed in Art. 17.3.

Since the cylinder passes through the prism, there are two separate loops of intersection and therefore two sets of numbers could be used. In this example one set of numbers, duplicated by prime (′) numbers for the far side of the prism have been used.

The consecutive points on the loop of intersection are established in the profile view of the cylinder and prism since the complete loop is shown in this view. The two loops appear in the Horizontal view of the given objects.

Critical Points on the line of intersection are located at points of abrupt change in direction and/or points where the visibility changes. Also the edge elements in each view of each solid intersect the other solid at critical points. Note that a cutting plane passed through the center line of a cylinder will cut the edge elements in the adjacent view of the cylinder. For example in Fig. 17-3, points 1, 4, 7, and 10 are the critical points on the cylinder and points 5 and 9 locate critical points on the prism. When finding the line of intersection, be sure that all critical points have been included.

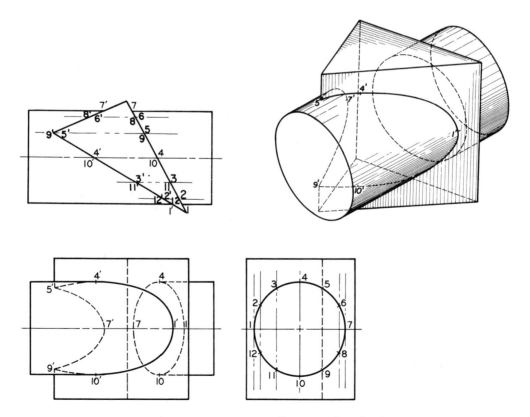

Fig. 17-3. *Intersection of Prism and Cylinder.*

SURFACE INTERSECTIONS

Referring again to Fig. 17-3, the cutting planes containing points 1 and 7 are called *Limiting Planes* since the loop of intersection does not extend beyond these two planes.

When a curved surface intersects a plane or another curved surface, additional cutting planes must be used to establish a sufficient number of points, called *Intermediate Points*, in order to establish a smooth line of intersection. In Fig. 17-3, points 2, 3, 6, 8, 11, and 12 are the intermediate points selected for this problem.

Study the visibility of the line of intersection and remember if a point on the curve is visible in any view, it must be the intersection point of two visible elements. *Note the termination of the edge elements* at their respective piercing points.

17.6 Intersection of Two Cylinders

A horizontal and vertical cylinder intersect in a continuous single curve in Fig. 17-4. From the pictorial drawing as well as an analysis of the orthographic views, it may be observed that some of the elements on each surface are continuous between the bases, while other elements are interrupted at their piercing points.

Problem. To find the curve of intersection of two cylinders. See Fig. 17-4.

Procedure:
1. Locate critical points at the extremities of each center line of both solids.
2. Select the necessary intermediate points, *numbering the points in sequence*.
3. Pass cutting planes, locating and numbering the points in the front projection where corresponding elements intersect.
4. Outline the curve using the numbering system to determine visibility.

17.7 Numbering System for Intersecting Solids

Since the line of intersection is a closed loop, start at any point and place numbers in sequence around the end view outline of one solid until reaching the point where this solid protrudes outside the other solid. At this point reverse the direction of travel around the loop, but continue to apply the numbers in sequence to the loop until returning to the original starting point. Refer to Fig. 17-4.

Note that only one number is placed at a reversal point since there is only one pair of intersecting elements at that point. All other points require two numbers, one on the near side and one on the far side.

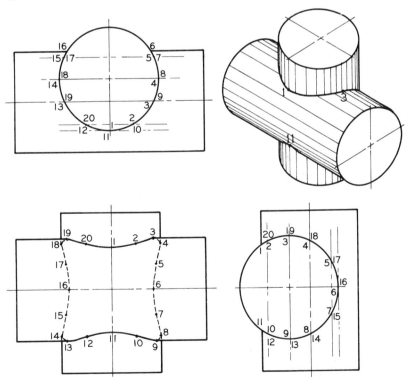

Fig. 17-4. *Intersection of Two Cylinders.*

17.8 Intersection of a Cylinder and Pyramid

Problem. To determine the line of intersection of a vertical cylinder and a triangular pyramid or tetrahedron. See Fig. 17-5.

Procedure:

1. Number the points of intersection in the contour view of the cylinder.

2. Pass planes which appear as edges in the contour view (top) of the cylinder and at the same time cut elements on both solids. All planes cutting elements on a pyramid must pass through the vertex. Point 2 is more accurately located by cutting the line 2X with a vertical plane parallel to the base line of the pyramid.

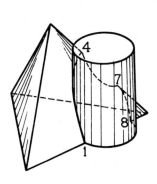

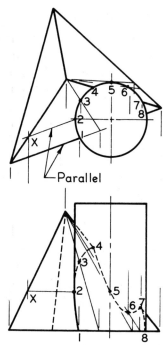

Fig. 17-5. Intersection of a Cylinder and Prism.

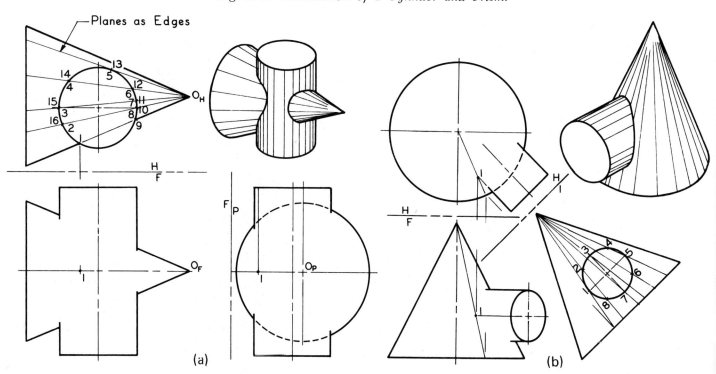

Fig. 17-6. Intersections for Student Completion.

SURFACE INTERSECTIONS

17.9 Intersection of Cylinder and Cone

Fig. 17-6 is partially complete, the points of intersection having been numbered and point 1 located in the adjacent views. The student should plot more points to complete the curve of intersection and determine the proper visibility of the curve and the edge elements.

17.10 Intersections of Other Surfaces

The line of intersection for any two surfaces can be found by passing planes that cut lines on both surfaces, locating points where the coplanar lines intersect, and joining the points found by consecutive cutting planes.

Two other combinations of intersecting surfaces are shown in Figs. 17-7 and 17-8. The student should be able to follow the solution in each problem from the construction that has been included.

17.11 The Sphere

The sphere, probably the most common double curved surface, is generated by a circle rotating about its diameter. Every orthographic view of a sphere is a circle whose outline on any adjacent view is a diameter line. Study

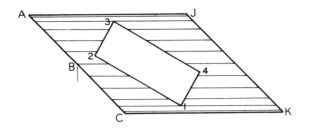

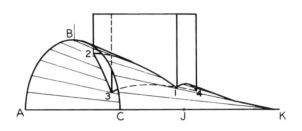

Fig. 17-8. *Intersection of a Prism and Conoid.*

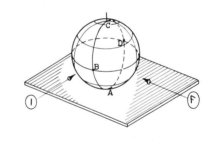

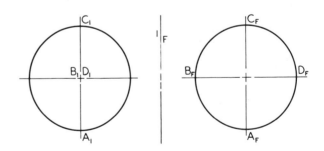

Fig. 17-9. *Great Circle on a Sphere.*

Fig. 17-9 which shows the F projection of the circle $A_F B_F C_F D_F$ and its adjacent view $A_1 B_1 C_1 D_1$. A cutting plane passing through a sphere cuts a great or small circle on a plane of projection parallel to the cutting plane.

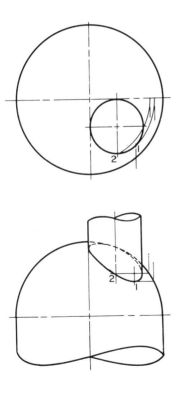

Fig. 17-7. *Intersection of a Cylinder and Hemisphere.*

17.12 Locating a Point on the Surface of a Sphere

The location of the projections of a point on the surface of a sphere is an important problem related to navigation. The shortest distance or path between two points on the sphere is along the great circle passing through these two points.

Problem. To locate an adjacent view of a point lying on the surface of a sphere. See Fig. 17-10.

SPACE ANALYSIS. *A point on the surface of a sphere lies on either a great circle or a small circle cut by a plane.*

Procedure:
1. Pass the edge view of a cutting plane K, parallel to the adjacent plane of projection, through the given horizontal view of point X.
   Plane K cuts a small circle which appears in true size in the front projection.
2. Locate the front projection of X on this true size circle.

Note that there are two possible positions of X_F. If X is visible (on top half of the sphere) in the given view, X_F is the correct F projection. If X is hidden, X_F' is correct. Both solutions are shown since visibility of the given point was not specified.

17.13 Line Piercing a Sphere

Problem. To find the piercing points of a given oblique line AB with the surface of a sphere. See Fig. 17-11.

SPACE ANALYSIS. *The piercing points of a line piercing a sphere are located at the intersection of the line and the circle cut by a plane that contains the line.*

Procedure:
1. Pass a plane appearing as an edge and containing the given line AB. Note that the cutting plane appears as an edge in the H view and cuts a small circle.
2. Construct an auxiliary elevation view showing the small circle in true size.
3. Locate the points of intersection of line AB and the true size small circle.
4. Project the points (P_1 and O_1) to the H and F views of the sphere and line.
5. Determine the correct visibility of the piercing points in the H and F views.

Note that P_H is visible; O_H hidden; P_F hidden; O_F visible. Why?

Fig. 17-11c shows a solution by rotation for the above problem. The procedure is left to the student.

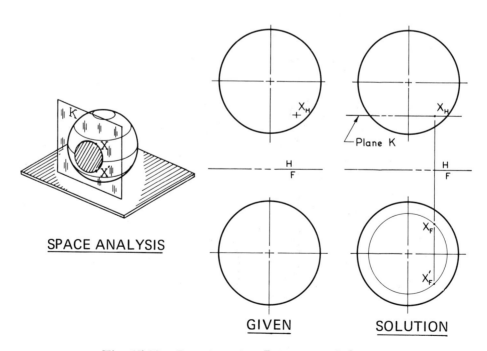

Fig. 17-10. *Location of a Point on a Sphere.*

SURFACE INTERSECTIONS

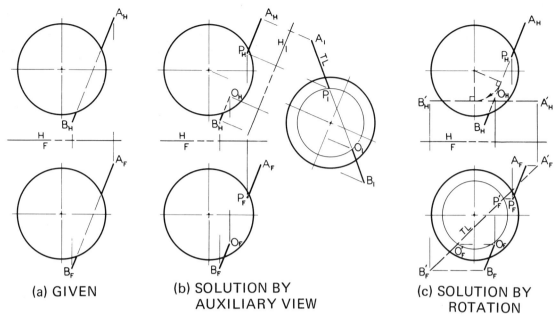

Fig. 17-11. *Line Piercing a Sphere.*

17.14 Plane Intersecting a Sphere

Problem. To find the projection of the line of cut formed by a plane intersecting a hemisphere, Fig. 17-12.

SPACE ANALYSIS. *The points on the line of cut between a plane and a sphere are located in the view showing the intersecting plane as an edge. The true size of the cut is found in the true size view of the intersecting plane.*

Procedure:

Study the method of construction shown in Fig. 17-12. The critical points of the line of intersection are lettered.

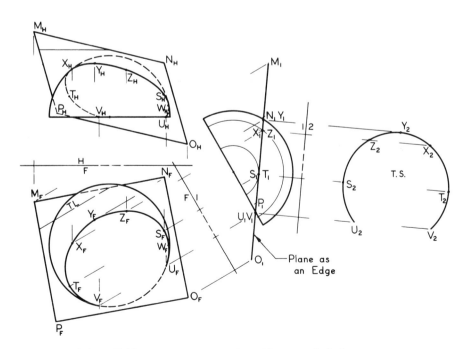

Fig. 17-12. *Intersection of a Plane and Sphere.*

17.15 Applied Problems

Problem 1. Determine the line of intersection between two sloping control panels. Panel R. has a slope angle of 45° and a base length (AB) of 3 feet, while panel S has a slope angle of 60° and a base length (BC) of 2 feet. The angle between AB and BC equals 150°. The vertical height of the panels is 2 feet. See Fig. 17-13.

How can a true size and shape pattern be drawn for each of the panels?

Problem 2. It has been found necessary to connect a 9 inch outside diameter vertical pipe to a 12 inch outside diameter 90° elbow on a condensing system. The center lines of the pipe and the elbow coincide. See Fig. 17-14.

Determine the shape of the outside line of intersection so that the end of the pipe can be prepared for attachment to the elbow.

How can the inside intersection line be found so that the proper size hole can be cut in the elbow?

Examination of the welded drain pipe illustrated in Fig. 17-15 shows several intersection lines which must be accurately located before the welding can be done.

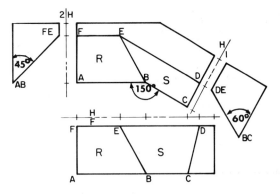

Fig. 17-13. *Intersection of Planes.*

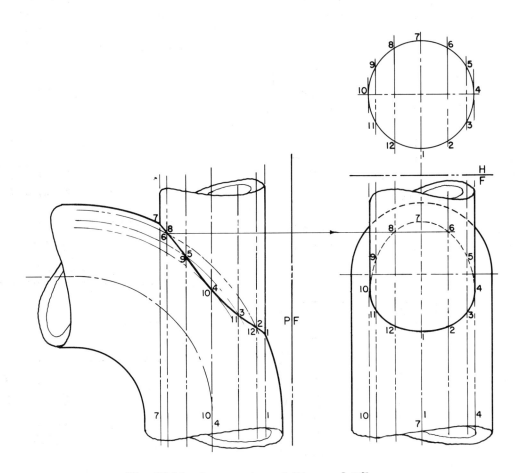

Fig. 17-14. *Intersection of Pipe and Elbow.*

SURFACE INTERSECTIONS

Fig. 17-15. *Intersections (Courtesy Crane Co.)*

PROBLEMS

1. Fig. 17-16. Given the top and right-side views of two prisms, draw the front view. Include the line of intersection and indicate correct visibility.

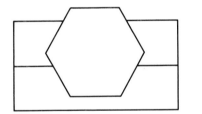 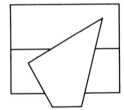

Fig. 17-16.

2. Fig. 17-17. Determine the line of intersection of the two given cylinders. Show correct visibility.

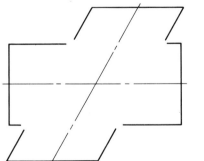 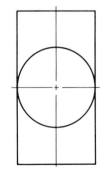

Fig. 17-17.

3. Fig. 17-18. A conical feed hopper with open top is intersected by a horizontal circular pipe and a square downspout as shown. Determine the lines of intersection of the three shapes. Show correct visibility.

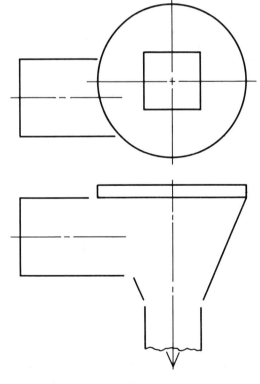

Fig. 17-18.

4. Fig. 17-19. The larger diameter pipe carries exhaust gases to the atmosphere from several diesel engines inside a building. The smaller pipe is the lead-in from the exhaust of one of the engines. Locate the missing lines of intersection. Show correct visibility.
5. Two pieces of 10 in. outside diameter steel tubing (AB and MN) with 1/2 in. wall thickness are to be welded together to fulfill the following specifications: AB is horizontal and normal to the front plane of projection. The centerline of MN intersects and makes an angle of 45° to centerline of AB at point N.
   a. Show the development of a pattern for scribing the cut on the tube MN.
   b. Show a pattern for scribing the opening in the tube AB.
6. Figure 17-20. Determine the line of intersection between the prism and the sphere. Show correct visibility.
7. Delineate the surface of (a) a hyperbolic paraboloid; (b) a conoid; and (c) a cylindroid. Intersect each surface with a cylinder of any desired diameter and position and show two views of the resulting line of intersection.

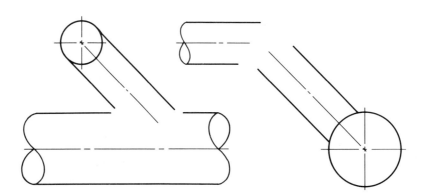

Fig. 17-19.

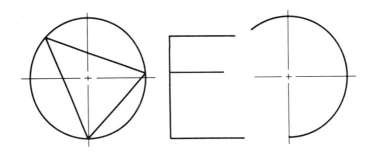

Fig. 17-20.

unit 18

Surface Developments

18.1 Developments

Many industrial drawings require the inclusion of development details in order that patterns can be made in the desired shapes to be cut from sheet metal.

A sheet metal designer should have a broad knowledge of the methods of constructing varied types of developments because of rapid advances in methods of seaming, folding, rolling, or pressing sheet metal during fabrication. This unit will discuss only the principles involved in the development of simple geometric shapes. Many practical considerations, a few of which are mentioned in Art. 18.3, will be omitted in the interest of brevity.

18.2 Developability

A *Development, Stretchout, Pattern,* or *Layout* is an area formed by laying out the true size and shape of a desired surface on a plane.

Surfaces composed of plane or single-curved areas are said to be *truly developable* since these surfaces can be unrolled into a plane thus forming a surface development.

Warped Surfaces are *not truly developable* since no two consecutive elements can form a plane. A double-curved surface is also nondevelopable since the surface contains no straight line elements. However, patterns for warped or double-curved surfaces can be laid out on a plane by approximation. Usually this approximate pattern is sufficiently accurate for practical purposes if the material used to manufacture the desired shape is somewhat flexible since the surfaces are "formed" by stretching or shrinking the approximated area into the desired surface.

18.3 Development Practices

Developments are usually laid out with the *inside surface up* because most bending machines are designed to bend the material with the markings folded inward. Inside-up also provides a surface upon which the scribe marks, punch marks, etc., may be made as guides in folding or rolling the surface. These guide marks will not be visible when the product is completed.

When pattern material of considerable thickness is bent to fit corners, the metal stretches on the outside and contracts on the inside while maintaining a constant length on a line known as the *neutral axis of bend*. Therefore, bend allowances, depending on the kind and thickness of material and the method of bending, must be considered when size accuracy is important. Tables of bend allowances or equations for their computation can be found in engineering handbooks.

In actual sheet metal work extra material must be provided for seams and joints to lock the component parts of the pattern to each other. Many

SURFACE DEVELOPMENT

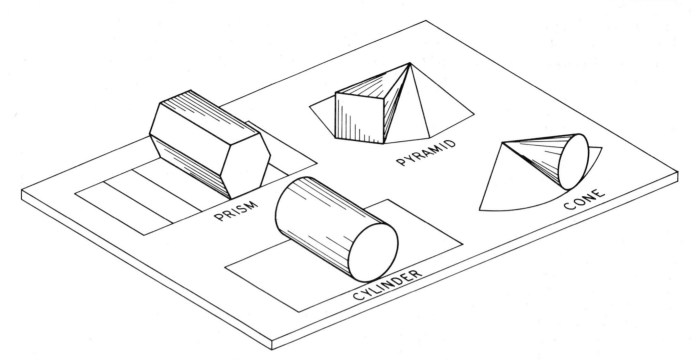

Fig. 18-1. *Developments or Stretchouts.*

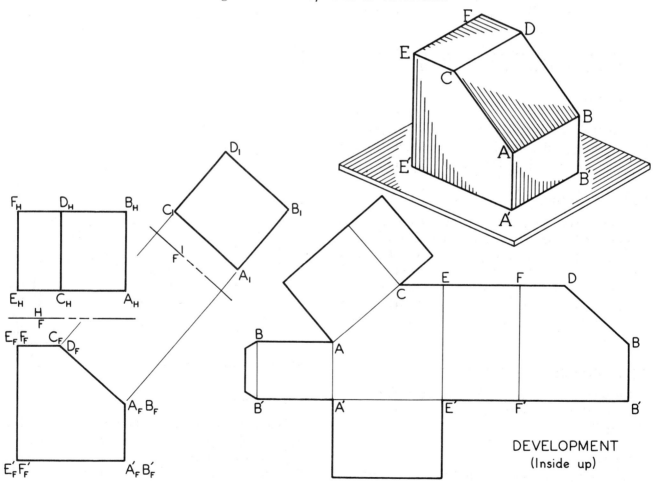

Fig. 18-2. *Development of a Right Prism.*

SURFACE DEVELOPMENTS

methods of locking or seaming can be found in handbooks concerning sheet metal work.

The development is ordinarily *split on the shortest element* in order to keep the joints or seams as short as possible. This minimizes joint failure, decreases fabrication cost, and usually provides a more economical pattern area.

PRINCIPLE. *All measurements laid out on a pattern must be in true length.*

18.4 Prism

A *right prism* is composed of three or more lateral faces perpendicular to its end bases. The development of these lateral faces will be rectangular in shape. See Fig. 18-1. If the bases are parallel, the faces of an *Oblique Prism* are parallelograms. Its development will consist of a series of adjacent parallelograms found by using the true lengths of the sides and diagonals (triangulation, see Art. 18.10). The oblique prism can also be developed by unrolling the surface about its girth line. See Art. 18.5.

Problem. To make a development of a truncated rectangular prism shown in Fig. 18-2.

Procedure:
Lay off adjacent areas of the lateral surfaces in true size. Notice that all corners are identified by letters, that the development is inside-up, and the seam is located on the shortest element.

18.5 Right Section

A Right Section is the area cut by a plane perpendicular to the axis of the object. For instance, the right section of a surface of revolution is a circle. See Fig. 18-3. The right section or girth line on a prism or surface of revolution forms a straight line on the development of the surface equal in length to the perimeter of the right section.

18.6 Cylinder

Problem. To develop the lateral surface of a circular cylinder with sloping bases shown in Fig. 18-3.

Procedure:
1. Establish a sufficient number of elements on the surface by dividing the circumference of the right section into a series of chords. Use a minimum of 12 equally spaced elements.
2. Number in sequence the elements on the right section.
3. Lay off a straight line equal in length to the circumference. This true length can be found graphically by stepping off the previously established short chords. It can also be found mathematically and then subdivided graphically into the proper number of elements.
4. Number the elements on the developed girth line in a sequence such that the inside surface is up.
5. Transfer the true lengths, starting with the shortest element as found in the Front view, to the corresponding elements on the development.
6. Draw a smooth curve through the ends of the elements.
7. Sharpen lines and label the development as necessary. Indicate side up.

If the cylinder is a solid, the true size and shape of the bases, found in auxiliary views, can be attached to the development of the lateral surface.

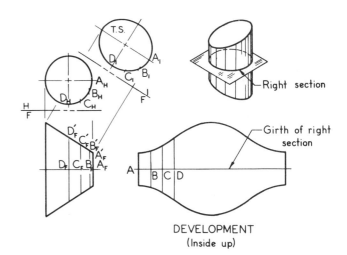

Fig. 18-3. *Development of a Right Circular Cylinder.*

18.7 Pyramid

The pyramid is a polyhedron whose surface consists of a polygonal base and adjoining triangular planes meeting at a common point called the vertex. The axis of a regular pyramid passes from the vertex to the center of the base. The axis of a *Right Pyramid* is perpendicular to the base plane while in an *Oblique Pyramid* the axis forms an acute angle with the base plane.

Note that all elements of a pyramid are in true length on the development and meet at a common point, the vertex.

Problem 1. To develop the lateral surface of a right-rectangular pyramid shown in Fig. 18-4.

Procedure:
1. Determine true lengths of the elements by rotation. The true length of only one element must be determined since all elements of a *regular right pyramid* are of equal length.
2. Swing an arc, whose radius is equal to the true length element, about a selected vertex point.
3. Lay off the true lengths of the bases of the sides on the arc.
4. From the extremities of each base line draw a straight line through the vertex point.
5. Sharpen the lines and add any necessary labels. Indicate which side up.

Problem 2. To develop the complete surface of a truncated oblique pyramid shown in Fig. 18-5.

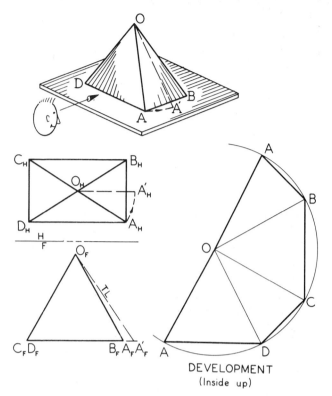

Fig. 18-4. *Development of a Right Rectangular Pyramid.*

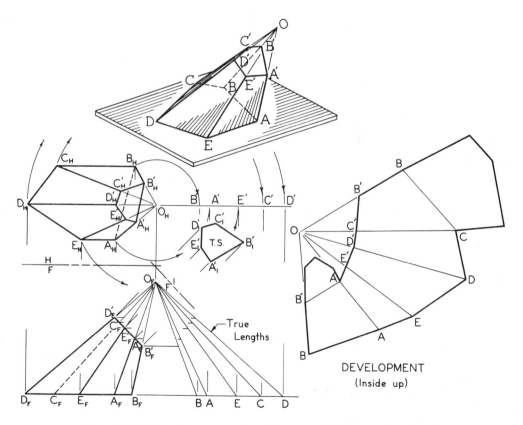

Fig. 18-5. *Development of a Truncated Oblique Pyramid.*

SURFACE DEVELOPMENTS

Procedure:

1. Determine the true lengths of the critical elements by means of rotation. Note that the elements of an oblique pyramid vary in length.
2. Swing an arc, whose radius equals the true length of the shortest element OB, about a selected vertex point O.
3. From point B swing an arc equal to the true length base line BA.
4. From point O swing an arc equal to the true length of element OA until it intersects the arc BA. This intersection fixes point A.
5. Use the same procedure to locate the remaining base points E, D, C, and B. Draw straight lines BA, AE, etc.
6. Find the true lengths of the truncated elements. Since the Front view and the true length diagram are aligned, points A′, B′, etc., can be transferred to the true length diagram thus giving the desired true lengths.
7. From O on the development lay off the true length distances OB′, OA′, etc., and draw the straight lines B′A′, A′E′, etc.
8. Attach the top plane and the base, using their longest sides for the most economical area. The true size of the base is shown in the Horizontal view. The true size of the top plane must be found in an auxiliary view.
9. Sharpen lines of the development and label.

18.8 Cone

Problem. To develop the surface of a right circular cone which has been cut by a vertical plane. See Fig. 18-6.

Procedure:

1. Triangulate the surface of the cone. Minimum of 12 triangles. See Art. 18.10.
2. Plot the shape of the cut in the Front view. Note that the Front projection of point C must be located by rotation.
3. Plot the true shape of the cut in an auxiliary elevation view.
4. From a selected vertex point swing an arc whose radius equals the true length of the edge element of the cone.
5. On this arc, step off the short chord lengths for the bases of the triangles and draw the element lines of the surface.
6. On the development locate the points of cut, A, B, C. The true length of the elements containing points B and C are found by rotation. Chord distance for point A is found in the Horizontal view.
7. Sharpen lines and label.

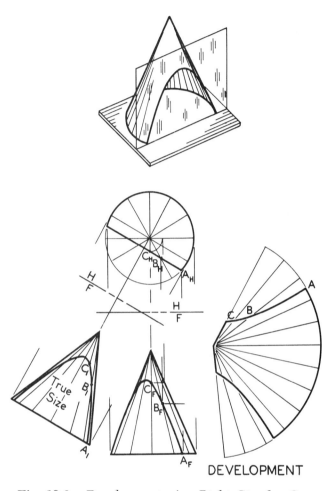

Fig. 18-6. *Development of a Right Circular Cone.*

18.9 True Length—Right Triangle Method

When true lengths are determined by rotation, the great number of lines superimposed on a given view of an object may cause confusion. In Fig. 18-5, the lines were revolved away from the views to avoid this confusion.

True lengths of lines may be conveniently found by the *Right-Triangle Method*. If a right triangle is constructed, Fig. 18-7, so that one leg equals the difference in elevation of the ends of the given line AB and the length of the other leg is equal to the horizontal projection of this line, $A_H B_H$, then the length of the hypotenuse of this triangle will be equal to the true length of AB.

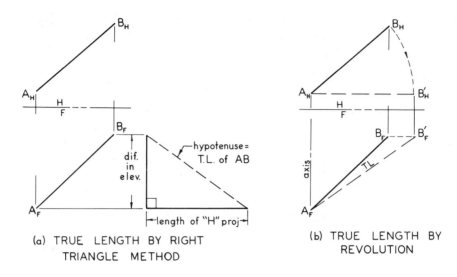

Fig. 18-7. *True Lengths.*

Fig. 18-8. *Geodesic Dome.*
(Courtesy WOI, Iowa State University)

18.10 Triangulation

The process of dividing a surface into a series of triangular areas is called *Triangulation*.

When triangulation is used in conjunction with the development of a surface, the true size triangles are assembled in sequence to form the pattern for the desired surface. One common example is the Geodesic Dome illustrated in Fig. 18-8. Many warped and double curved surfaces can be approximated in this manner.

18.11 Oblique Cone

Problem. To develop the complete surface of a truncated oblique elliptical cone shown in Fig. 18-9.

Procedure:

1. Triangulate the surface of the cone.
2. Find true lengths of the elements by the Right Triangle Method. The altitude of the triangle is the perpendicular distance from the vertex, O_H, to the base plane of the cone as seen in the Horizontal projection. The distances on the base of the cone are found by laying off the lengths of the elements shown in the Front view of the cone.
3. On the true length diagram locate the true lengths of the elements on the truncated portion of the cone. The Front view of the cone and the true length diagram are aligned. Since the front plane of the cone, parallel to the

SURFACE DEVELOPMENTS

back plane, appears as an edge, it will also appear as an edge in the true length diagram cutting the elements to true length.

4. Construction of the pattern, following the method used in Art. 18.7, Problem 2, is left for the student. Study Figs. 18-5 and 18-9.

18.12 Transition Pieces

A connector between two dissimilar openings is called a *Transition Piece*. The surface of this connector may be composed of a combination of planes, single curved surfaces, and in some cases, warped surfaces. These types of connectors are quite common in heating and ventilating systems and for skin surfaces on aircraft, ships, and automobiles.

18.13 Tanget Plane Convolute

The *Tangent Plane Convolute* is the envelope of all positions of a plane tangent to two curved line directrices lying in different planes. The successive tangent positions of the plane define the location of consecutive straight line elements on the surface. Fig. 18-10.

A smooth connecting surface between two dissimilar curves may be a tangent plane convolute which is a developable surface.

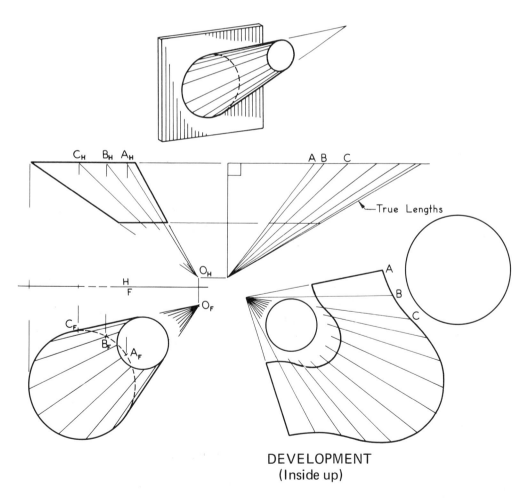

Fig. 18-9. *Development of a Truncated, Oblique Elliptical Cone.*

18.14 Tangent Plane Convolute with Parallel Bases

Problem. To delineate and develop a convolute surface lying between two dissimilar curves in parallel planes as shown in Fig. 18-10.

This surface can be represented by a series of straight line elements connecting the curved-line directrices, each element lying in a plane tangent to the two curves.

Procedure:

1. At any point X on the ellipse construct a line MN tangent to the ellipse, by bisecting the exterior angle formed by lines from the focii through point X. (See Appendix, page 341 to locate focii.)
2. Draw a line JK, parallel to MN, tangent to the circle. Locate the point of tangency, Y.
3. Draw the element XY which lies in the tangent plane JKMN.
4. In a similar manner locate enough other elements to fully delineate the surface.
5. Triangulate the area formed by each pair of successive elements.
6. Using the right triangle method determine the true lengths of the elements, VW, XY, ZA, etc., and the diagonals WX, YZ, AB, etc.
7. Lay off triangles 1, 2, 3, 4, etc., in sequence to form the pattern.
8. The actual construction of the layout has been left for the student.

Note that the method outlined here provides an approximate development. For a true development construct a model of the convolute and roll the surface pattern onto a plane.

18.15 Transition Connectors

Figs 18-11 and 18-12 illustrate two common types of transition connectors frequently used in heating or ventilating installations.

A study of the surfaces shows that the connectors consist of planes and cones which must be developed in proper sequence. In each case the cones are triangulated and the true lengths of the elements are found by use of the right triangle method.

If the connector is symmetrical, as in Fig. 18-11, it is possible to construct a half development which can be turned on its center line to form the other half of the pattern.

The importance of specifying the "side-up" of a development is very important in a nonsymmetrical surface of this type, since if folded incorrectly, the connector will not fit the openings.

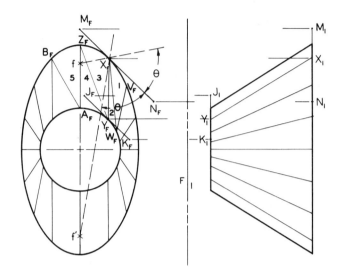

Fig. 18-10. *Development of a Tangent Plane Convolute.*

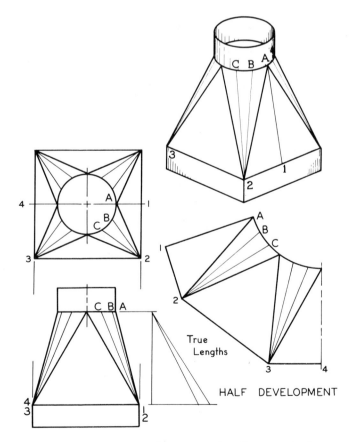

Fig. 18-11. *Development of a Symmetrical Transition Pipe.*

SURFACE DEVELOPMENTS

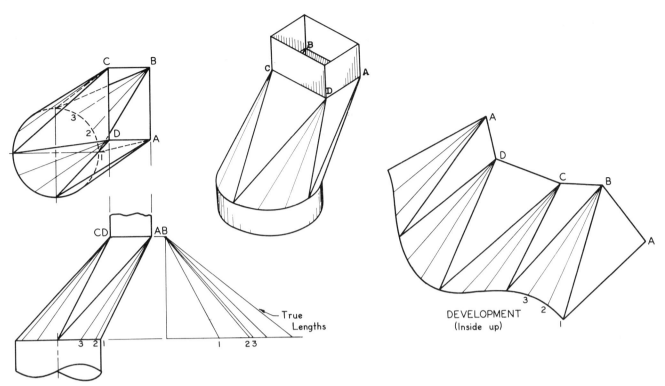

Fig. 18-12. *Development of an Offset Transition Pipe.*

18.16 Development of a Sphere

ZONE OR POLYCONIC METHOD

Problem. To make an approximate development of the surface of a sphere. See Fig. 18-13.

PRINCIPLE. *A spherical surface can be approximated by dividing the surface by a series of parallel planes. Each of these zones may be developed as the frustum of a cone.*

Procedure:

The construction for determining the vertex of the cone is shown in Detail A of Fig. 18-13. Note that the edge element of the cone, partially inside and partially outside the surface of the sphere, is perpendicular to a radial line on the sphere so as to approximate the length of a segment of the great circle.

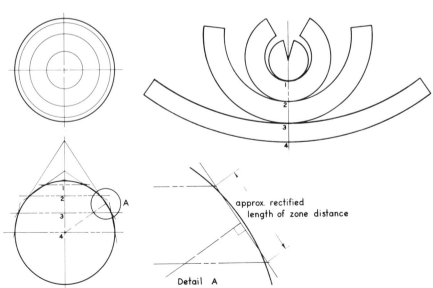

Fig. 18-13. *Development of a Sphere, Zone Method.*

SEGMENT OR POLYCYLINDRIC METHOD

The sphere, shown in Fig. 18-14, is developed by the segment method, often called the orange peel or gore method. The sphere is divided by vertical meridian planes into equal gores, usually 12, and each arc is replaced by its chord. Parallel cutting planes, equally spaced on the circular contour, are located in the front view. A true size view of one of the gores, developed in alignment with the horizontal view, represents the approximation of the surface lying between two of the vertical cutting planes. A partial pattern is shown in the illustration.

18.17 Applied Problems

This article will discuss the development of the patterns for the two applied problems whose required intersections were previously completed in Unit 17.

Problem 1. Construct the true size patterns for the sloping control panels of problem 1, Art. 17.15. See solution in Fig. 18-15.

SPACE ANALYSIS. *The true size view of a plane is found on an image plane parallel to an edge view of the given plane.*

Procedure:

1. Construct edge views of planes R and S.
2. Construct true size views on image planes parallel to the edge views.
3. Sharpen lines and label as necessary.

Problem 2. Construct a true size pattern showing the contour of the cut to be made on the 9 inch pipe of Problem 2, Art. 17.15. Solution in Fig. 18-16.

Procedure:

1. Lay off a straight line equal in length to the circumference of the 9 inch pipe.
2. Graphically divide this line into 12 equal parts and construct perpendicular element lines. See

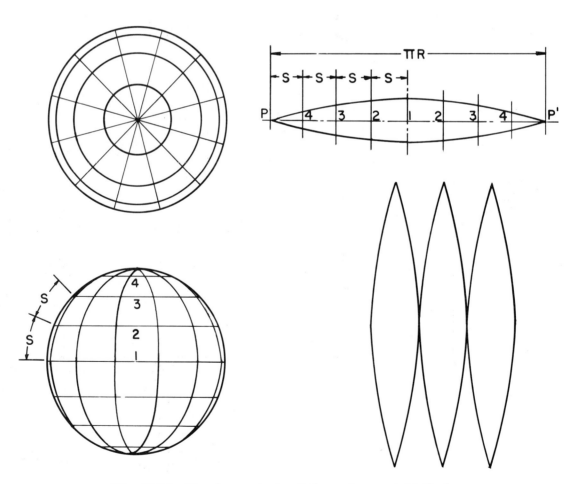

Fig. 18-14. *Development of a Sphere, Segment Method.*

SURFACE DEVELOPMENTS

Appendix, page 340 for graphical division of a line.
3. Starting with the shortest elements, number 7, number the element lines as shown.
4. Transfer the true lengths of the elements, shown on the Front view, to the pattern.
5. Draw a smooth curve on the pattern.
6. Sharpen lines and label.

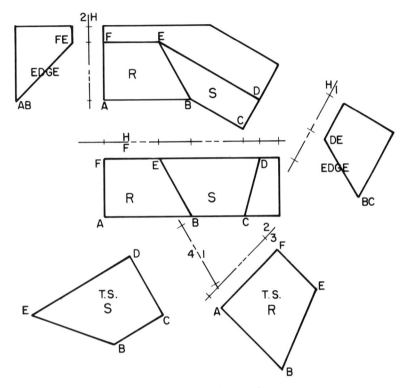

Fig. 18-15. *Control Panel Pattern.*

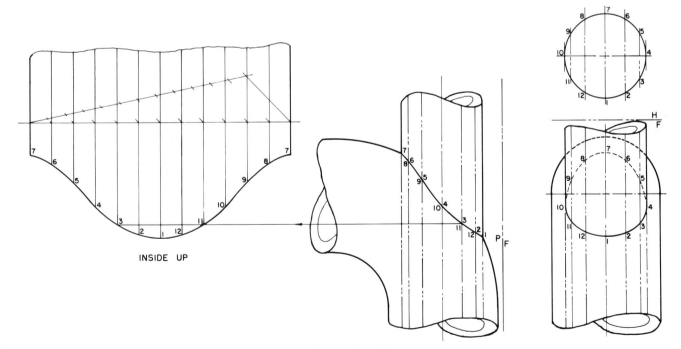

Fig. 18-16. *Pipe Pattern.*

PROBLEMS

1. Make a pattern development of each of the following figures in the order given: 18-17 (a, b, c, d), 18-19, 18-24.
2. Fig. 18-18. Complete the top view and develop the given pyramid.
3. Develop the lateral surfaces of the following figures: Fig. 18-20, 18-21, 18-22. Name and classify each surface. Show intersection of cutting plane (c.p.) on the development of Fig. 18-22.
4. Fig. 18-23. It is desired to find a scale area for the layout (development) of the convolute surface for a portion of a fuselage included between the circle, 3'-0 dia. at station 1 and the ellipse, 8'-0 major, 6'-0 minor axis at station 5. Use 24 elements. Show true shape of the ribs at stations 2, 3 and 4 evenly spaced between stations 1 and 5.

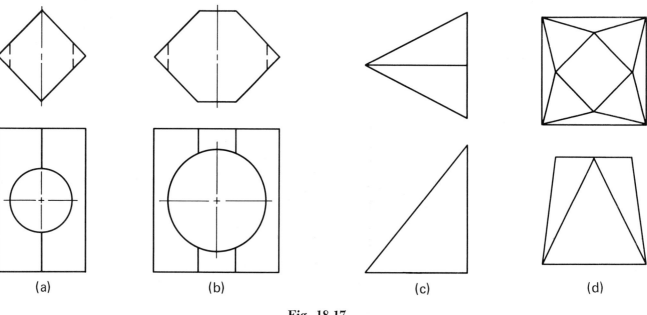

Fig. 18-17.

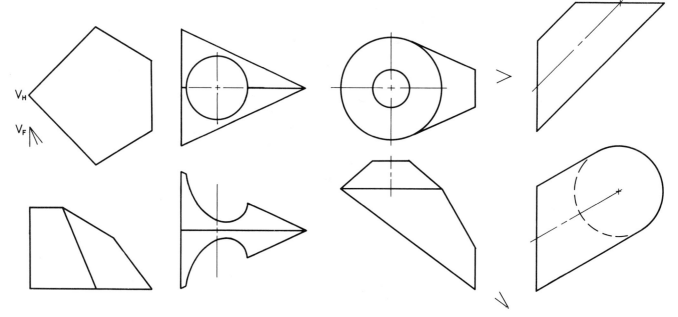

Fig. 18-18.

Fig. 18-21.

SURFACE DEVELOPMENTS

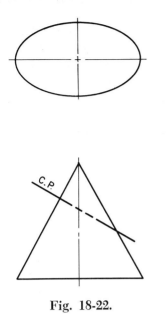

Fig. 18-22.

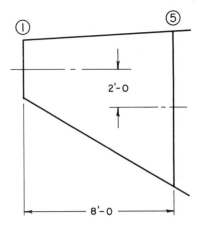

Fig. 18-23.

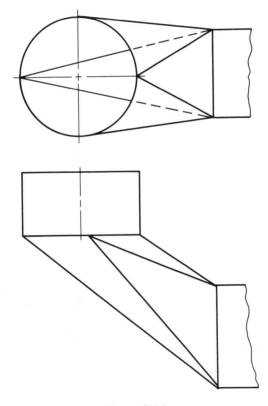

Fig. 18-24.

unit 19

Fasteners-Threaded

19.1 Classification of Fasteners

The types of fasteners of interest to the engineer can be generally classified as *threaded, removable,* and *permanent.* This unit will treat only threaded fasteners.

Threaded Fasteners. The common fasteners whose function depends on screw threads include bolts and nuts, studs, cap screws, setscrews, and other miscellaneous forms of screws. The following articles describe and show the standard methods of delineation and specification of these fasteners. Fig. 19.1 shows a variety of threaded fasteners.

19.2 Screw Threads

These definitions, abridged from the American National Standards (ANSI) relate to Fig. 19-2, the true projections of an internal and external screw thread.

Fig. 19-1. *Threaded Fasteners. (Courtesy Industrial Fasteners Institute)*

STANDARD FASTENERS—THREADED

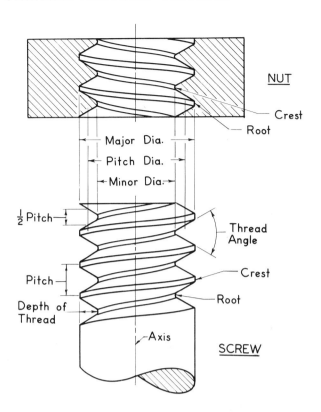

Fig. 19-2. *Screw Thread Nomenclature.*

A *Screw Thread* is a uniform ridge in the form of a helix on the external or internal surface of a cylinder (straight thread) or a cone (taper thread).

Major Diameter is the largest diameter on a screw thread.

Minor Diameter is the smallest diameter on a screw thread.

Pitch is the distance from a point in one thread to the corresponding point on the next adjacent thread measured parallel to the axis.

Pitch Diameter is the diameter of an imaginary, concentric cylinder cutting the threads so that the width of the cut thread metal and width of separating space are equal.

Lead is the distance a point on a thread advances axially in one revolution.

Crest (external thread) is the point on the screw thread at the greatest distance from the axis.

Crest (internal thread) is the point on the thread nearest to the axis.

Root (external) is the point on the screw thread nearest to the axis.

Root (internal) is the point on the screw thread at the greatest distance from the axis.

Depth of Thread is the radial distance (measured perpendicular to the axis) between the crest and root of a screw thread.

19.3 Standard Thread Forms

Cross-sections of the most common thread forms are shown in Fig. 19-3.

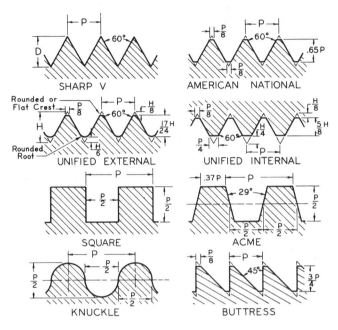

Fig. 19-3. *Common Thread Forms.*

Sharp V is a type of thread used on small diameters for adjustment purposes. However, because of the difficulty encountered in producing the sharp roots, it is infrequently used. A slight dulling of the crests and roots produces the American National thread form.

American National Thread, once called the United States Standard (USS) is still used in this country although the Unified thread has replaced it for the most part.

Unified Thread is an international standard screw thread form adopted in 1948 by the United States, Canada, and Great Britain. It is interchangeable with the American National thread for the same diameter, series, and class. Note the rounded root and crest on the external and the wider crest on the internal unified thread.

Square Thread is a power transmission thread, the faces are at right angles, or nearly so, to the axis. A slight taper on the thread face assists disengagement of a mating member.

Acme Thread is a modification of the square thread as seen in Fig. 19-3. The Acme has the advantages of being easily cut, stronger, and more easily disengaged. It is likewise used to transmit power.

Knuckle Threads, usually formed of sheet metal, are rolled or molded. An example can be found on the base of a lamp bulb or on screwed bottle tops.

Buttress Threads are used to transmit power only in one direction or to withstand shock. They are found in the breechblocks of large guns and in jacks and other mechanisms that withstand thrust in one direction.

19.4 Right and Left-Hand Threads

A right-hand thread advances as it is turned clockwise, while the left-hand thread advances with counterclockwise rotation. A thread specification is understood to be right-hand unless left-hand (LH) is a part of the specification. Fig. 19-5 shows detailed representation of both RH and LH threads.

19.5 Multiple Threads

A *single thread* has a pitch equal to the lead. Refer to Art. 19.2 for definition and to Fig. 19-4 for illustration.

A *double thread* will advance a distance of twice the pitch in one revolution. The thread consists of two adjacent helices winding around the cylinder.

A *triple thread* will advance a distance of three times the pitch in one revolution.

Multiple threads are used for faster motion of the nut along the bolt when great power or holding is not required. Illustrations of multiple threads are shown in Fig. 19-4.

19.6 Thread Representation

Screw threads may be drawn in *true projection,* by *detailed representation,* by *schematic representation,* or by *simplified thread symbols.* True projection is seldom used except for illustration purposes as in Fig. 19-2.

Detailed Representation shows the thread form drawn with proper pitch and lead. Straight lines connect the crests and roots as shown in Fig. 19-5, although the connecting lines are usually omitted on internal threads. This representation is *infrequently* used due to the time and effort required for drawing.

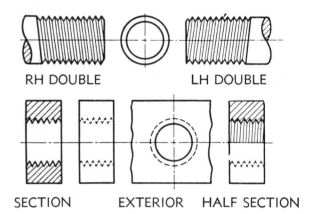

Fig. 19-5. *Detailed Thread Representation.*

Schematic Representation is shown in Fig. 19-6 for external, internal, section, and end views. While some companies feel this form to be more realistic, the simplified representation is used more than all others.

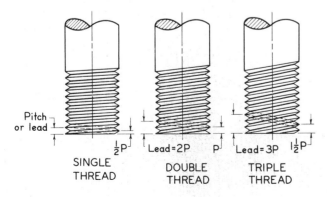

Fig. 19-4. *Multiple Threads.*

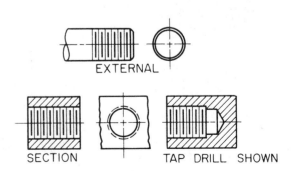

Fig. 19-6. *Schematic Thread Representation.*

Simplified Representation can be drawn more quickly than can the other types of symbols. The drawings in Fig. 19-7 should be studied carefully with respect to the solid and hidden outlines on the several views.

Square Threads are represented in external, internal, end view, and by symbols in Fig. 19-8.

Acme Threads. The external and internal representations of an Acme thread is shown in Fig. 19-9. In drawing the Acme, one-half pitch is measured on a line midway between the crest and root of the thread. Slant lines are drawn through the equally spaced points on the pitch line to determine the crest and root widths. Refer to Fig. 19-3 for an enlarged Acme form.

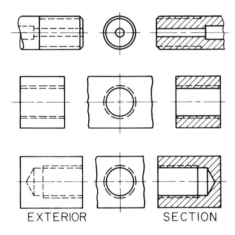

Fig. 19-7. *Simplified Thread Representation.*

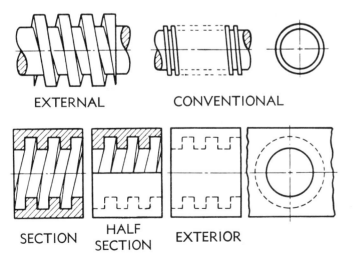

Fig. 19-8. *Square Thread Representation.*

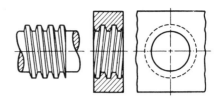

Fig. 19-9. *Acme Thread Representation.*

19.7 American Standard and Unified Thread Series

The various diameter-pitch combinations of the standard thread series covered by the Unified and the American Standard are tabulated in the Appendix, page 351. This is an important table to which the student will need to refer many times.

Coarse Thread Series is used for most general purpose screw threads. They are designated as UNC (Unified Coarse).

Fine Thread Series are used where special holding conditions require a finer thread such as for automotive and aircraft work. A fine thread does not loosen under vibration as readily as a coarse thread. The designation is UNF although machinists often refer to it as SAE.

Extra Fine Thread Series is used for highly stressed applications where the engagement is limited. Examples of extra fine thread are found on thin-walled tubes, ferrules, and couplings. The designation is UNEF.

8-Pitch Thread is a uniform pitch series for large diameters requiring threads of medium pitch. Originally intended for high pressure joints, it is now widely used as a substitute for the coarse thread series on diameters larger than 1 inch. The designation is 8UN.

12-Pitch Thread is a uniform pitch series for diameters requiring threads of medium fine pitch. Originally intended for boiler practice, it is now used as a continuation of the fine thread series for diameters larger than 1 1/2 inches. Designation is 12UN.

16-Pitch Thread is a uniform pitch series for large diameters requiring threads of fine pitch. It is suitable for adjusting collars, retaining nuts, and also serves as a substitute for the extra fine series for diameters larger than 2 inches. Designation 16UN.

Special Thread Series involve combinations of diameter, pitch, and length of engagement other than those in the standard series. They are designated UNS. Special threads should not be used unless standard series is unsuitable for the purpose intended. There are also other constant pitch series for certain diameter ranges as 4UN, 6UN, 20UN, 28UN, 32UN.

The American Standard series have the same designation as the Unified series except to omit the "U." Examples: NF, NEF, 12N, etc.

19.8 Thread Classes

Classes of thread are distinguished from each other by the accuracy or precision used to manufacture the thread. Classes 1A, 2A, and 3A apply to *unified external threads*, and Classes 1B, 2B, and 3B apply to *unified internal threads*. Classes 1, 2 and 3 apply to both external and internal American National form of threads.

Classes 1A and 1B are used for free assembly and disassembly with a minimum of binding, and where low cost of production is important.

Classes 2A and 2B are general purpose threads for bolts, nuts, and screws with normal applications in the mechanical field.

Classes 3A and 3B are made with precision to produce a tight fit. They are recommended only where the higher cost is warranted.

Acme Thread Classes are standard in only one diameter-pitch series shown in the table in the Appendix, page 362. There are two types of Acme threads, *General Purpose* and *Centralizing*. As the name implies, General Purpose is the more common class which has some clearance for free movement. The Centralizing is tighter to maintain alignment. The diameter-pitch relation is the same.

19.9 Thread Specifications

A screw thread is designated on a drawing by a note with leader and arrow pointing to the thread. The minimum of information required in all notes is a specification in sequence of 1) the nominal diameter or size number, 2) number of threads per inch, 3) thread series symbol, and 4) the thread class number or symbol, supplemented optionally by the pitch diameter limits. Unless otherwise specified, threads are right-hand and single lead; left-hand being designated by the letters LH following the class symbol. Double or triple lead threads are designated by the words DOUBLE or TRIPLE preceding the pitch diameter limits. The use of fractional values or the decimal equivalents for nominal diameters is optional. If decimal equivalents are used for size call-out, they should be shown to four places, omitting the fourth place only if it is a zero. For number sizes, three place decimals are used.

Examples:

1/4-20 UNC-2A-LH
PD 0.2164-0.2127 (optional)
1/4 = diameter; 20 = number of threads per inch; UNC = unified coarse series; 2A = class of fit (external); LH = left-hand; PD = pitch diameter limits.

Check the following specifications in the Appendix, pages 351 and 362, for accuracy and identification.

.375-16 UNC-2A   .4375-20 UNF-3A
.500-13 UNC-2A   1 7/8-8 UN-2A
.190-32 UNF-2B   7/8-4 1/2 SQUARE-LH
1"-5 ACME-2G (1" major diameter, 0.20 pitch, class 2 general purpose)

For internal threading it is customary to give the tap drill diameter followed by the thread specifications: #7 (.2010) DRILL, .250-20 UNC-2B. This specification should be referred to the location view of the tapped hole. Tap drill diameters are obtained from the thread tables.

19.10 Threaded Fasteners

A threaded fastener is a removable device used to hold two or more parts together. Tables giving sizes, head and thread specifications are found in the Appendix.

A Bolt or Through Bolt is a partially threaded cylinder with an integral head on one end. It is used to secure two or more parts by engaging the bolt with a threaded nut, usually hexagonal or square, thus gripping the parts together as the nut is tightened. See Fig. 19-10a and Art. 19.11 for specifications.

A Cap Screw is similiar to the bolt except that it is used without a nut. Two parts may be held by passing the cap screw through a clearance hole in one piece and tightening into a threaded hole in the second part. See Fig. 19-10b. Cap screws are distinguished by several styles of head shapes including hexagon, socket, flat, fillister, and round. The Appendix table, page 354, gives details of several styles of cap screws.

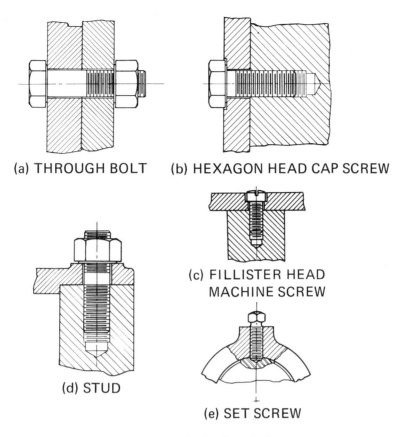

Fig. 19-10. *Standard Threaded Fasteners.*

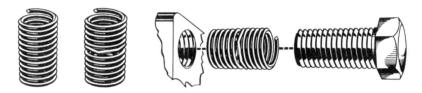

Fig. 19-11. *Heli-Coil Inserts.*

Specification Example: .500-13 UNC-3A x 1.25 long, fillister head cap screw.

A Machine Screw resembles a cap screw except for its limitation to small sizes. An example is shown in Fig. 19-10c. Refer to the table on page 355 in the Appendix for details of sizes, types of heads and other specifications. Machine screws regularly have sheared ends, not chamfered, and fine or coarse threads with a class 2 fit.

Specification Example: #10(.190)-24 NC-2 x .75 long, round head machine screw.

A Stud is a cylinder threaded on both ends. When one end of the stud is tightened into a threaded hole as shown in Fig. 19-10d, a nut is screwed on the outer end after the stud passes through a clearance hole in the second piece. Studs are specified by nominal diameter, length, thread specifications, and length of threads. Usually a detail drawing of the stud is shown on which complete specifications are given.

A Setscrew is shown in Fig. 19-10e as typically used to prevent turning motion between two assembled parts. The table and illustrations on page 361 in the Appendix give details of standard sizes, heads, points and threading.

Specification Example: .250-20 UNC-2A x 1.0 slotted, headless, cone point setscrew.

Thread or Heli-Coil Inserts are shaped like a spring where the wire corresponds to the screw threads. See Fig. 19-11. The inserts are made from hard steel or bronze and provide a lining for threaded holes in soft metal or plastic.

Miscellaneous Screws can be purchased in many styles for special purposes. Specifications are obtained from manufacturers' catalogs. A partial list of miscellaneous screws and bolts:

Wood Screws	U-Bolts
Lag Screws	Eye Bolts
Thumb Screws	Expansion Bolts
Drive Screws	Wing Nuts
Screw Hooks	Collar Screws
Turnbuckles	Shoulder Screws
Stove Bolts	Hanger Bolts
Carriage Bolts	Step Bolts
Plow Bolts	Hook Bolts

19.11 Standard Hexagon and Square Head Bolts and Nuts

American Standard Bolts and Nuts are specified by their series and finish. ANSI B 18.2.1–1965 and ANSI B 18.2.2–1965 give full information concerning sizes and tolerances for all bolt and nut standards. Drawings from these standards are shown in the Appendix, pages 352, 353.

Series. American Standard Bolts and Nuts may be classified in two series, *Heavy* or *Regular*.

Hexagon Head Bolts and Nuts, and Square Nuts (not Bolts) are standard in the *Heavy Series* which, as the name implies, is used for heavy work where larger bearing and wrench surfaces are needed.

The *Regular Series* is for general use and includes *both Hexagon and Square* headed bolts and nuts.

Finish. Specifications for bolts and nuts indicate *Unfinished* or *Finished*. Previously semi-finished was also included but the latest standard has combined some categories so that a fastener is classified as finished if it is finished to any degree.

Square head Bolts, square nuts, hexagon head bolts, and flat hex nuts may be *Unfinished*. All Hex Nuts are *Finished* except the flat hex which has neither washer face nor chamfer.

The latest standard combines *finished hex head bolts* and *hex head cap screws* with the same specifications, into a common classification of *Hex Cap Screws*. Also, the previous heavy semi-finished and heavy finished hex head bolts have been combined into a series called *Heavy Hex Screws*.

The one particular feature which characterizes a *Finished* bolt or nut is the machined or formed "washer face" on the bearing surface of the bolt head or nut. The washer face is about .031 thick and 1.5 times the bolt diameter. Some nuts have chamfered corners to produce the circular bearing surface.

Threads on *Unfinished Bolts and Nuts* are *Coarse, Class 2A*. The *Hex Cap Screw series* (finished) may have *Coarse, Fine, or 8-Pitch Threads, Class 2A*. Finished Nuts have Coarse, Fine and in some cases 8-Pitch Threads, Class 2B.

Length of Bolts and Threads are given in the tables, pages 352, 353.

Basic sizes and Proportions for drawing may be taken from the tables, or in general for Regular Hex and Square Bolts and Nuts:

W (across flats) = 1.5D (bolt dia.)
H (head height) = .67D;
T (nut thickness) = .88D

For Heavy Hex Bolts and Nuts, and Heavy Square Nuts:

W = 1.5D + .125; H = .67D; T = D.

Specification of Bolts and Nuts. The specifications and their order for American Standard bolts and nuts include the following:

Bolt diameter	Finish
Thread	Head shape
Length of bolt	Name

If not otherwise stated, bolt and nuts are considered to be *Regular Series* and *Finished*.

Examples:
Bolts: .250-20 UNC-2A x 1.5 Heavy Hex Screw
1/2-13 UNC-2A x 2 Square Bolt
Nuts: .250-20 UNC-2B Heavy Hex Nut
1/2-13 UNC-2B Square Nut

19.12 Plain and Lock Washers

While washers cannot be considered as threaded fasteners, they are used with bolts and cap screws and therefore their specifications should be included with the threaded devices.

American Standard Plain Washers should be specified by giving inside diameter, outside diameter, and thickness. The table on page 356 gives the range of sizes.

STANDARD FASTENERS—THREADED

American Standard Lock Washers are now available in the preferred series as Regular and Extra Duty. To specify give nominal size and series. For example, .625 Regular Lock Washer. A table of standard sizes is on page 357 of the Appendix.

19.13 Special Locking Nuts and Devices

Special Nuts and Devices are used to keep the nut from loosening on the bolt or the cap screw from its threaded hole. The following list is given without explanation and includes special devices whose details and specifications may be obtained from manufacturers' catalogs.

Slotted Nut	Elastic Stop Nut
Castle Nut	Set Screw
Split Nut	Cotter Pin (See Art. 20.3)
Jam Nut	Plastic Insert in thread

The *Jam nut* is a relatively thin nut used as a lock by tightening against the gripping nut. The table, page 358 in the Appendix, gives Jam nut specifications.

PROBLEMS

1. Make an extensive classified listing of various types of threaded fasteners. Write correct standard specifications for each.
2. Sketch enlarged pictorial views of short segments of the common thread forms.
3. Illustrate with instrument drawings or freehand sketches all the standard methods of thread representation for a given thread form.
4. With freehand sketches or instrument drawings illustrate a use for each of the following fasteners. Write complete specifications for each.
   a. Bolts, washers, and nuts.
   b. Cap screws (several head shapes) and lock washers.
   c. Set screws (with various combinations of heads and points).
   d. Special and miscellaneous screws and locking devices.
5. A hex nut is 3/4" thick. Its mating bolt is 5" long with coarse thread.
   a. Write complete specifications for both bolt and nut.
   b. What is the length of thread on the bolt?
   c. How many revolutions are required to advance the nut along the full length of the thread?
6. Two pieces of metal, each 1/2" thick, are to be fastened together by a 3/8" hex-head bolt, lock washer, and nut.
   a. What is the minimum standard-length bolt required if 2 threads of the bolt must extend beyond the nut.
   b. Write specifications for the fasteners.
7. a. How far will a 3-inch diameter, triple NC thread advance into a mating nut in two revolutions?
   b. A one-inch diameter shaft and its nut have standard square threads. If the nut is kept stationary and the shaft rotated at 184 rpm, how fast does the shaft move in inches per minute?
   c. How many complete turns will be made on a 5/16 UNC thread when the screw advances 3 1/2 inches?
8. What is the thickness of a 1/2 in. American Standard Lock Washer?

unit 20

Fasteners - Removable and Permanent

20.1 Definitions

Removable Fasteners are those other than threaded fasteners that can be removed by force or pressure. Many depend on the friction of metal contact or on a taper for holding power. Included in this classification are such items as keys, splines, pins, and springs.

Permanent Fasteners do not allow the components to be readily separated. Riveting, soldering, brazing, gluing, and welding are examples of fixed or permanent fasteners.

Part I

REMOVABLE FASTENERS

20.2 Keys and Keyways

This type of fastener is used to prevent relative motion between wheels and shafts or similar machine parts. The key is a piece of metal which is placed in a groove called a keyseat cut in the shaft. A portion of the key when seated extends outside of the shaft diameter to fit into a slot or keyway in the hub.

Square and Flat Keys are made from cold finished stock, not machined. See Fig. 20-1 for their shape and function. Tables for keys are given on page 373 in the Appendix.

Specification Examples: 1/4 Am. Nat. Std. Sq. Key x 1 1/2 long. 5/16 x 1/4 Am. Std. Flat Key x 2 long.

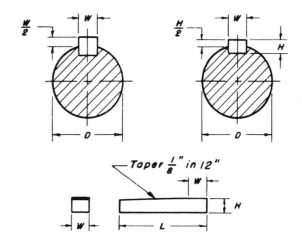

Fig. 20-1. *Square and Flat Keys.*

The Pratt and Whitney Key is similar to a flat key except it has round ends and is placed in a milled keyseat with two-thirds of its height in the shaft and one-third extending into the keyway in the hub. Pratt and Whitney keys are specified by number which fixes the cross section size as well as the overall length. Tables for keys are given on page 373. See Fig. 20-2.

Specification Example: #12 Pratt and Whitney Key.

Woodruff Keys are widely used semi-circular disks, made to fit into a keyseat milled in the shaft by a Woodruff cutter. The keyway in the hub

FASTENERS—REMOVABLE AND PERMANENT

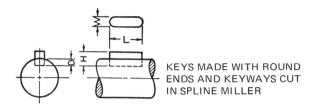

Fig. 20-2. *Pratt and Whitney Keys.*

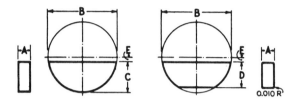

Fig. 20-3. *Woodruff Keys.*

is similar to that for a flat key. They are specified by number, a table for which is shown in the Appendix on page 374. See Fig. 20-3.

Specification Example: #606 WDF Key.

Gib-Head Keys are tapered on their upper surfaces. They may be driven into place to form a very secure fastener. Tables for square and flat gib-head keys are shown on page 373. Specifications are identical to square and flat keys except Gib-Head noted. See Fig. 20-4 for shape.

Specification Example: 1/4 x 3/16 Gib-Head Key x 1 1/2 long.

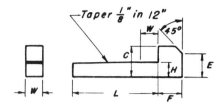

Fig. 20-4. *Gib-Head Keys.*

Splines. Design frequently requires a positive, high strength connection to prevent rotation between a shaft and its related member. A device called a spline, equivalent to multiple keys spaced uniformly around the shaft and mating hub, provides such a connection. Fig. 20-5 not only shows the shape of splines, but also shows a method of dimensioning. Further information can be found in ANSI Y14.2-1958 and in manufacturers' catalogs.

Keyways are generally specified as to width and depth. The proper method for dimensioning keyseats and keyways for interchangeable manufacture is shown in Fig. 20-6. They are frequently specified by note.

Example: KEYWAY FOR .250 x .188 FLAT KEY.

Pratt and Whitney keyseats are dimensioned by a note: KEYWAY FOR #7 P & W KEY, or by giving the width, depth and length of the seat.

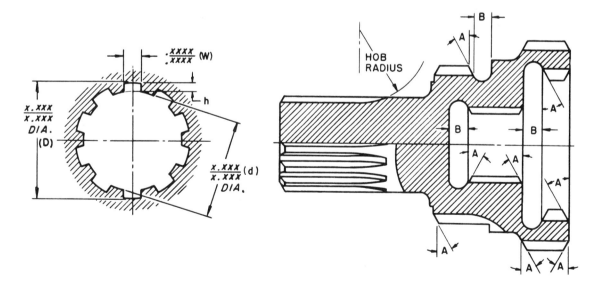

Fig. 20-5. *Splines.*

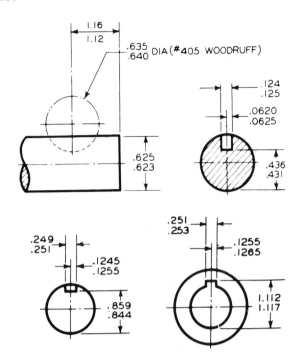

Fig. 20-6. *Keyways and Keyseats.*

Woodruff Key slots may be specified: MILL FOR #608 WOODRUFF KEY or dimensioned as shown in Fig. 20-6.

20.3 Pins

Taper pins are used for such light work as fastening hubs or collars to shafts as shown in Fig. 20-7. Details of standard taper pins may be found on page 360 of the Appendix.

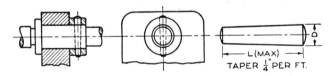

Fig. 20-7. *Taper Pins.*

Dowel pins are used especially to keep two adjacent pieces aligned in a prescribed position. Although sometimes straight, the pins are usually tapered and made of hardened steel.

Cotter pins are used principally as a locking device for slotted or castle nuts. Sometimes the cotter pin is placed in a drilled hole outside of a standard nut which will prevent loss but allow some loosening of the nut. It is also inserted in a hole in a shaft to retain the wheel on a child's wagon or tricycle. Cotter pins are specified by their diameter and length. See Fig. 20-8 and the Appendix, page 360 for available sizes.

20.4 Springs

Helical Springs are formed by winding spring wire on a cylinder. Tension springs or compression springs are respectively formed by winding the helix coils close together or more widely apart. While the detailed method of drawing (similar to screw threads) may be used, much time can be saved by single line representation. The information needed to purchase springs of given specification is shown on the ANSI drawings, Fig. 20-8.

Flat Springs are, as the name implies, made by predesigned bending and heat treating flat spring steel. Their special shapes are usually designed for the particular job requirements. Automobile leaf springs are a widely used type of flat spring.

Part II
PERMANENT FASTENERS

20.5 Rivets

Rivets, as permanent fasteners, have been widely used in structural steel assembly, boiler work and other sheet metal fabrication. Holes are punched or drilled in the parts to be fastened. The rivet is inserted in the hole, usually red hot, and pressed or hammered until the head is formed and the rivet tightened in the hole. Fig. 20-9 shows standard heads and rivet joints. The use of rivets is also illustrated in Unit 23 on Structural Drawing.

Riveting has gradually decreased in use as high strength bolts and welding have replaced the older method of permanent fastening.

20.6 Adhesives

Productive research in the field of adhesives has developed some extraordinary bonding strengths for joining metals as well as ceramics, wood, and plastics. Adhesives will bond almost all materials in any combination of thickness and type. The greatest problem is the selection of the best adhesive at the lowest cost from the thousands of compounds that meet desired adhesion and

FASTENERS—REMOVABLE AND PERMANENT

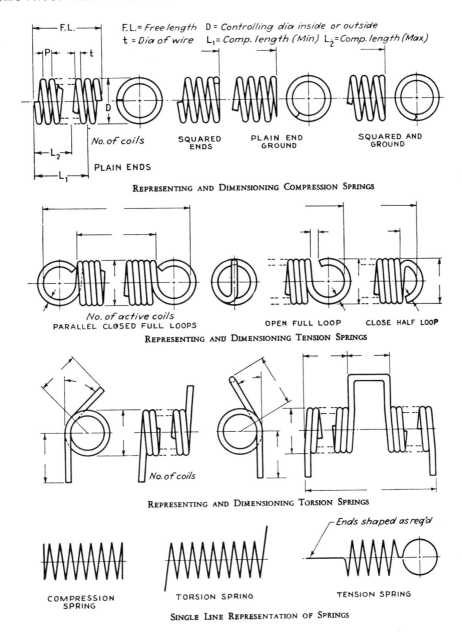

Fig. 20-8. *Standard Spring Representation.*

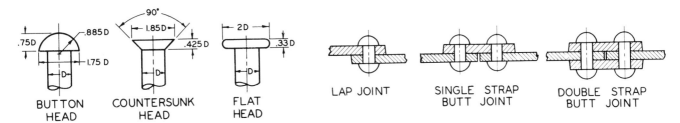

Fig. 20-9. *Standard Rivet Heads and Joints.*

strength requirements. Some of the factors that must be considered in selecting the best adhesive for a job include:
1. Thickness, chemical analysis, and texture of the surface to be bonded.
2. Strength, conditions of temperature, moisture, color, toxicity, and electrical properties required of the bond.
3. Necessary equipment, time required, temperature, pressure, and form of the adhesive to be applied.
4. Cost limitations.

The epoxy and phenolic compounds are among many combinations of products used for adhesives. Information and specifications may be obtained by reference to manufacturers' catalog files or an Assembly Directory and Handbook.

20.7 Welding

Welding, an important permanent fastener, is a process of uniting metallic parts by localized heating to a degree which will allow the parts to fuse; or by compressing or hammering the parts together with or without prior heating.

Only an outline of the part that welding plays in the industrial world will be presented. Obviously, any process that involves physics, chemistry, and metallurgy cannot be explained in a few paragraphs. However, the student should gain a basic understanding of the several welding processes, why and where they are used, and a knowledge of the representation of the standard welding symbols on a drawing. Much of the material and illustrations are based on ANSI Y32.3-1959. Other good references on welding include publications of the American Welding Society, New York City, and Lincoln Electric Co., Cleveland, Ohio.

20.8 Welding Processes

The common types of welding processes may be classified as *Gas, Arc,* and *Resistance Welding.* Other types less frequently encountered are thermit, atomic hydrogen, and induction welding.

Gas Welding. A *nonpressure* process, gas welding uses a mixture of gases, usually oxygen and acetylene, which burn at a very high temperature, about 6000°F, to produce the heat for welding. By means of the gas torch, the metal along the line of weld is heated to a molten state and "puddled" with metal from a filler rod.

Arc Welding. This *nonpressure* process unites two pieces of metal by the intense heat, about 6000°F, formed by sustaining an arc between the work and a metal welding rod used as an electrode. The metal at the desired localized area and the end of the rod are melted by the arc. Small globules from the rod fuse with the molten metal of the work to form the weld.

Resistance Welding. Sometimes referred to as *pressure* welding, Resistance welding occurs when the work is heated to the fusing point by a strong electrical current. Due to the higher resistance to the flow of current at a joint, the metal can be brought to fusion heat by controlling the amount of current. When the proper temperature is reached, pressure is applied to the metal to form a welded joint. Several forms of resistance welding are briefly described:

Resistance Spot Welding joins two or more sheets of material at a concentrated point or "spot" by passing a high current between two pointed electrodes which exert pressure upon the sheets.

Resistance Seam Welding is a variation of spot welding whereby rollers are substituted for the pointed electrodes. As the rollers revolve pressing the work between them, many spot welds are made by a rapid series of high current pulsations.

Flash or Upset Welding joins pieces that are held slightly apart, allowing an arc to pass between them. This arc or "flash" heats and prepares the ends to be welded, and at the proper instant the pieces are united by pressing them together.

Projection Welding, applicable to sheet metal, requires projections or spots to be embossed on one piece to form a path for the flow of current between the parts. Current flows through the projection until, at the desired temperature, the sheets are pressed together to form a welded joint similar to a spot weld.

20.9 Types of Welds

The basic welded joints are illustrated in Fig. 20-10 with types of weld appropriate for each.

Gas and Arc welds may be classified into Fillet, Plug or Slot, Arc Seam or Arc Spot, Backing or Melt-Thru, and Edge or Corner Flange. Groove welds include the Square, V, Bevel, U, J, Flare-V and Flare Bevel. The shapes of these welds and their symbols are shown in Fig. 20-11.

The four basic *Resistance welds* have been described in Art. 20.8. Their symbols can be found in Figs. 20-11 and 20-12.

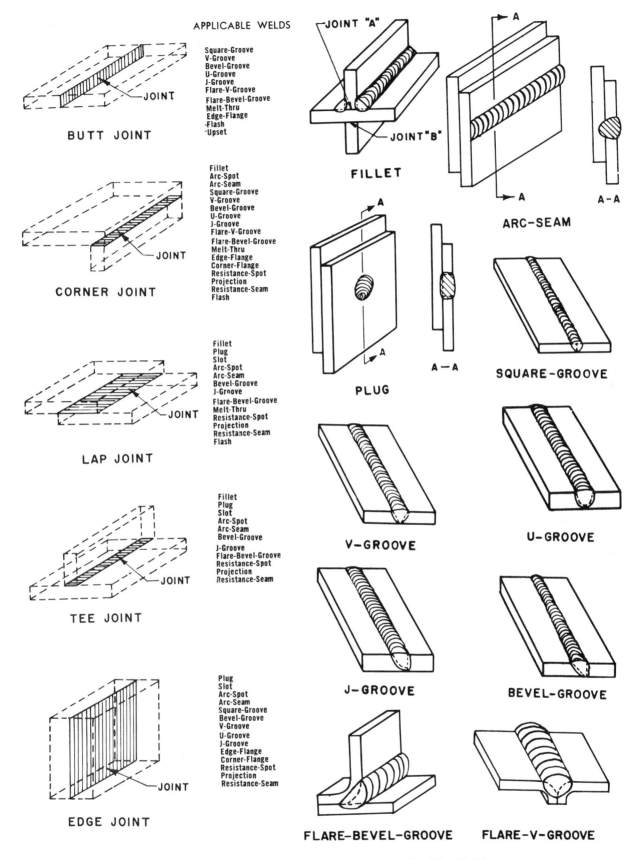

Fig. 20-10. *Basic Weld Joints and Applicable Welds.*

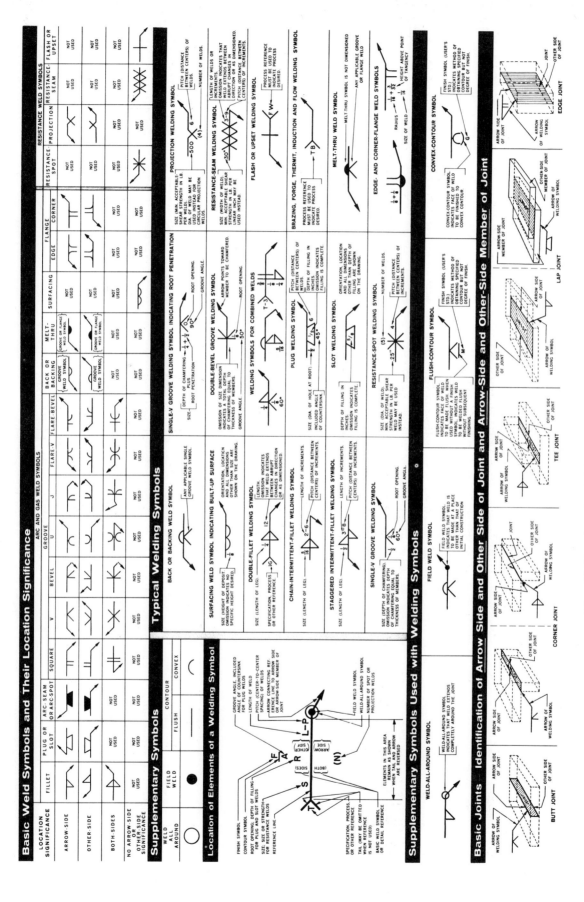

Fig. 20-11. *Standard Weld Symbols.* (ANSI Y32.3)

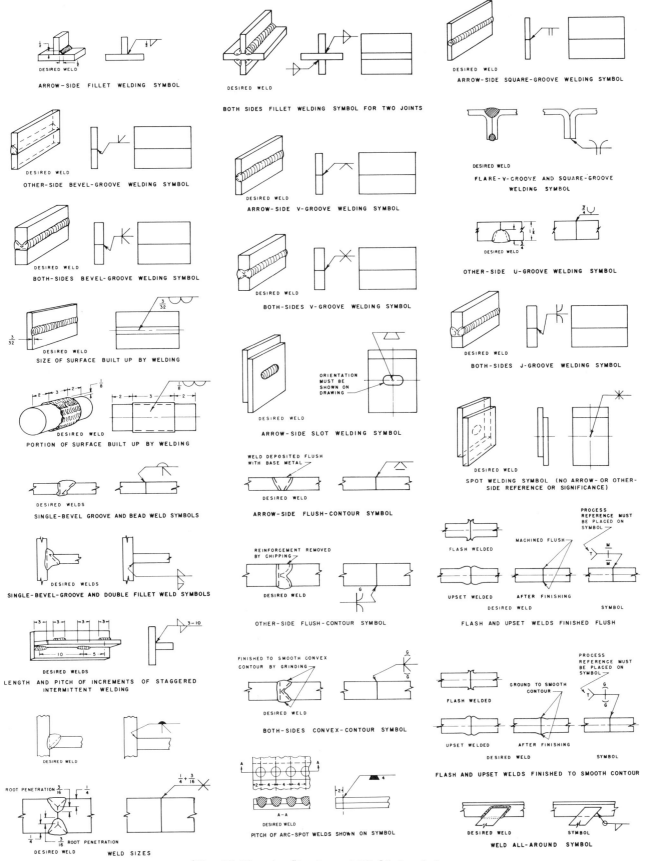

Fig. 20-12. *Application of Weld Symbols.*

20.10 Welding Symbols

The Summary of Standard Welding Symbols, approved by the American Welding Society, is shown in Fig. 20-11. Important information contained in this summary merits careful attention by the student. Especially noteworthy are the shape of the different welding symbols; the location of the elements of the welding symbol; the position of the arrow with respect to the shaft, or reference line; the weld all-around symbol; the field weld symbol; and the methods for specifying the size, length, pitch, groove angle, root opening, weld finish, and position of the weld on arrow or other side.

20.11 Application of Welding Symbols

At one time it was sufficient to place a note on the drawing, TO BE WELDED, leaving the responsibility of type, size, and quantity of welding to the shop. Today's highly specialized manufacturing processes often require control of a welded product by specifications and symbols applied to the drawing by the designer and the draftsman. Fig. 20-13 is included just for kicks.

HEY BOSS!!!
I got this part done.

Fig. 20-13. *Welding*

The decision to manufacture an article as a mild steel weldment instead of a grey iron casting is often dependent upon the required strength, the quantity, and the weight limitation of the article. Mild steel is some 2 1/2 times as rigid as cast iron and about four times stronger. Provided the weldment is of good design, its cost for a small number of pieces is considerably less than for comparable castings. Housings, covers, gear cases, links, cranks, and yokes are typical parts that can be produced economically by welding. There has also been a steady increase in the use of welding for the connection of structural steel members and fabrication of structural beams and girders.

Fig. 20-14 shows a Bracket made from stock material. *Note that each component part of the weldment is completely dimensioned so that it can be cut to size before fabrication.* Also, note the material list which includes proper stock identification for each piece.

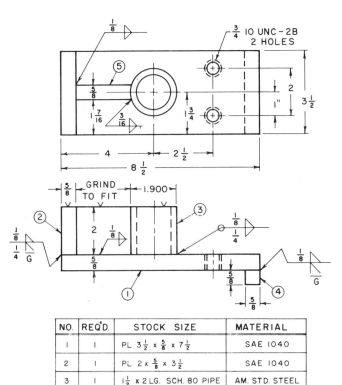

WELDED BRACKET
ONE REQUIRED
SCALE: $\frac{1}{4}$ SIZE
BREAK ALL SHARP EDGES

Fig. 20-14. *Weldment Drawing.*

PROBLEMS

1. With freehand sketches or instrument drawings, illustrate a use for the following. Write complete specifications for each.
   a. Three different keys
   b. A spline
   c. Taper pin
   d. A tension spring
2. a. What is the width of an American standard flat key for a 3-inch shaft?
   b. What is the nominal key size for a #404 Woodruff key?
   c. What is the width, length, and height of a #15 Pratt and Whitney key?
   d. What is the length of a #5 Am. Std. Taper Pin?
   e. What is the maximum length of a 3/16 cotter pin?
3. List as many industrial applications as possible where adhesives might be used for fasteners. Check the library.
4. Make two view orthographic sketches of the five basic joints for welding.

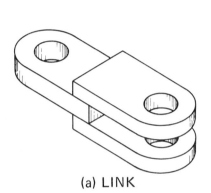

(a) LINK

(b) PIPE SADDLE

(c) LINK

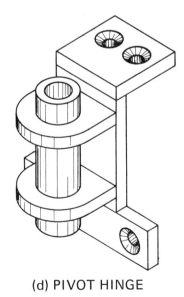

(d) PIVOT HINGE

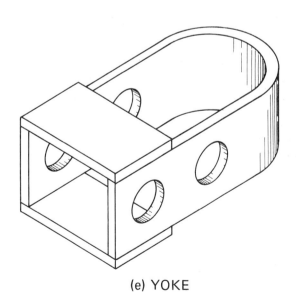

(e) YOKE

Fig. 20-15.

a. Apply an appropriate arc weld symbol to each joint giving all weld size designations.
b. Apply an appropriate resistance weld symbol to each joint giving all weld size designations.
5. Make welding drawing of the objects shown in Fig. 20-15, a, b, c, d, e. Dimension each completely as a welded part. Make a materials list similar to the one shown in Fig. 20-14.
6. Sketch and dimension details 2 and 4 in Fig. 26-15 as welded parts. Include a complete list of stock specifications.
7. Redesign the structural assembly, Figs. 23-17 and 23-19 for welded construction. Utilize plug welds in place of rivets, and continuous and intermittent fillet welds as necessary.
8. Design a completely welded steel table for the welder's own use. Include as an integral part of the table, a circular metal flux box; trough or rack for holding weld rods; cylindrical or structural steel legs; and any other desirable features.

unit 21

Manufacturing Processes and Measurements

Part I
PROCESSES

21.1 Introduction

It might be legitimately asked why a unit on processing appears in a text on engineering graphics. There are several very sound reasons for its inclusion.

Although the engineer will never be even a mediocre machinist, he must be familiar with shop processes since these processes will influence his design limitations or freedom. Every part must be "manufacturable." This knowledge of how components of a mechanism may be made will also influence his decisions regarding dimensioning, surface finish, and other specifications.

Since few university engineering curricula permit time for shop courses, knowledge must be gained by an early general acquaintance and continued voluntary observation and study of processes wherever possible.

It is hoped that in studying this brief treatment of processing, the student will realize the great strides being taken in this challenging field.

21.2 Fundamentals

Processing involves altering the shape of a piece of stock material to obtain a single component of some required configuration. There are basically four ways to obtain a desired shape of a part:

1. *Shaping;* to form without adding or losing material.
2. *Machining;* removing unwanted surplus material from solid stock.
3. *Molding;* forming the part from molten material.
4. *Welding;* combining pieces of stock material into a whole. This process is discussed in Unit 20.

The chart in Fig. 21-1 shows these processes grouped for easier reading.

21.3 Shaping Processes

Forming is a process where solid material is subjected to stresses great enough to cause a permanent desired deformation or flow of the material.

Extrusion is a process using a press comparable to an oversized toothpaste tube in which a plastic material like hot metal is squeezed through a shaped die or nozzle as in Fig. 21-2. It is primarily applicable to long items of constant cross section.

Rolling of steel and aluminum plates, bars, and structural shapes is done by passing hot ingots between appropriately contoured rolls. Hot rolling results in some surface oxidation and a rela-

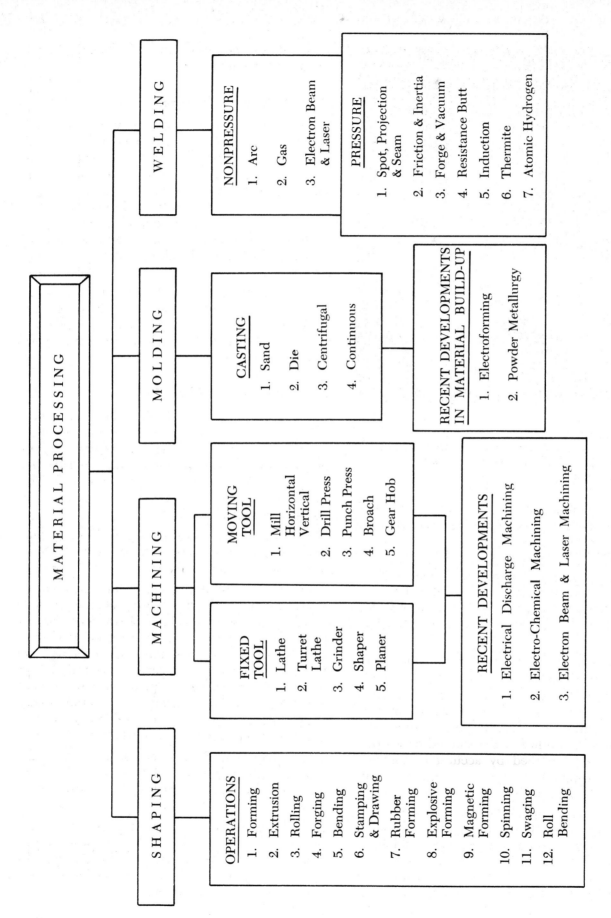

Fig. 21-1. *Production Processes.*

MANUFACTURING PROCESSES AND MEASUREMENTS

Fig. 21-2. *14,000 Ton Extrusion Press. (These 26 ft. aluminum alloy extrusions are main upper rib chords for the Boeing 747 jet.) (Courtesy Aluminum Company of America.)*

tively soft metal condition. Cold rolling produces a higher quality surface finish on a product of high strength.

Forging is an important process for obtaining high strength, tough parts for such items as connecting rods, crank shafts, and high quality tools. Here, a drop forging press with two formed dies mate to shape a preheated blank into a finished shape.

Bending of sheet metal and plate for mass production is usually accomplished on a press brake.

Stamping and Drawing of complex shapes such as car body members, wing tanks, wing root fairings, small parabolic reflectors, pots, washing machine drums, and other double-curved or warped surfaces is accomplished by accurately contouring two mating parts of a punch and die to stamp or draw the desired shape.

Other Methods of shaping material are available including: Rubber and explosive forming, magnetic forming, spinning, swaging (cold forging), and roll bending, just to mention a few. A complete discussion here would be impossible and inappropriate. For reference, several good texts on these subjects are listed in the bibliography at the end of this unit.

21.4 Machining Processes

Machining is the removal of unwanted material to obtain a desired contour by any one of several methods. In the past, this removal of material has been almost entirely limited to use of a sharp tool moving over the work surface to shear or carve away a little material at a time. Recent space age developments permit economical machining of extremely hard materials or unusual shapes by electrical discharge, chemical, and laser or electron beams.

Mechanical machining devices may be separated into two broad classes; those in which the work moves past a fixed-position tool or cutter and those in which the work sits on a stationary bed and the tool passes over it.

21.5 Machining—Fixed Tool

The *Lathe*, Fig. 21-3, is the oldest, most versatile, and most common machine tool. Cutting is accomplished by turning the generally cylindrical work into the sharp tool which reduces the outside or inside diameter of the material. The tool can be moved across the work to contour, taper or thread the cylinder.

A *Turret Lathe* is one having multi-sided tailstocks, programmed to perform several tasks automatically or semiautomatically.

The *Grinder* produces highly finished surfaces by means of an abrasive wheel rotating at high speed. Grinders are in general of two types, namely, *surface grinders* and *cylindrical grinders* or a combination known as a universal machine. External or internal curved surfaces are finished on cylindrical grinders while plane or flat surfaces are finished on surface grinders. This process may also be done on a lathe using a high speed fine grinding wheel against the work. See Fig. 21-4.

A *Shaper*, Fig. 21-5, is a machine where the tool is pushed over the stationary work which automatically moves a small distance after every stroke.

A *Planer*, usually used for large work, has the reverse action of the shaper, i.e., the work moves while a tool similar to that of a shaper advances automatically. The work carriage may have a back and forth or a rotary motion.

220 MANUFACTURING PROCESSES AND MEASUREMENTS

Fig. 21-3. *Engine Lathe. (Courtesy South Bend Lathe, Inc.)*

Fig. 21-4. *Cylindrical Grinding. (Courtesy Norton Company)*

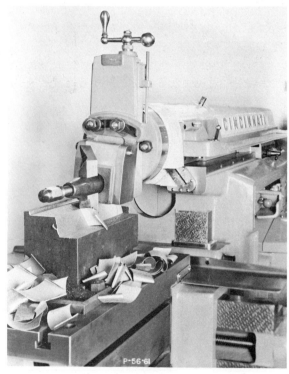

Fig. 21-5. *Shaper. (Courtesy Cincinnati Shaper Company)*

21.6 Machining—Moving Tool

Many different types of milling machines, Fig. 21-6, are available for general use and specialized tasks.

Fig. 21-6. *Milling a Helical, Cast Iron Rotor Blade. (Courtesy Cincinnati Milling Machine Co.)*

A *Horizontal Milling Machine*, Fig. 21-7, has a cutting wheel or wheels mounted on a horizontal arbor and turning at a relatively slow speed while the work moves slowly under the arbor by a synchronized drive. Long, straight contours and planes without undercuts can be machined with very good surface finish.

Vertical Milling Machines, Fig. 21-8, use a variety of special shank-type tools to cut straight or complex contours as the work moves under it. Surface finish is generally very good.

A *Punch Press* forms holes of simple or complex shapes in relatively thin material, where hole finish is not critical.

A *Drill Press* is used for drilling through holes, as well as blind holes. It can also be used to counterbore, countersink, and spot face.

A *Gear Hobber* is used to machine gears from rough blanks. The hob is a multi-toothed device similar to a mill cutter but contoured precisely to shape the gear teeth.

A *Broach*, Fig. 21-9, is a multi-toothed machining device which is pushed or pulled through a pilot hole in the work. Each tooth removes a little more material than the previous one, completing the machining process in one stroke.

21.7 Machining—Recent Developments

Electrical Discharge Machining uses an inexpensive low quality tool as an anode and the hard material to be worked as a cathode. Sparks jumping the anode-cathode gap erode the cathode and chips are carried away by a non-conductive fluid. Since there is no contact the anode (tool) is not worn by the process and may be used indefinitely.

Electro-chemical Machining uses chemicals and electrical energy to obtain the same result as with Electrical Discharge. In this case the liquid is conductive and the electrical current densities are extremely high, up to 5000 amps./sq. in.

Electron Beam and Laser Machining respectively use electrons and light focused optically or magnetically onto the native material to create holes. In these methods, no tool is used; there is no wear, hence, the potential for accuracy is great.

21.8 Molding

Another basic way to form a single part is by melting the metal or plastic and pouring the liquid into a mold. There are several casting techniques but all have some things in common. It should be recognized that casting is primarily a high volume production process, since the cost of the model and mold represent a relatively large investment.

The detail drawing of the part is used by a skilled model maker to produce a pattern which is used to form the cavity into which liquid is poured. The molten material will solidify and shrink slightly as it cools to room temperature.

Sand Casting has been used for many centuries to produce bronze, iron and steel castings using

Fig. 21-7. *Milling Shaft Keyways.*
(Courtesy Cincinnati Milling Machine Company)

Fig. 21-8. *Contour Milling an Aluminum Wing Strut Using a Tracer. (Courtesy Cincinnati Milling Machine Company)*

Fig. 21-9. *Broaching Hexagonal Holes.*
(Courtesy duMont Corporation)

sand as a mold. A two-part flask, consisting of a cope and drag containing a pattern, is filled with special compacted sand. The pattern is removed by separating the cope and drag, leaving a cavity into which molten metal is poured after the cope and drag are reassembled.

Die Casting is a process used where good finish and high speed production is required. Liquid material is injected under pressure into the polished steel die and after a short initial cooling, is extracted mechanically. Many plastic and metal alloy parts are die cast for fuel pumps, washing machines, small engine blocks, and countless other items.

Centrifugal casting is often used to produce large pipes and other suitable shapes. The mold is rotated to force the material outward against the walls of the cavity.

Continuous Casting is a process largely used for pouring ingots directly from the steelmaking furnace into a vibrating cavity with a temporary bottom. The shrinkage and weight of the liquid metal keeps it sliding down the cavity to emerge as a solid ingot which can be cut off as necessary for easy handling.

21.9 Recent Methods of Material Buildup

Electroforming is a recent development in material buildup technique. Here, an extremely heavy detachable plating is built up by normal plating methods. This is a very advantageous technique for obtaining complex shaped thin shells.

Powder Metallurgy is another modern technique of buildup. A pulverized metal is compressed in a mold to the required shape and heated until the individual particles reach the point of fusion. At this point they become sintered together to form a microscopically porous but strong product.

21.10 Automation

The field of production automation is expanding explosively. Highly specialized and ingenious techniques have been applied to processing and feedback control. All of the processes discussed have been automated to some degree using mechanical switching, tracing, and numerical control depending on the requirements. Numerical control of machines is discussed briefly in Unit 25.

21.11 Job Analysis

One of the reasons for the discussion of materials, processes, operations, and machines in this unit is to give the student an idea of where and how a particular part is produced. A knowledge of how a part is made should be a guide in its design and show the importance of giving clear and complete specifications upon a detail drawing.

Large industries have production departments which are concerned with production efficiency; cost, speed, and quality control. The detail drawing of a new design or part is thoroughly analyzed before its production is approved. Operation sheets are prepared which list each job required for production together with the machine, bench, special tools, etc., for each step. This careful study will often reveal where some change of design, material, or finish will produce a satisfactory product at a lower cost. The drawings may then be returned to the drafting room for reworking or revision. A production engineer, no matter in what field, must be able to read and understand drawings in addition to being familiar with design, materials, and production processes.

21.12 Conclusions

The processing field is a vast and fast moving one. Space exploration and the nuclear age have created new demands for exotic materials of ultra-pure quality which require new methods for their manufacture and processing. It is a field for creative men with a desire for challenging work.

It is also necessary for every design engineer to keep up to date on processing advancements for his employer's interests and his own technical improvement, both in knowledge and in ability to improve his product designs. Several texts are available, and periodicals carry many frequent articles relating to this field.

Part II

MANUFACTURING MEASUREMENTS

21.13 Historical

Ever since man developed the ability to cultivate land and build, there has been a demand for ever increasing accuracy in measurement. One of the earliest recorded measurements was the cubit, the length of a man's forearm from elbow to fingertip. Many of these early measurements were similarly determined by Royal decree, with the result that the English system of measurement grew rather chaotically. Hence the foot was literally the length of a man's foot; the inch, the length from the end of the thumb to the first joint; the yard, the distance from the nose to the tip of the outstretched fingers. As recently as the last century, counties were often determined by the distance a man could ride horseback in one day.

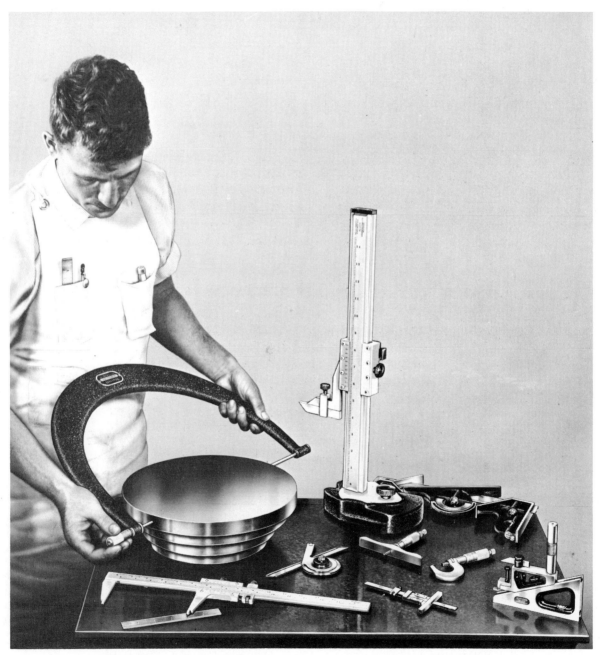

Fig. 21-10. *Measuring Devices. (Courtesy L. S. Starrett Company)*

The English system has been standardized considerably by government action and is the system used in manufacturing the vast majority of the world's products. Still it is criticized by advocates of the metric system, which is admittedly far more rational. Even so, conversion to the metric system in the United States is not likely in the foreseeable future due to its unfamiliarity and the prohibitive cost of conversion.

Everyday measurements in the English system employ fractional (1/2, 1/4, 1/8 . . .) parts of an inch; engineering measurements demand more accuracy. American industries are gradually converting to the decimalized inch, which is used throughout this text.

21.14 Linear Measuring Devices

There are many measurements important to engineering design: Mass, surface roughness, hardness, radioactivity, thermal and electrical conductivity, etc. Of these, geometrical measurement is most important to graphical representation and is discussed in this unit.

Sample Measurements. Measurement of sample parts can be accomplished by several methods, Fig. 21-10, depending on the geometry. A few of the many devices available are listed here.

1. *Vernier Calipers*, Fig. 21-11, have sliding jaws to fit the sample and a principal scale which indicates the nominal distance between jaws. The vernier is a short adjacent scale which is divided into several parts and is used to interpolate between smallest subdivisions of the principal scale. The vernier is used by initially reading the principal scale at the zero index of the vernier scale. If no principal graduation aligns with the index, the reading of the next smaller graduation is noted. To this, the number on the vernier scale graduation which most closely aligns with a graduation on the principal scale is added. Fig. 21-12 shows several settings on vernier scales. Application of the method given will permit the reading of each measurement.

2. *The Micrometer* was invented in the seventeenth century, and is the oldest and still most common precision measuring instrument. The operation of the micrometer, shown in Fig. 21-13, is based on the advance made by a threaded spindle revolving in a stationary nut. One revolution advances the spindle one pitch of the thread, and a partial turn advances the spindle proportionally. A series of scribed lines on the sleeve and barrel of the micrometer makes possible readings to thousandths or even ten-thousandths of an inch.

3. *Adjustable gages* require setting the gage to fit a feature and measuring its setting with a micrometer or vernier. Included in this classification are Expansion Gages, Snap Gages and Calipers.

4. *Go or No-Go Gages* are accurately made devices which either will or will not fit a feature. Typical examples are shown in Fig. 21-14.

5. *Dial Indicators* display amplified changes which are sensed by a mechanical, electrical, or fluid pressure circuit.

6. *Optical devices* employ graphical displays for interpretation and comparison. Optical flats, Fig. 21-15, employ light refraction patterns to detect surface flaws. Optical Comparators, like the thread gage in Fig. 21-14, require visual comparison of product contour with a known standard. Magnification creates highly reliable contour checking as in Fig. 21-16.

7. *Johannsson Blocks* are sets of precision steel blocks, combinations of which can measure nearly any dimension. First produced near the end of the nineteenth century by Carl Johannson, a Swedish machinist, these blocks were so carefully made that each was accurate to within a few millionths of an inch. Precision block sets are not used to make measurements, but to set or check gages, and are now available to any shop that can afford them.

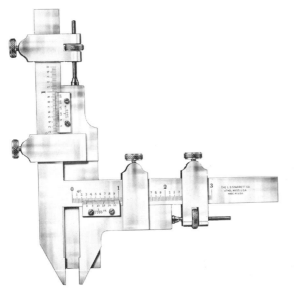

Fig. 21-11. *Vernier Calipers. (Courtesy L. S. Starrett Company)*

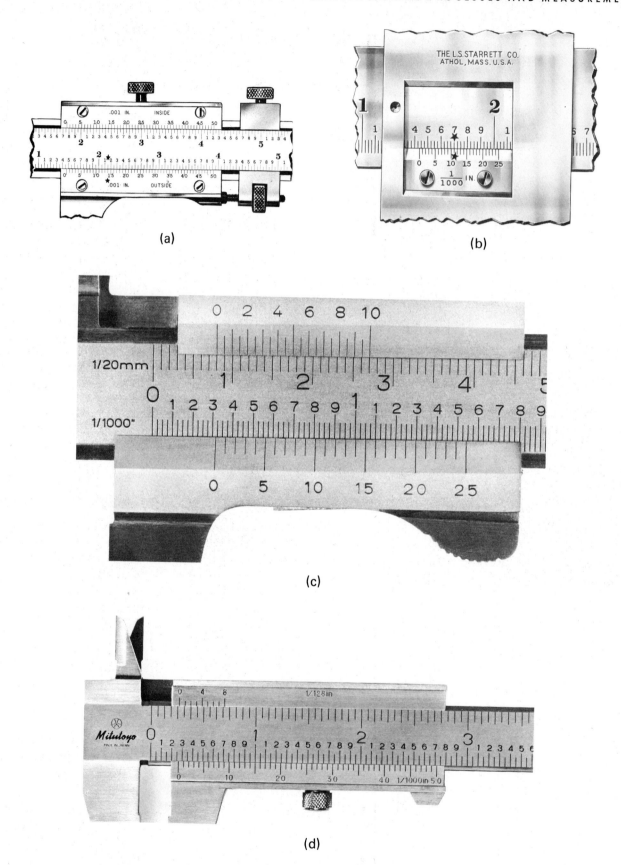

Fig. 21-12. *Vernier Scale Settings.*

MANUFACTURING PROCESSES AND MEASUREMENTS

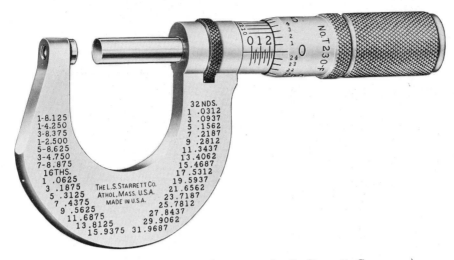

Fig. 21-13. *Micrometer. (Courtesy L. S. Starrett Company)*

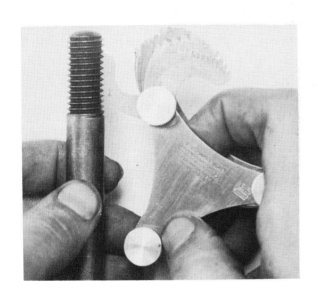

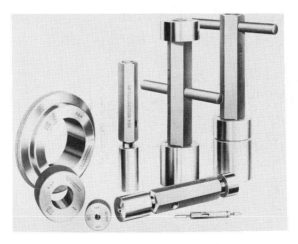

Fig. 21-14. *Thread Gage, Caliper Gages, Plug Gages, Ring Gages. (Courtesy Brown and Sharpe)*

21.15 Special Devices

Specialized measuring techniques based on known physical phenomena are common in industry today. These are often used in mass production where sample checking is difficult and continuous methods are needed. Examples of these techniques include:

1. Attenuation, dispersion, and reflection of sound, X-rays, or high energy beams are used to measure thickness and detect voids or imperfections. The laser measured the distance to the moon within 6 inches using the target placed by astronauts Armstrong and Aldrin.
2. Light reflection characteristics can be used to determine surface quality by sensing with a phototube.
3. Electrical effects such as resistance or inductance can be used to obtain readings indicating thickness or roughness when used to detect movement of a mechanical follower.

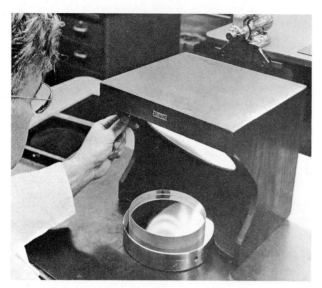

Fig. 21-15. *Optical Flat. (Courtesy Van Keuren Company)*

BIBLIOGRAPHY

De Barr and Oliver, *Electro-Chemical Machining.* MacDonald

De Garmo, E. P., *Materials and Processes in Manufacturing.* MacMillan

Forming Aluminum. Aluminum Company of America

Lindberg, R. A., *Materials and Manufacturing Technology.* Allyn & Bacon

Metals Handbook. American Society for Metals

Stainless Steel Fabrication. Allegheny-Ludlum Steel Corporation

Steel Castings Handbook. Steel Founders' Society

Welding Aluminum. Kaiser Aluminum Corporation

Welding Handbook. American Welding Society

Fig. 21-16. *Projection-type Optical Comparitor. (Courtesy Bausch and Lomb)*

PROBLEMS

1. Make a list of various measuring devices such as scales, micrometers, gages, etc., in their order of relative accuracy, indicating the range of accuracy possible with each. Opposite each gage or instrument give its purpose or use.
2. Construct a chart listing the basic shop machines, their operating characteristics, primary purpose, and other possible operations performed on each machine.
3. Make a list of common machine operations and list, in order of probability, the machine most likely to be utilized for that operation on a production basis.
4. With freehand sketches illustrate the following: counterboring, countersinking, counterdrilling, spotfacing, taper cutting, tapping, turning, facing, broaching, and cutting a key seat for a Woodruff Key.
5. After necessary supplementary study of information from reference sources, write a concise description article on:

    a. Jigs and fixtures   d. Heat treatment
    b. Job analysis   e. Pattern shop and foundry
    c. Superfinishes   f. Standard steel stock

6. List in sequence with any necessary notes the production process steps required to produce one of the items shown in Figs. 12-24, 20-14, or 26-14.

Note: Much of the material and many illustrations in this unit have been supplied through the courtesy of the National Machine Tool Builders, from "Machine Tools Today."

unit 22

Dimensioning for Production

22.1 Introduction

General size and location dimensioning of *single parts* was studied in Unit 12; the problem of dimensioning *mating parts* is discussed here.

When one part must fit another, the sizes of the mating parts must be controlled so that they will function as intended. For example, if a rotating shaft is to fit a bearing the shaft must be small enough to revolve freely yet large enough to prevent rattling or knocking. In the days of small shops where one master machinist made all the parts and fitted them by hand, it was unnecessary to give dimensions to thousandths of an inch; a note describing the type of fit required was sufficient. Mass production, however, requires that interchangeable parts be produced on different machines with different tools by personnel with limited skill. Obviously not all these parts are identical, so it becomes apparent that accurate specification of size is required and in addition, how much variation from that size can be tolerated before the part becomes a reject.

The methods of determining these accurate size specifications is the subject of this unit. Admittedly, a knowledge gained through designing experience is helpful for proper size specification of mating parts. However, certain common practices, established by the experience of other designers, can be applied to determine proper fits for economical production. Moreover, the methods discussed here are not limited to hardware but can be applied to electrical circuits, chemical mixtures, etc.

22.2 Definitions of Terms

The following definitions are abridged from ANSI B4.1-1967.

A *Shaft* on a mechanism is any piece of material or protrusion whose motion or position is constrained.

A *Hole* on a mechanism is any hollow formed in a part whose purpose is to constrain the motion or position of the shaft.

Size is a designation of magnitude.

Nominal Size is the measurement used for general identification. In Fig. 22-1 the link and clevis throat are both nominally 1 inch wide.

Basic Size is the theoretical size from which the limits for a dimension are derived by application of allowances and tolerances. It is an exact expression of the nominal size. Hence, the link and clevis have a basic size of 1.000 inches.

Actual Size is the measured size of an individual part. Other links may be slightly larger or smaller than the one shown in Fig. 22-1, but still be acceptable.

Limits of Size are the specified maximum and minimum sizes of a part, commonly called *Limits*,

DIMENSIONING FOR PRODUCTION

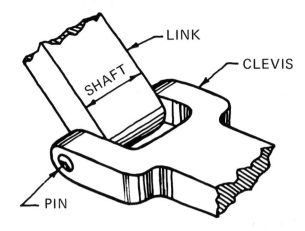

Fig. 22-1. *Link and Clevis Assembly.*

Any link or clevis whose actual size falls outside the limits will be a reject.

Maximum Material Limit (MML) is that condition which leaves the *most* material on the part after processing. It describes either the *largest* link or the *smallest* clevis.

Minimum Material Limit (mml) is that condition which leaves the *least* material on the part after processing. It describes either the *smallest* link or the *largest* clevis.

Tolerance is the total permissible variation in actual size of a part. It is the difference between the limits of either the link or clevis. In Fig. 22-1, assume a tolerance of .004 inches. The largest acceptable link will be .004 inches larger than the smallest link.

Design size is that size of a given part to which the tolerance is applied to give its limits. In the *unilateral system* the design size is the maximum material limit. In the *bilateral system*, the design size is often the average material condition.

Basic Hole System is a method for computing the limits of mating parts in which the design size of the hole is the basic size. The Basic Hole System is frequently used for computing limits because tooling economies can be realized if the size within the selected limits can be produced by a standard tool (reamer, broach, etc.). In Fig. 22-1:

Basic Size = Minimum Clevis = 1.000 = Clevis Design Size.

Therefore, Maximum Clevis = 1.004.

Allowance is the prescribed or intended difference in dimensions of mating parts. It is the *minimum clearance* (positive allowance) or *maximum interference* (negative allowance) between mating parts. Mathematically,

Allowance = Minimum Hole − Maximum Shaft.

Referring again to Fig. 22-1 and using an allowance of +.003 inches (clearance) in this equation,

+.003 = 1.000 − Maximum link, or Maximum link = .997. Applying the definition of tolerance again,

Minimum link = .997 − .004 = .993

These calculations determine the dimensions for this pair of parts using the Basic Hole System, and are shown in Fig. 22-2.

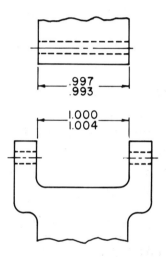

Fig. 22-2. *Dimensioned Link and Clevis.*

Basic Shaft System is a method for computing the limits of mating parts in which the design size of the shaft is the basic size and the allowance determines the hole size. The Basic Shaft System is recommended if there is a particular advantage as, for example, when a standard size of shafting can be used without modification, or several fits are necessary on the same shaft. Using this system:

1. Link Design size = basic size = maximum link = 1.000
2. Minimum link is one tolerance smaller, = .996

Applying the definition of allowance,

3. Allowance = minimum clevis − maximum link
   Minimum clevis = 1.000 + .003 = 1.003
4. Maximum clevis is one tolerance larger = 1.007

Note that allowance and tolerance are the same in either basic hole or basic shaft system.

Clearance Fit occurs between mating parts having limits of size so prescribed that a clearance always results in assembly. This is the type of fit used to match the clevis and link just discussed, where the allowance, or the difference between maximum material conditions is positive.

Interference Fit occurs between mating parts having limits of size so prescribed that an interference always results in assembly. In this case, both the allowance and the loosest fit are negative.

Transition Fit occurs between mating parts having limits of size so prescribed that either a clearance or interference may result during assembly. In this type of fit, the allowance is negative but the loosest fit is positive.

22.3 Unilateral and Bilateral Tolerancing Systems

A *Unilateral System* of tolerancing is the type used in the preceding article. It allows variation in only one direction from the design size. See Fig. 22-2. This method of specifying a tolerance is often useful to a machinist when a critical size is being approached as material is removed in a machining operation.

A *Bilateral System* of tolerancing allows variation in two directions from a design size. While bilateral variations, Fig. 22-3c are generally used for *location* dimensions or for dimensions that can be allowed to vary in either direction, some firms have adopted it universally.

Applying the bilateral basic hole system to the example of Art. 22.2, Fig. 22-1, the following steps are taken:

1. Clevis Design Size = Basic Size = Average Material Condition = 1.000
2. Total tolerance = .004, varies both ways from design size,
   Therefore: Clevis = 1.000 ± .002
3. Allowance = minimum clevis − maximum link
   Maximum link = .998 − .003 = .995
4. Link Design size = Average link = .993
   Therefore: Link = .993 ± .002

Similar calculations are used in the Bilateral Basic Shaft System.

22.4 Expressing Allowable Variation

Various expressions are used to state the amounts of permissible variation for dimensions indicated on the drawing. The expressions recommended by ANSI Y14.5-1966 are described as follows:

1. *Maximum and Minimum Limits* of size are specified. The numerals should be arranged in the following ways:
   a. For size dimensions, the numeral representing the *maximum material condition* is placed above and the numeral representing *minimum material condition is placed below*. Fig. 22-3a.
   b. For location dimensions, the high limit is always placed above the low limit when shown between witness lines, and the low always precedes the high limit when given in note form. Fig. 22-3b.
2. *A Combined Plus and Minus Sign* followed by a single tolerance numeral. This method is generally followed if the plus and minus variations are equal. Fig. 22-3c.
3. *Two Tolerance Numerals* are specified, one plus one minus. This form of expression is necessary if the plus variation differs from the minus variation. Fig. 22-3d.
4. *One Limit Only* should be used as follows:
   a. For a unilateral variation where the other direction is zero. Fig. 22-3e.
   b. For a condition where one limit is not important, MIN or MAX is often placed after the numeral given. Fig. 22-3f.
5. *Expressions of True Position* are specified with a note indicating the tolerance, as in Fig. 22-3g and discussed in Art. 22.14.

22.5 Selection of Tolerances and Allowances

In the examples previously given, tolerance, allowance, and basic size have been specified. In a design project, this selection is a decision to be made by the designer. In some cases tolerance and allowance can be determined using empirical design equations. Many companies determine proper fit limitations by test and research in order to formulate tables which are reliable for conditions of manufacture of their products. No one allowance and tolerance table is comprehensive enough so that it satisfactorily covers all possible combinations of materials and conditions.

DIMENSIONING FOR PRODUCTION

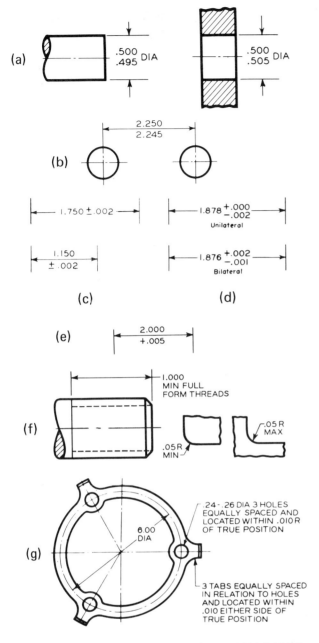

Fig. 22-3. *Expressing Limits (ANSI Y14.5-1966)*

In choosing the class of fit for manufacture, the engineer should keep in mind that cost usually increases exponentially with the accuracy required, and that no finer class of fit should be chosen than the functional requirements actually demand. The length of engagement, lubrication, load characteristics, and type of mating materials also play an important part in the selection of the class of fit for a piece of work.

Preferred Basic Sizes are recommended basic standards which should be used to reduce the large number of possible limit combinations. Fewer basic sizes reduce the necessary supply of raw material and tools. Preferred basic sizes up to 8.00 inches are shown in Fig. 22-4.

0.010	0.10	1.00	3.00	5.00
0.012	0.12	1.20	3.20	5.20
0.016	0.16	1.40	3.40	5.40
0.020	0.20	1.60	3.60	5.60
0.025	0.24	1.80	3.80	5.80
0.032	0.30	2.00	4.00	6.00
0.040	0.40	2.20	4.20	6.50
0.05	0.50	2.40	4.40	7.00
0.06	0.60	2.60	4.60	7.50
0.08	0.80	2.80	4.80	8.00

Fig. 22-4. *Preferred Basic Sizes. All values are in thousandths of an inch. (ANSI B4.1-1967)*

Tolerances and Allowances in preferred increments are a logical complement to preferred basic sizes. For general purposes, the American National Standards recommend that tolerances and allowances should be chosen from the range of values shown in Fig. 22-5.

0.1	1	10	100
...	1.2	12	125
0.15	1.4	14	...
...	1.6	16	160
...	1.8	18	...
0.2	2	20	200
...	2.2	22	...
0.25	2.5	25	250
...	2.8	28	...
0.3	3	30	...
...	3.5	35	...
0.4	4	40	...
...	4.5	45	...
0.5	5	50	...
0.6	6	60	...
0.7	7	70	...
0.8	8	80	...
0.9	9	...	...

Fig. 22-5. *Preferred Tolerances and Allowances. All values are in thousands of an inch. (ANSI B4.1-1967)*

22.6 Classification of Fits

Because research, experience, and cost are factors governing the choice of allowance and tolerance for a particular job, a general chart has been devised to assist the student in selecting reasonable values, Fig. 22-6. It should be noted that while the class of fit depends upon its purpose or use, the amount of tolerance or allowance for a particular class increases within prescribed limits as the size increases.

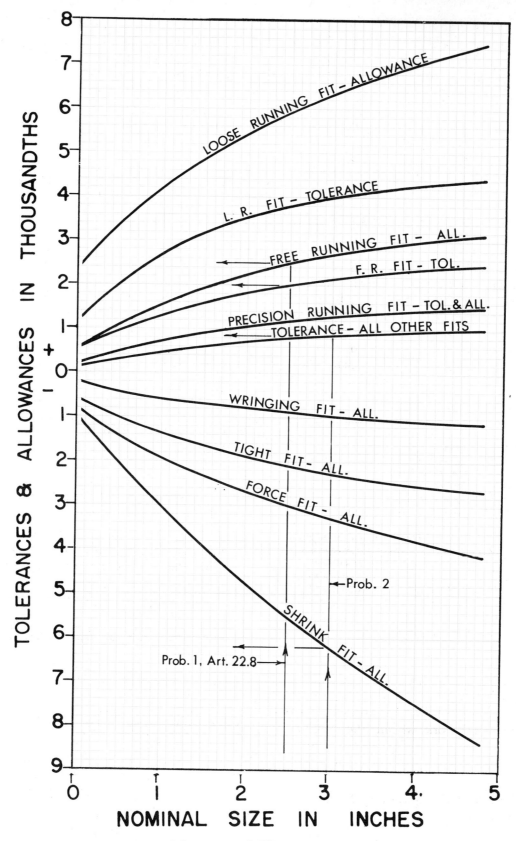

Fig. 22-6. *Tolerances and Allowances for Specified Fits.*

Nominal Size Range Inches Over — To	Grade 4	Grade 5	Grade 6	Grade 7	Grade 8	Grade 9	Grade 10	Grade 11	Grade 12	Grade 13
0 — 0.12	0.12	0.15	0.25	0.4	0.6	1.0	1.6	2.5	4	6
0.12 — 0.24	0.15	0.20	0.3	0.5	0.7	1.2	1.8	3.0	5	7
0.24 — 0.40	0.15	0.25	0.4	0.6	0.9	1.4	2.2	3.5	6	9
0.40 — 0.71	0.2	0.3	0.4	0.7	1.0	1.6	2.8	4.0	7	10
0.71 — 1.19	0.25	0.4	0.5	0.8	1.2	2.0	3.5	5.0	8	12
1.19 — 1.97	0.3	0.4	0.6	1.0	1.6	2.5	4.0	6	10	16
1.97 — 3.15	0.3	0.5	0.7	1.2	1.8	3.0	4.5	7	12	18
3.15 — 4.73	0.4	0.6	0.9	1.4	2.2	3.5	5	9	14	22

Fig. 22-7. *Standard Tolerance Grades. All values are in thousandths of an inch. (ANSI B4.1-1967)*

22.7 Manufacturing Limits

American National Standard tolerances are listed under Grades 4 to 13, as shown in Fig. 22-7, to provide a suitable range from which approximate tolerances for holes and shafts can be selected to permit the use of standard gages. The standard tolerances have been so arranged that for any grade they represent approximately similar production difficulties throughout the range

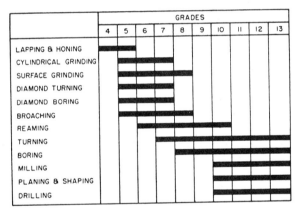

Fig. 22-8. *Grades for Machining Processes. (ANSI B4.1-1967)*

of sizes. For larger sizes refer to ANSI B4.1-1967 from which these tables have been taken.

Fig. 22-8 has been provided to indicate the machining processes which may normally be expected to produce work within the tolerances indicated by the grade given in this Standard. This information is intended merely as a guide.

22.8 Estimating and Computing Limit Dimensions

The proper limit dimensions for any set of mating parts can be determined by applying the information discussed in Arts. 22.5, 22.6, and 22.7.

Problem 1. A heavily loaded shaft is to have a *free-running fit* with a bearing whose nominal size is 2 1/2 inches. Determine the bearing and shaft limit dimensions. Use only three decimal places for economy. Fig. 22-9.

1. Basic Size = 2.500
2. Tolerance (Fig. 22-6) = .002
   Preferred tolerance (Fig. 22-5) = .002
3. Allowance (Fig. 22-6) = +.0025
   Round to .003 to permit free play.
   Preferred allowance (Fig. 22-5) = +.003
4. Use Basic Hole System
   Basic size = minimum hole = 2.500
   Maximum hole = 2.500 + .002 = 2.502
5. Allowance = minimum hole − maximum shaft
   Maximum shaft = 2.500 − .003 = 2.497
6. Minimum shaft = 2.497 − .002 = 2.495

Little difficulty will be encountered in determining limit dimensions if the student will arrange his data and computations *systematically.*

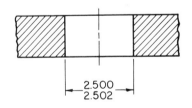

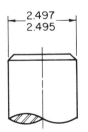

Fig. 22-9. *Basic Hole Dimensioning.*

Problem 2. A steel wheel is to be shrunk-fit onto a 3-inch axle. Determine the limit dimensions for each part. Use the basic shaft system and 3 place accuracy. Fig. 22-10.

1. Basic Size = maximum shaft = 3.000
2. Tolerance (Fig. 22-6 and Fig. 22-5) = .001
3. Allowance (Fig. 22-6 and Fig. 22-5) = −.006
4. Maximum shaft = 3.000
   Minimum shaft = 3.000 − .001 = 2.999
5. Minimum hole = maximum shaft + allowance
   Minimum hole = 3.000 + (−.006) = 2.994
6. Maximum hole = 2.994 + .001 = 2.995

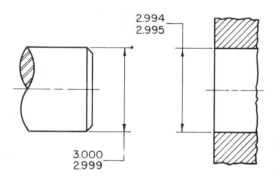

Fig. 22-10. *Basic Shaft Dimensioning.*

22.9 Limit Dimensions From ANSI Tables

Some 30 years of work and the combined experience of many men went into developing tables on pages 363-366. These tables give a series of standardized dimensions, to the nearest ten-thousandth of an inch, for several types and classes of fits on a unilateral Basic Hole System. These dimensions are such that the fit produced by mating parts in any one class will produce similar performance throughout the range of sizes.

Five types of fits are specified and each is subdivided into several numbered classes. These recommended fits, designated by letter symbols for study and discussion purposes only are not to be included on production drawings; only the limit dimensions determined by use of these tables are specified.

RC: *Running and Sliding Fits,* shown on page 363 are intended to provide suitable running performance, with allowance for lubrication, throughout the range of sizes.

LC: *Locational Clearance Fits,* page 364, are intended for parts that are normally stationary, but which can be freely assembled or disassembled.

LT: *Locational Transition Fits,* page 365, are a compromise between clearance and interference fits, and applied where accuracy of alignment is important, with a small amount of clearance or interference permissible. These are also used for selective assembly.

LN: *Locational Interference Fits,* page 365, are used where accuracy is of prime importance, and for parts requiring rigidity and alignment with no special requirement for bore pressure.

FN: *Force and Shrink Fits,* page 366, constitute a special type of interference fit, normally characterized by maintenance of constant bore pressures throughout the range of sizes.

Graphical description of each class of fit is shown below each table.

Use of the tables is summarized in the statement above each:

Limits for hole and shaft are applied to the basic size to obtain limits of size for the parts.

Problem 1. A punch and die set is to have a running and sliding fit, RC2. Nominal Size = 1 inch. From page 363, the die (hole) limits are 0 and +0.5 thousandths; the punch (shaft), −0.7 and −0.3 thousandths.

Therefore, the die size is:

$$\frac{1.0000}{1.0005} \text{ or } 1.0000 \pm \begin{matrix}.0005\\.0000\end{matrix} \text{ or } \begin{matrix}1.0000\\+.0005\end{matrix}$$

the punch size is:

$$\frac{.9997}{.9993} \text{ or } .9997 \pm \begin{matrix}.0000\\.0004\end{matrix} \text{ or } \begin{matrix}.9997\\-.0004\end{matrix}$$

Problem 2. A steel gear is to have a Heavy Drive fit, FN3, onto a 3-inch diameter shaft. From page 366, the Hole limits are −0 and +.12 thousandths; the shaft, +3.0 and +3.7 thousandths.

Therefore, the hole size is:

$$\frac{3.0000}{3.0012}, \text{ or } 3.000 \pm \begin{matrix}.0012\\.0000\end{matrix}$$

the shaft size is:

$$\frac{3.0037}{3.0030}, \text{ or } 3.0037 \pm \begin{matrix}.0000\\.0007\end{matrix}$$

DIMENSIONING FOR PRODUCTION

22.10 Adapting ANSI Tables for Basic Shaft System

Again, the Basic Shaft System (Art. 22.2) is recommended if there is a particular need, such as adapting a design to accomodate purchased standard shafting.

A study of the ANSI Tables will show that in the *Hole Limits* column the numbers are *always* 0.0 and +X so that the hole dimensions are equal to and slightly larger than the basic size, which agrees with the definition of Basic Hole System, Art. 22.2.

To convert the tables to the Basic Shaft System, 0.0 and −X are needed in the *Shaft Limits* column. This can be done without affecting the class of fit by adding the negative of the number corresponding to the largest shaft to each of the four figures.

Problem. Determine the limits dimension for a 1 inch RC4 shaft and hole using B. S. system (RC4S).

From page 363, the largest shaft is .0008 undersize.

Add +.0008 to each figure.

Limits: hole $\frac{1.0008}{1.0016}$ shaft $\frac{1.0000}{.9992}$

22.11 Cumulative Tolerances

An effort should be made to avoid chain dimensioning when limits are being used. If the summation of the minimum values in a chain of dimensions is compared with the summation of the maximum, the total tolerance of the chain has increased by the product of the number of individual dimensions times their individual permissible variation. This cumulative error may be eliminated by using base line dimensioning. Fig. 22-11 shows good practice for precision locations.

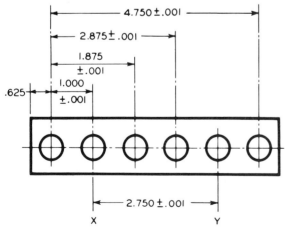

Fig. 22-11. *Dimensioning to Prevent Tolerance Accumulation Between X and Y. (ANSI Y14.5-1966)*

22.12 Tolerance Control Symbols

Positional or Form tolerances are sometimes controlled by the symbols shown in Fig. 22-12.

GEOMETRIC CHARACTERISTIC SYMBOLS			
	Characteristic		Symbol
Form Tolerances	For Single Feature	FLATNESS	▱
		STRAIGHTNESS	—
		ROUNDNESS (CIRCULARITY)	○
		CYLINDRICITY	⌭
		PROFILE OF ANY LINE (1)	⌒
		PROFILE OF ANY SURFACE (1)	⌓
	For Related Features	PARALLELISM (2)	∥
		PERPENDICULARITY (SQUARENESS)	⊥
		ANGULARITY	∠
		RUNOUT (3)	↗
Positional Tolerances		TRUE POSITION	⊕
		CONCENTRICITY (4)	◎
		SYMMETRY (5)	≡
SYMBOLS FOR MMC AND RFS			
Modifier			Symbol
(MMC) MAXIMUM MATERIAL CONDITION			Ⓜ
(RFS) REGARDLESS OF FEATURE SIZE			Ⓢ

Fig. 22-12. *Geometric Tolerance Symbols. (ANSI Y14.5-1966)*

The geometric symbols are shown in a box, Fig. 22-13, which includes the tolerance and often a modifier. These control symbols are either added as a note or referred to the feature by a leader from the box.

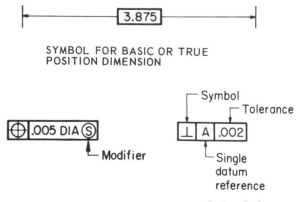

Fig. 22-13. *Feature Control Symbols. (ANSI Y14.5-1966)*

22.13 Positional Tolerances

In the past, holes and other features have been located by rectangular or polar coordinates, shown predominately with individual tolerances. The engineering intent can often be expressed more precisely if locations are given as *True Position*, with tolerances stating how far actual positions can be displaced from the true positions. There are two methods for applying positional tolerancing.

1. The following or a similar form of note is used for features located by dimensions to their axes or where the location may be allowed to vary in any direction. 6 HOLES LOCATED AT TRUE POSITION WITHIN .005 DIAMETER. Radius may be used instead of diameter.

2. The following is a typical form of note for features located by dimensions to a center plane or to one surface of the feature. 6 SLOTS LOCATED AT TRUE POSITION WITHIN .005 TOTAL.

Fig. 22-14 illustrates *True Position Dimensioning* both in *note* and *symbol* form.

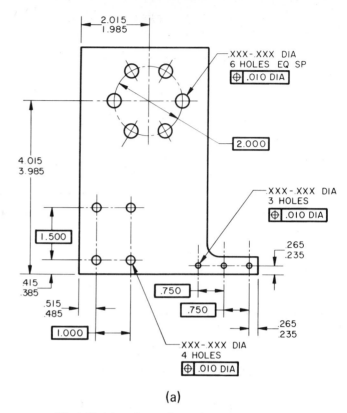

Fig. 22-14. *True Position Dimensioning.*
(a) *Hole Pattern Locations by Coordinate Limits.*

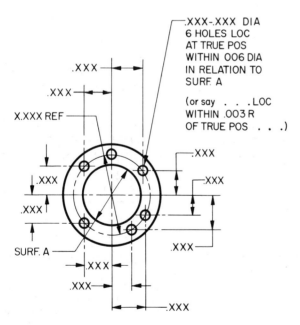

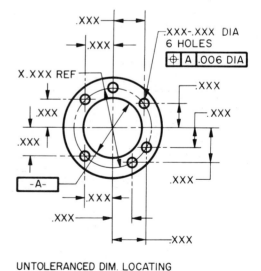

Fig. 22-14. *True Position Dimensioning.* (b) *Round Flange with Datum Bore.*
(ANSI Y14.5-1966)

DIMENSIONING FOR PRODUCTION 239

| DRAWING CALLOUT | | INTERPRETATION |

SPECIFYING ANGULARITY

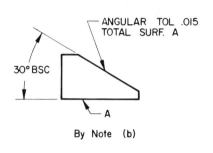
By Note (b)

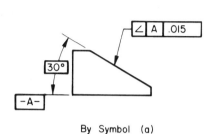
By Symbol (a)

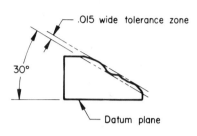

The surface must be within the specified tolerance of size and must lie between two parallel planes (.015 apart) which are inclined at specified angle to the datum plane.

SPECIFYING PARALLELISM

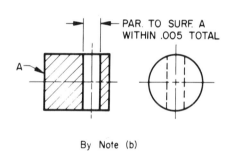
By Note (b)

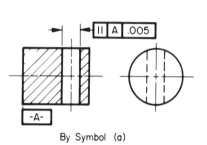
By Symbol (a)

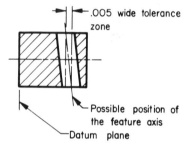

The feature axis must be within the specified tolerance of location and must lie between two planes (.005 apart) which are parallel to the datum plane.

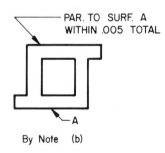
By Note (b)

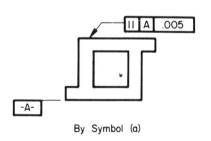
By Symbol (a)

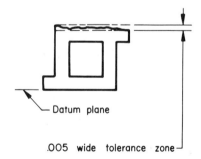

The surface must be within the specified tolerance of size and must lie between two planes (.005 apart) which are parallel to the datum plane.

Fig. 22-15. *Tolerances of Form.* (ANSI Y14.5-1966)

22.14 Geometrical Tolerances

Tolerances of Form state how far actual surfaces are permitted to vary from the perfect geometry implied by drawings. Expressions of these tolerances refer to straightness, flatness, parallelism, squareness, angular displacement, symmetry, concentricity, roundness, etc. Geometrical tolerances are to be observed regardless of the actual finished sizes of the features concerned. The following are typical expressions of tolerances of form.

STRAIGHT WITHIN .003 TOTAL

FLAT WITHIN .003 TOTAL

PARALLEL TO A (centerline or surface) WITHIN .005 TOTAL

CONCENTRIC TO A WITHIN .003 F.I.R. (full indicator reading)

ROUND WITHIN .003 ON DIA.

When tolerances of form are not specified on a drawing, it is commonly understood that the actual part will be acceptable if it is within the dimensional limits given, regardless of form variations.

Fig. 22-15 shows examples of *Geometrical Tolerancing*.

22.15 Surface Quality

A means for specifying the quality of finish on a surface has been standardized, since there may be a great deal of variation in the finish due to the machining operation or to the skill of the workman. The following definitions and explanations are abstracted from ANSI B46.1-1962.

Roughness is the height of the irregularities caused by the action of the machine tool.

Microinch, one-millionth of an inch, is the unit of measurement of roughness.

Waviness, the undulation or wave in a surface, is caused by a faulty machine or machine operation, warping, or strain.

Lay is the predominant direction of machine tool marks.

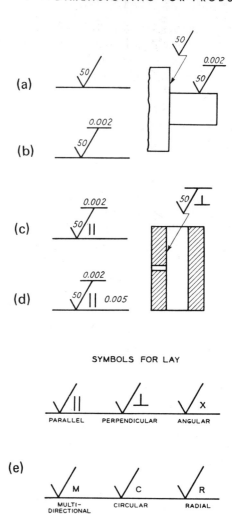

Fig. 22-16. *Surface Quality Symbols.* (ANSI B46.1-1962)

A surface with a specified finish is labeled with a finish symbol having the general form of a check mark, Fig. 22-16, so that the point of the symbol is either on the line indicating the surface or on a leader pointing to the surface.

The simplest form of roughness symbol specifying roughness height only is used when the width of roughness or direction of tool marks is not important, see Fig. 22-16a. This height may be either maximum peak to valley height, or average deviation from the mean. The numerical value is always placed above the short leg of the symbol.

Where it is desired to specify waviness height in addition to roughness height a straight horizontal line is added to the top of the symbol with the numerical value of waviness height indicated as shown in Fig. 22-16b. If the nature of the preferred lay is to be shown, in addition to these

DIMENSIONING FOR PRODUCTION

two characteristics, it will be indicated by the addition of a combination of lines as shown in Fig. 22-16c and e. The parallel and perpendicular part of the symbol indicates that the dominant lines on the surface are parallel or perpendicular to the major boundary line of the surface in contact with the symbol.

The complete symbol including the roughness width placed to the right of the lay symbol, is shown in Fig. 22-16d.

The use of only one member to specify the height or width of roughness or waviness indicates its maximum value. Any lesser degree of roughness will be satisfactory. When two numbers are used, separated by a dash, they indicate the maximum and minimum permissible values.

22.16 General Notes

When the same instructions apply to more than one dimension on a drawing, it is standard practice to avoid repetitions by using a general note on the drawing. Where applicable, a note may be preceded by the phrase: UNLESS OTHERWISE SPECIFIED.

The following list of general notes are typical:
LINEAR TOLERANCES .xxx
BREAK SHARP EDGES
FILLET RADII .xxx MAX (OR MIN)
THREAD DEPTHS ARE MINIMUM FULL THREAD
DIMENSIONS APPLY BEFORE (OR AFTER) PLATING

PROBLEMS

1. Write, in your own words, an accurate but concise definition of the following terms: (a) nominal size; (b) actual size; (c) allowance; (d) tolerance; (e) clearance fit; (f) transition fit; (g) interference fit; (h) basic hole system; (i) basic shaft system; (j) unilateral tolerance; (k) bilateral tolerance; (l) cumulative error.

2. Construct a table similar to the one shown in Fig. 22-17. From the information given, supply the missing numerical values in the spaces provided. Note that both unilateral and bilateral tolerance systems are shown. For further exercise in computing limit dimensions, allowances, etc., substitute suitable values for

Nominal Size	Allowance	Tolerance	Basic Hole System		Basic Shaft System	
			Hole	Shaft	Hole	Shaft
¾	+0.0035	0.0004				
3						
1½			1.500 ±.002	1.493 ±.002		
½	−0.0005	0.0003				
2		0.004		1.994		
1¾			1.750 ±.001	±.001	1.749 ±.001	±.001
1½	Zero	0.0003				
¼			0.254	0.241		
1¾			±.0003	1.7501 ±.0003	±.0003	

Fig. 22-17.

the one given and determine corresponding results.

3. Fig. 22-18. Dimension the punch and die to avoid cumulative tolerance error. Use limit values only (basic hole system).

   Specifications: Nominal size of notches—3/4 in. vertical, 1/4 in. and 1/2 in. horizontal;

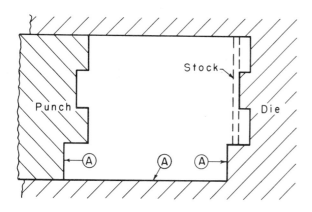

Fig. 22-18.

zero allowance on surfaces marked "A"; all other mating surfaces: +0.005 allowance, 0.003 tolerance (each dimension); stock thickness, 0.012.

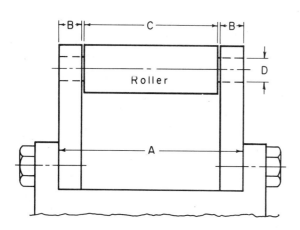

Fig. 22-19.

4. Fig. 22-19. Determine values for A, B, and C on the given assembly so that the roller will have a minimum of 0.004 end play. Determine values for D (hole) and D (shaft) from the given information.
   Specifications:
   Dimensions A: Nom. = 6; Tol. = 0.002
   B: Nom. = 1/2; Tol. = 0.002
   C: —   Tol. = 0.002
   D: Nom. = 1/2; Tol. = 0.002;
      Allow. = +0.002 (use basic hole)

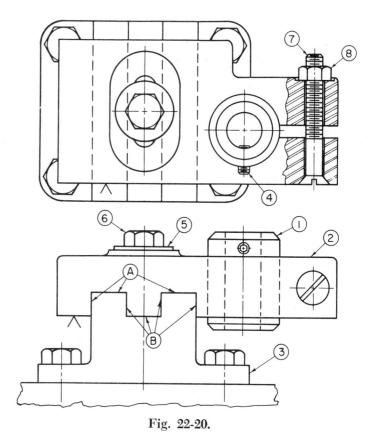

Fig. 22-20.

5. Fig. 22-20. Sketch and dimension completely the nonstandard parts of the mill jig shown. Scale the assembly drawing for all nominal sizes. Scale shown is 3" = 1'-0.
   Fit Specifications:

   a. Between parts 1 and 2 before cap screw (part 7) is tightened (basic shaft system), +0.003 allowance, 0.002 tolerance (each part).

   b. Between parts 2 and 3 (basic hole system) —between each mating surface marked B, +0.002 allowance. Use 0.001 tolerance on each limit dimension.

   c. Part 4 has 20 unified threads per inch.

   d. Part 6 is a 1 inch cap screw, 4 1/2 in. long, which slides through the slot in part 2, and screws into part 3. It has 8 National coarse threads per inch.

   e. Part 7 has 16 unified threads per inch.

DIMENSIONING FOR PRODUCTION

6. Sketch a shaft and hole similar to Fig. 22-10 for each of the following cases. Determine the limit dimensions from the ANSI tables and apply to the sketches. Label each.
   a. Nominal Size 1 inch, using RC7, free running fit.
   b. Nominal Size 1/2 inch, using RC3, precision running fit.
   c. Nominal Size 1 1/2 inches, using FN1, light drive fit.
   d. Nominal Size 2 inches, using FN4, force fit.
7. Figs. 22-1 and 22-2. Determine the correct limit dimensions for the 3/8 diameter pin and its holes in the link and clevis. The pin should fit tightly into the clevis hole, and the link should rotate freely on the pin. What system (B.H. or B.S.) should be used? Why?

unit 23

Piping and Structural Drawing

23.1 Introduction

Engineering is interdisciplinary since engineers must use each other's work in conjunction with their own. The electrical engineer may need high voltage distribution towers; the structural engineer, a pipe column for joist support. The mechanical engineer may wish to install heavy equipment on a roof, and the hydraulic engineer may need an electrically driven centrifugal pump. In every case the competent and creative engineer is expected to be capable of recognizing, interpreting, and solving basic problems involving the design function in fields other than his own. The man without some ability in this way is bound to be "low man on the totem pole."

It is the purpose of this unit to acquaint the student with the basic elements of piping and structural components; their applications, standards, and graphical representation.

Part I
PIPE AND PIPING SYSTEMS

23.2 Piping

The practicing engineer often encounters problems which involve the handling of liquids, gasses or solids. He may be called upon to select the proper piping, tubing, and fittings to be used in the design of machines, process plants, power plants, water systems, transportation lines, or structures.

The intended purpose determines the material to be used in the manufacture of the pipe. The more common materials include iron, copper, and aluminum alloys, but concrete, glass, plastics, rubber, and many other materials are also used.

Individual pipes are connected by fittings and valves to form a circuit in which the direction and amount of flow can be controlled. Pipes are also extensively used for structural purposes such as staging, handrails, flag poles, and simple framing.

Pipe and tubing specifications have been standardized as to composition, size, threads, etc., through the efforts of such organizations as the American Water Works Association (AWWA), American Gas Association (AGA), American Petroleum Institute (API), American National Standards Institute (ANSI) and others. Specifications for special fittings, valves and other piping products found in manufacturers' catalogs should be used to supplement the tables found in the Appendix.

23.3 Pipe Materials and Sizes

Carbon steel, stainless steel, some alloys, and aluminum pipe are commercially available in three weights or schedules: Standard (schedule 40), Extra Strong (schedule 80) and Double Extra Strong (schedule 160). Other schedules are available on a limited basis. These weights are specified by nominal diameter and wall thickness, details of which are shown in the Appendix, page 368. The use of schedule numbers is gradually replacing the older strength designation. The schedule number required for a specific application is determined by the formula

$$\text{Sch. No.} = (1000)\left(\frac{P}{S}\right)$$

where P = working pressure, psi, and S = allowable stress in the pipe, psi. "S" values for various materials can be found in any of several handbooks. Note on page 368 that the nominal diameter does not exactly correspond to any measurable size on the pipe. For sizes 12 inches and under the nominal size is close to the inside diameter (I.D.), while for sizes 14 inches and larger the nominal size is exactly the outside diameter (O.D.)

Since the outside diameter for all weights of a given size pipe remains constant, it is possible to interchange fittings between different weights of a given diameter pipe. The inside diameter of heavier pipe decreases as its wall thickness increases. This is illustrated in Fig. 23-1. Notice that the O.D. of three-quarter-inch pipe is over one inch.

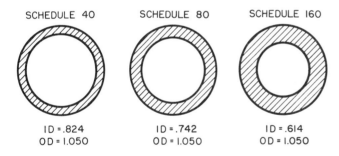

Fig. 23-1. *Comparison of Three-Quarter Inch Steel Pipes.*

Cast-Iron Pipe is available in sizes from 3 to 84 inches in diameter. The larger sizes are used principally for sewer pipes, gas lines, and water mains. Flanged and bell-and-spigot joints are most commonly used on cast-iron pipe.

Copper, Brass, and Wrought-Iron Pipe are used for the same purposes as steel pipe and have the same outside diameters, although wall thickness is slightly different than that of corresponding sizes of steel pipe. Copper and brass pipe is available in two standard weights, regular and extra strong. Wrought iron pipe is available in regular, extra strong, and double extra strong weights.

Tubing, both flexible and rigid, is available in many materials and sizes up to 12 inches in diameter. Copper water tube is available in four standard weights: K, L, M, and DWV, in order of decreasing wall thickness. Its outside diameter is consistently one-eighth inch greater than its normal size.

Copper refrigeration tube is specified by its outside diameter and wall thickness.

Aluminum, steel, titanium, etc., tubing are available in a wide range of diameters, wall thicknesses, and shapes, as shown in the Appendix, page 372. Specifications for tubing should include the outside diameter, wall thickness, material, type, and ratings required if any. When specifying fittings, indicate the type of tubing since fittings are often not interchangeable.

Example: 37° flared tube bulkhead union, stainless steel, for 1 1/4 O.D. x 0.065 wall, type 304 S.S. cold drawn seamless tubing.

23.4 Pipe Joints

Pipes are connected most commonly by screwed, flanged, bell and spigot joints, or by welding. See Fig. 23-2. Flared, crimped, and soldered joints with appropriate fittings and adapters are used to join thinwall tubing.

23.5 Pipe Threads

American Standard (National) pipe thread is similar in form to the American Standard thread described in Unit 19 except that the pipe thread is tapered 1/16 inch per inch, measured on the diameter. The taper insures a tight joint and fixes the entry distance of the pipe into the fitting. The external thread is known as the male thread while the internal is the female thread. Fig. 23-3 shows the details of a pipe thread.

American Standard straight pipe thread, used for couplings, fittings, and other pressure tight joints, has the same form as the taper and the same number of threads per inch for a given diameter. The table on page 368 in the Appendix gives the number of threads per inch for different diameters.

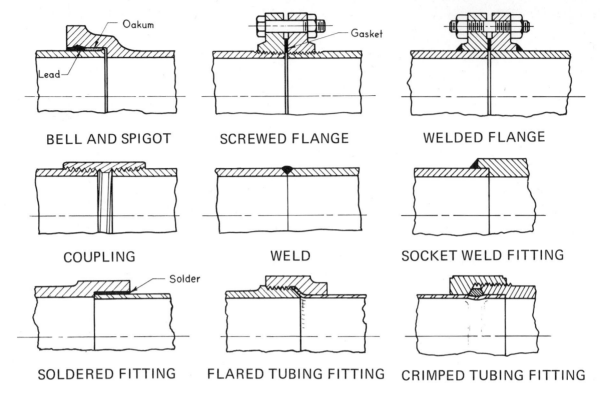

Fig. 23-2. *Pipe Joints.*

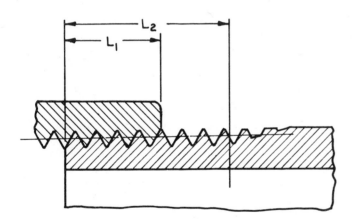

L_1 = LENGTH OF HANDTIGHT ENGAGEMENT
L_2 = LENGTH OF EFFECTIVE THREAD

Fig. 23-3. *American National Taper Pipe Thread.*

Either schematic or simplified representation may be used to show pipe threads on a drawing. The taper may be shown but is generally omitted as in Fig. 23-4.

A pipe thread is specified by giving its diameter and form, for example: 1"-NPT, specifies a 1"

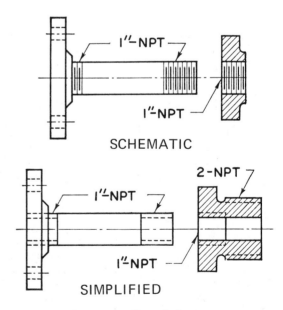

Fig. 23-4. *Pipe Thread Representation.*

nominal diameter, National, Pipe, Taper. The 1"-NPS specifies a 1" nominal diameter, National, Pipe, Straight. The specification 1" Am. Std. Pipe Thd. is a common usage.

23.6 Pipe Fittings

Many different types of fittings are used in the installation of pipe lines, the most common being screwed or flanged, although welded fittings are available for steel and aluminum. Soldered and flared fittings are usually used on brass and copper pipe and tubing.

Screwed Fittings, Fig. 23-5, shows an assortment of common plumbing fittings. Complete details of these and others may be found in company catalogs. The table on page 370 gives dimensions for the items most often used.

Flanged Fittings of cast iron or forged steel are often used for pipe connections, especially in the larger sizes. The usual fittings are commercially available in sizes from 1" diameter. The table on page 371 shows the detail for fittings in common use.

Couplings, consisting of short cylinders with internal threading, are used to connect pipes in pressure tight joints. The usual coupling is threaded through with right-hand National pipe straight threads. Reducing couplings in standard sizes can be purchased.

Unions are used as the closing link in screwed pipe circuits. A common union of the type shown in Fig. 23-6 consists of three principal parts tightened together either against a gasket or a ground joint. Flanged systems do not require unions since any two flanges screwed or welded on the pipe ends may, after inserting a gasket, be drawn together by bolts.

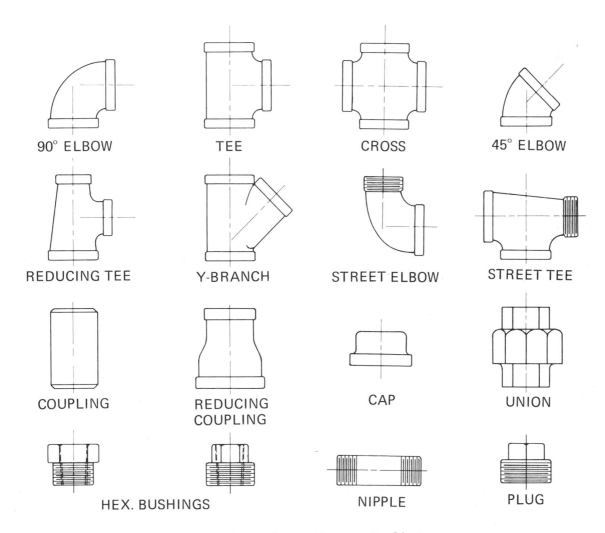

Fig. 23-5. *Screwed Pipe Fittings—Double Line.*

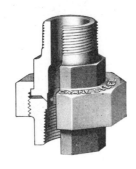

Fig. 23-6. *Screwed Unions. (Courtesy Crane Co.)*

23.7 Valves

Valves of many different types and styles are used to regulate the flow in pipes. Some of the more common ones are:

Globe	Angle	Pressure-regulating
Gate	Safety	Butterfly
Check	Expansion	Control
	Needle	

Gate Valves offer less restriction to straight-line flow than other types and are extensively used as shut-off valves for liquids. Fig. 23-7a illustrates a gate valve.

Globe Valves are principally used to throttle or regulate quantity of liquid or vapor flow. The manner of operation is shown in Fig. 23-7b. Control valves are similar to globe valves but are opened and closed automatically by some control signal.

Check Valves prevent reversal of liquid flow. A swing check is shown in Fig. 23-7c.

The overall length of these valve types is tabulated on page 369. See manufacturers' catalogs for use and diagrams of other valves.

23.8 Pipe and Fitting Representation

A majority of piping layouts are shown by means of single-line representation, either in orthographic projection or in pictorial form. The pipe, regardless of diameter, is represented by single lines and the fittings and valves by standard single-line symbols, page 367 in the Appendix. Single-line representations, such as illustrated in Fig. 23-8, can be used since standard pipes and fittings suitable for almost any type of installation are available. Identifying letters are used on the pictorial to call out each part in the system for listing in a bill of materials.

A double-line drawing, more easily interpreted by the layman, is used for piping installations where the accuracy of clearness is important. Many companies manufacturing pumps and process equipment supply double-line drawings to insure correct installation by the purchaser. The fittings shown in Fig. 23-5 are double-line representations.

Many of the larger process industries using extensive piping installations construct accurate scale models like Fig. 7-1 showing all components

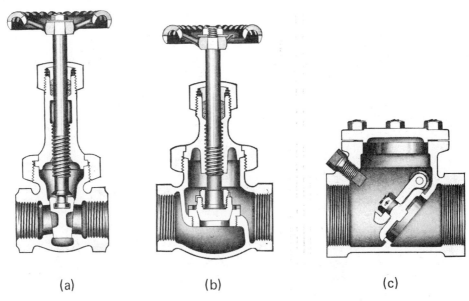

Fig. 23-7. *Common Valves in Section. Gate, Globe, and Check. (Courtesy Crane Co.)*

PIPING AND STRUCTURAL DRAWINGS 249

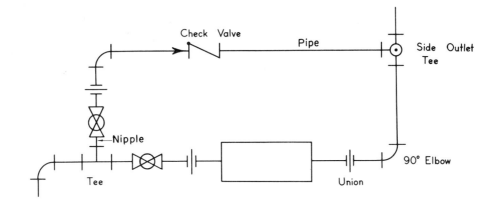

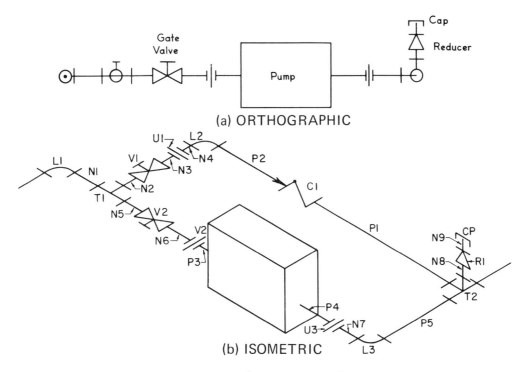

Fig. 23-8. *Single-Line Piping Layout.*

in place, thereby controlling required accuracy of location of the piping system and component parts.

Most fittings can be specified by name, size, and material However, reducing fittings such as tees, crosses, and laterals must have connection sizes properly indicated. In the case of a tee or lateral the straight, or flow direction is given with the larger of the two openings expressed first, followed by the dimension of the side outlet. Example: 1 x 3/4 x 1/2 tee, or 4 x 4 x 2 lateral. For a cross, the largest opening is given first followed by its straight line outlet, and then the 90° flow expressed in the same manner. Example: 4 x 3 x 2 x 2 cross.

23.9 Computing Pipe Lengths

Pipe is often cut and threaded in the shop and delivered, with necessary fittings, to the erection site. To specify the proper fittings and the necessary lengths and diameters of pipe for the installation, all pieces are identified on the dimensioned drawing from which the computations of lengths are made. Standard tables, found in the Appendix of this text, are used to provide dimensions and entrance lengths for pipes and fittings. Little difficulty should be encountered in this type of computation if the work is properly identified and kept in an orderly manner.

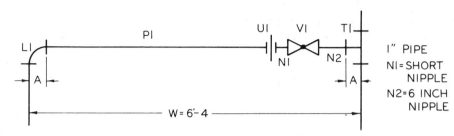

Fig. 23-9. *Screwed Pipe Length Computation.*

Problem 1. Given the schematic of a simple piping system, compute the length of pipe P1. See Fig. 23-9.

The overall lengths of the union and the globe valve, the distance A, and the thread engagement for tight fit are found in the piping tables in the Appendix, pages 368, 369, and 370.

1. Obtain values: A = 1.50; U1 = 2 3/8; V1 = 3 3/4; N1 = 2; Thread Engagement (E) = 11/16
2. Write the equation for total length, W.
   W = A + (P1 − 2E) + U1 + (N1 − 2E) + V1 + (N2 − 2E) + A
   W = 2A + P1 + U1 + N1 + V1 + N2 − 6E
   P1 = W + 6E − 2A − U1 − N1 − V1 − N2
   P1 = 76 + 4 1/8 − 3 − 2 3/8 − 2 − 3 3/4 − 6
   P1 = 63 or 5′-3

Problem 2. Given the schematic layout, determine the length of pipe section P2. System pipe size is 1 1/2; gasket thickness, 1/8; P1, 18 inches. Length, X1, face-to-face of flanges, is 6′-2. See Fig. 23-10.

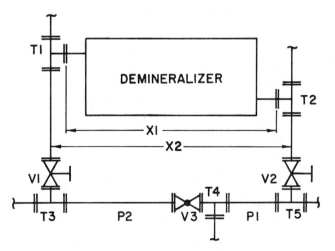

Fig. 23-10. *Flanged Pipe Length Computation.*

Face-to-face distances of fittings are found on pages 369 and 370.

1. Obtain values: Tees: A = 4, X1 = 74, V1 = V2 = 4 3/8, V3 = 5 1/2.
2. Write equations:
   X2 = X1 + 2A + 2(1/8)
   X2 = 74 + 8 + 1/4 = 82 1/4
   X2 = A + P1 + 2A + V2 + P2 + A + 5(1/8)
   P2 = 82 1/4 − 4A − P1 − V2 − 5/8
   P2 = 42 1/8 or 3′-6 1/8

Part II

STRUCTURAL MEMBERS AND STRUCTURES

23.10 Structures

Structural fabrication is complex and varied in scope. It involves the use of many kinds and shapes of cast iron, steel, aluminum, concrete, and other materials, each of which presents unique problems to the design engineer and construction worker. However, drafting room practices associated with the various structural fabrications are basically the same. Many fundamental principles of projection line work, and dimensioning used in machine drawings also apply to structural drawings.

Steel fabriciation includes punching, flame cutting, shearing, forging, bending, machining, fitting and aligning parts; and permanently fastening the assembly by bolting, riveting, or welding.

The information and illustrations that follow are a minimum outline of drafting room practices associated with structural steel fabrication. These standards are recommended by the American Institute of Steel Construction (AISC) and the American National Standards Institute (ANSI).

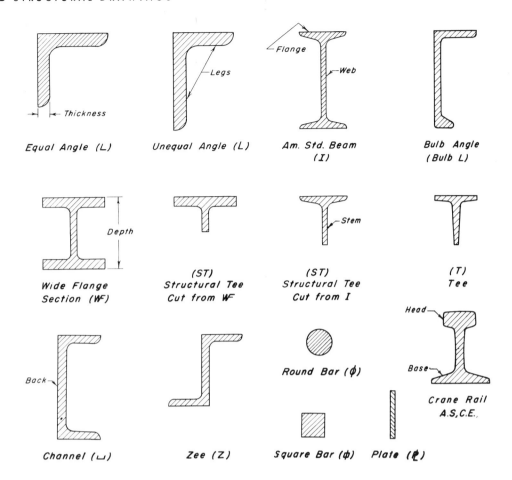

Fig. 23-11. *Rolled Structural Steel Shapes.*

23.11 Structural Steel Shapes

The illustrations in Fig. 23-11 show the principal structural steel shapes. Each is named by a description of its right-section appearance. The Steel Construction Handbook published by the American Institute of Steel Construction gives complete information about sizes, weights, and properties of all standard structural shapes. It also contains a great deal of other information pertinent to steel construction that is essential for the structural designer, such as stress formulas, deflection and load data, etc.

23.12 Specification for Rolled Structural Shapes

The standard specifications for rolled structural shapes are shown in the table in Fig. 23-12. Note carefully the order in which the symbol, size, weight, and length are specified for the various shapes.

23.13 Definitions

Many of the following terms are peculiar to the structural steel field of construction. A study of the pictorial drawing of a structural joint shown in Fig. 23-13 will graphically define these terms:

Lower Chord	Gage Distance
Upper Chord	Gage Line or Working Line
Purlin	Pitch
Gusset Plate	Working Point
Span	Edge Distance
Panel	Clearance
Bay	Rivets

Beam. A horizontal connecting member in a structure such as an I or wide flange beam.

Girder. A horizontal member of a structure built up of prefabricated structural shapes.

Truss. A rigid framework formed by structural members.

NAME	ORDER OF DESIGNATION	TYPICAL SPECIFICATION
Equal Angle	Symbol, leg, leg, thickness, length	L 4 x 4 x ½ x 9'-6
Unequal Angle	Symbol, long leg, short leg, thickness, length	L 3 x 2 x ⅜ x 11'-8
I Beam	Depth, symbol, wt./ft., length	15 I 75.0 x 42'-0
Bulb Angle	Symbol, web, flange, wt./ft., length	Bulb L 7 x 3½ x 17.1 x 9'-9
Wide Flange	Depth, symbol, wt./ft., length	10 WF 45 x 8'-3
Structural Tee	Symbol, depth, cut from wt./ft., length	ST 5 I x 15 x 10'-0
Tee	Symbol, flange, stem, weight, length	T 3 x 3 x 6.7 x 23'-3
Channel	Depth, symbol, weight, length	9 ⊔ 20.0 x 12'-5
Zee	Symbol, depth, flange width, weight, length	Z 6 x 3½ x 29.4 x 7'-7
All Bar Stock	Size, symbol, length	1 φ 8'-2
Plate	Symbol, width in inches thickness, length in feet if over 12 inches	℔ 8 x ⅜ x 1'-2
Crane Rail	Wt./yd., name, length of run in ft.	90 lb. A.S.C.E. rail x 142'-0

Fig. 23-12. *Standard Designations for Rolled Shapes.*

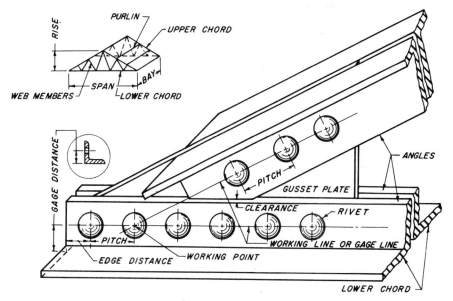

Fig. 23-13. *Graphical Definitions of Structural Terms.*

Column. A vertical member used to support parts of a structure.

Shop Rivets. Rivets to be installed in the shop. They are indicated on the drawing by an open circle equal to the diameter of the rivet *head*.

Field Rivets. Rivets to be installed at the place of final erection. They are indicated on the drawings by a filled-in circle equal to the diameter of the rivet *hole*.

Slope. The tangent of the angle between a member and the horizontal. It is indicated on the drawing by a triangle whose hypotenuse is parallel to the member, and whose longer leg is always 12 units in length.

23.14 Structural Rivets

Rivets are frequently used to join structural members, either directly or through gusset plates and clip angles. Rivets are made of carbon steel with a variety of head shapes. Rivet shanks range in size from 3/8 to 1 1/2 inches diameter. Fig. 23-14 shows rivet heads, riveted joints, and the conventional symbols for riveting with tables of minimum rivet spacing, edge distances and standard gage distances for angles.

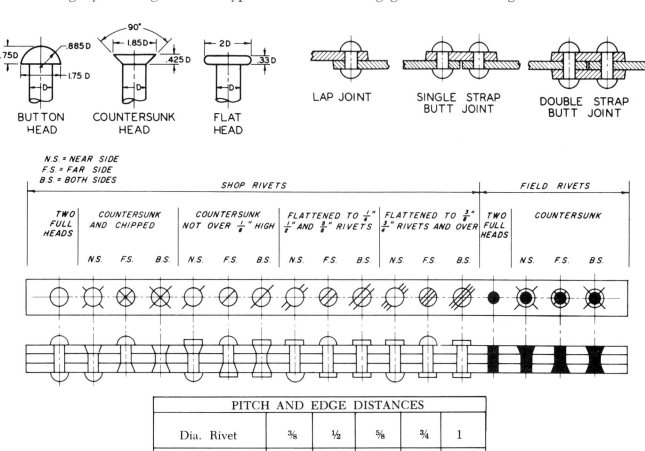

Fig. 23-14. *Rivet Symbols, Pitch and Edge Distances, and Gage Distances for Angles.*

23.15 Structural Welding

At present a large portion of structural work is either partially or completely welded, most commonly with fillet welds. Graphical weld specification corresponds to the standards described in Unit 20. The size and location of the welds are given on the design drawing.

23.16 High Strength Bolts

Steel bolts with high tensile strength have recently come into fairly wide use for connecting structural members. Bolts should be used with hardened steel washers under both head and nut, and set with proper tension by either a manual or compressed air torque wrench.

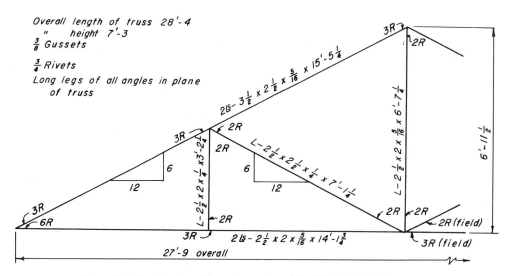

Fig. 23-15. *Design Drawing of a Modified Fink Truss.*

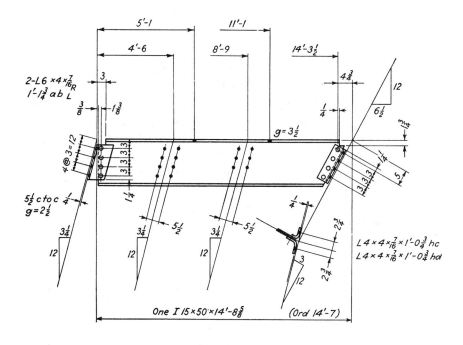

Fig. 23-16. *Shop Detail of a Beam. (ANSI Z14.1-1946)*

Each method of fastening—rivet, weld, or bolt—has particular advantages and disadvantages depending upon the conditions of the project. The noise of riveting is objectionable in some erecttion locations; while welding requires highly skilled workmanship. These disadvantages, perhaps, account for the increasing popularity of bolting. Bolts are removable and reusable; an obvious advantage if the structure is to be disassembled.

23.17 Types of Structural Drawing

Design drawings are single-line diagrams showing the structure shape, principal dimensions, and specifications for the structure. The layout or design drawing is used only in the drafting room as an aid in computations and the preparation of the shop drawings. An example is shown in Fig. 23-15.

Shop detail drawings are subassembly drawings showing detailed information necessary for shop fabrication of the structure. The various members to be fastened together in the shop are drawn in their assembled positions. A shop detail of a beam is shown in Fig. 23-16.

Erection drawings are used if a structure is too large to be completely assembled in the shop. These drawings are line diagrams, specially made for the erection crew, giving pertinent dimensions, identification marks, and sufficient notes to assure the proper assembly of the structure.

23.18 Detail Drawing of a Joint

A layout of a gusset plate connecting three structural members in a truss is shown in Fig. 23-17. The working lines are first laid in obtaining slopes and locations from the design drawing and members are drawn to the detail scale,

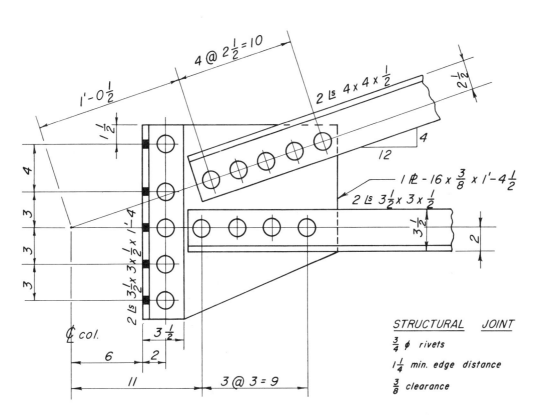

Fig. 23-17. *Structural Joint Design.*

using the working lines as gage lines. The clearance distance is laid off to limit the extent of the diagonal members. The first rivet in a member is located on the gage line and at least as far from the nearest sheared edge as the prescribed minimum edge distance. Other rivets are located at the pitch distance from the first rivet, usually three rivet diameters. The outlines of the gusset plate should be designed to avoid sharp corners protruding between members. The specifications for the gusset plate are given for the rectangle from which the plate will be cut. Several gusset plates are specified in Figs. 23-19 and 23-20.

23.19 Types of Trusses

Five general types of trusses frequently used in structures are shown in Fig. 23-18. Modifications of these, and others, can be found in structural design handbooks. An excellent reference for structural drafting is available in Structural Steel Detailing, published by the American Institute of Steel Construction.

23.20 Detail Drawings of a Truss

Because of the length of the members of a structure in relation to their transverse dimensions, two scales are generally employed for the detail drawing. The larger, called the detail scale, is used for the details around the joints which show the size of members, gusset plates, rivet locations, etc. The detail scale has the effect of enlarging the features at the joint in comparison to the rest of the drawing. The *working scale*, usually about half of the detail scale, shows the distances between points and the relationship between all the component parts of the structure without resorting to break lines. Scales frequently employed in structural detail drawings range from $1/2'' = 1'\text{-}0$ to $3'' = 1'\text{-}0$.

After selecting an appropriate working scale and detail scale, these steps of procedure may be followed.

1. Locate the working points and draw the working or gage lines for the lower chord, upper chord, and web members at the working scale.
2. Locate and draw the structural members relative to the gage lines using the detail scale.
3. Lay in clearance lines to limit the length of the members.
4. Locate the rivets observing pitch and edge distances.
5. Design gusset plates. (Art 23.18)
6. Add dimensions and shop notes. Sharpen all outlines showing the shop and field rivets or welds.

Figs. 23-19 and 23-20 illustrate truss details.

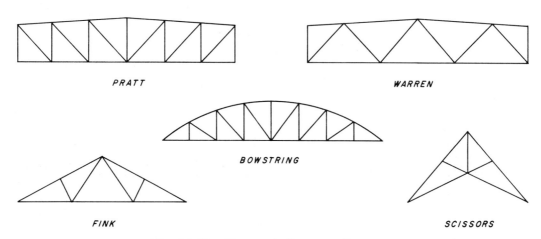

Fig. 23-18. *Types of Common Trusses.*

PIPING AND STRUCTURAL DRAWINGS

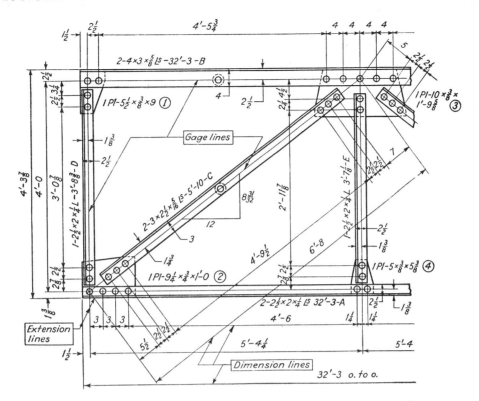

Fig. 23-19. Dimensioning a Truss Detail. (ANSI Z14.1-1946)

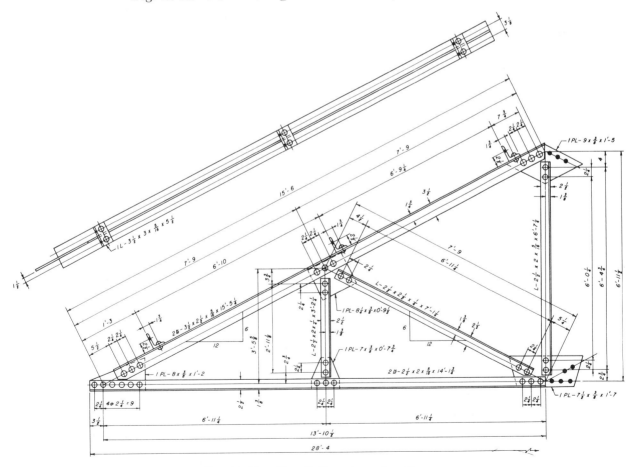

Fig. 23-20. Detail Drawing of a Truss.

23.21 Standards for Structural Dimensions and Notes

The most important dimensions for shop details are determined by the designer and indicated on the design sheets. Other dimensions are obtained from structural handbook tables or computed from the drawing. The following standards should be observed when placing dimensions and notes on the detail drawings. Refer to Fig. 23-20.

1. Rows of dimensions are placed similar to the standards for machine drawing with the shorter dimensions such as rivet pitch, edge distances, etc., nearer to the outline.
2. Dimension lines are continuous lines with the numerals placed above the line and near its midpoint.
3. Dimension lines should be spaced at least 1/2 inch from the object lines with 3/8 inch to 1/2 inch between successive rows of dimensions.
4. All dimensions larger than one foot are given in feet and inches except the width dimension of plates and the depth of structural members which are given in inches. A dash always separates feet and inches.
5. Feet marks (′) are included but inch marks omitted on structural drawings. Example: 1′-6; 3′-0 1/2.
6. Rivets and holes should be dimensioned center to center.
7. If a uniform edge distance is used, it may be given in a general shop note. Variations should be shown on the drawing.
8. Dimensions should be given to the center line of a beam and to the back of an angle or channel.
9. Staggered rivets should be dimensioned as if they were located on one gage line. If three or more pitch distances for a line of rivets are equal, they should be dimensioned thus: 4 @ 3 1/2 = 1′-2.
10. Slope triangles should be shown on inclined members with the unit 12 as the longest leg of the triangle.
11. Chain or complete series dimensioning is used on length dimensions to aid the fabrication shop and to avoid adding or subtracting values.
12. The size and length of the member is indicated by a specification near and parallel to the member.
13. Each member of a structure that is to be shipped separately for field erection is given a code mark with letter and number that appears both on the shop detail and the erection plan.
14. Painting instructions, size of rivets, hole sizes, edge distances, etc., are given in the front of general notes in a conspicuous place on the drawing.

PROBLEMS

1. After making a special study from reference sources, construct a table outlining significant and distinguishing characteristics of various kinds of piping and tubing.

2. Make freehand, double-line sketches of both screwed and flanged types of common pipe fittings. Opposite each, show its correct symbolic representation and specification.

3. From pictures and information in manufacturers' catalogues, sketch illustrations showing the basic features of commonly used valves.

4. The piping layout in Fig. 23-21 shows the proposed by-pass from A to B.
   a. Make a pictorial sketch of the piping system.
   b. Complete a bill of material for ordering all items needed for the installation.

Note: N_1 and N_4 are short nipples.
N_2 and N_3 are nipples of assumed or assigned length.

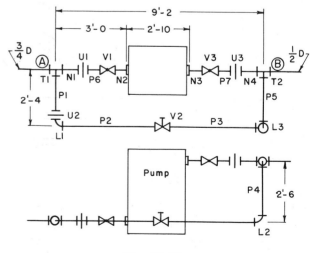

Fig. 23-21.

5. Two 1 inch pipe lines, "H" and "V," are part of a system going through an 8'-0 high room. "H" is horizontal, 2'-0 below the ceiling, 1'-0 from and parallel to the North wall; "V" is vertical, 2'-0 from the East wall and 6'-0 from "H" at its nearest point. On a well-proportioned isometric sketch show the shortest 1 inch connecting pipe "C" between "H" and "V," with a globe valve at the midpoint on "C" and the minimum required fittings on all three pipe lines. The viewing direction should be from the lower southwest corner of the room. Include enough of the pipes and the room to show the connector, the point where "H" pierces the east wall, and where "V" pierces the ceiling. Compute the total length of straight pipe needed for "C."

6. A three-foot pipe column must support a 1000 lb. load. If 20,000 psi stress is allowable, determine what size sch. 40 pipe is required assuming no buckling occurs.

7. Design one of the following items, with complete sketches or drawings, using structural members, pipe, or tubing, with plywood or similar surfaces if needed.

   workbench   chair
   child's vehicle   desk
   playground "monkey bar"   bookcase
        room divider

8. Compile a list of differences between structural and machine dimensioning. Explain these differences, why they are used, and illustrate each by an example.

Clearance distance 1/2
Gusset plate 1 thick
Rivets 1 1/2 diameter
Minimum edge distance 2

Number of Rivets:
3 in each diagonal member
5 in horizontal member
Detail scale: 1 1/2" " = 1'-0
Pitch = 3 rivet diameters

10. Fig. 23-23. The working line drawing shows the specifications for a steel roof truss. Additional specifications:

    Gage distances: 2 angles = 1 1/8; 2 1/2 angles = 1 3/8; 3 angles = 1 3/4; 3 1/2 angles = 2; 4 angles = 2 1/2.
    Clearance distance = 1/4.
    Gusset plate = 3/8 thick.
    Rivets = 3/4 x 2 1/4 minimum pitch.
    Minimum edge distance = 1 1/4.
    Three purlin clips: 4 x 3 x 3/8 x 6, evenly spaced.
    Shop paint red oxide.

    a. Make a completely dimensioned sketch showing the detail at joint B. Locate with respect to working point A.

    b. Make a detail drawing of the roof truss. Include: elevation view of one-half of the truss; top view of upper chord; complete shop details.

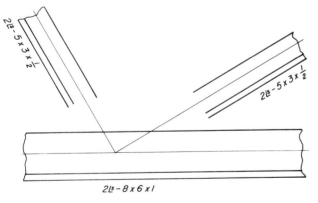

Fig. 23-22.

9. Fig. 23-22. Design and detail a gusset plate which will connect two members of a roof truss to the bottom chord. Complete the drawing using the following specifications:

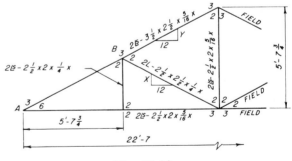

Fig. 23-23.

11. Using dimensions given in a structural steel handbook, draw half size dimensioned views of:
    a. Heaviest section 10 inch I
    b. Lightest section 15 inch channel
    c. Heaviest section 8 inch angle
    d. Lightest section 6 inch zee

12. Make a shop detail drawing of a scissors truss shown in Fig. 23-24.

Upper chord members	2	3 1/2 x 3 x 1/4
Lower chord members	2	3 1/2 x 3 x 1/4
Other members	2	2 1/2 x 2 x 1/4
Rivet pitch 2 inches		5/8 ⌀ shop rivets

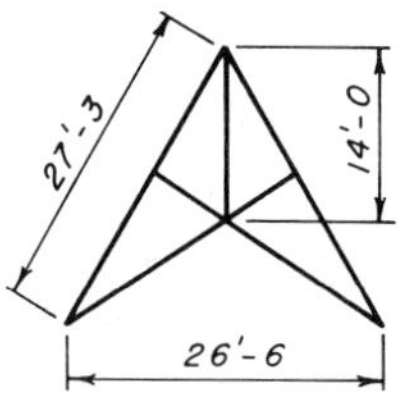

    Fig. 23-24.

13. You are required to prepare a *complete set of drawings* to enable the construction of a simple garage wtih workshop. The workshop will be located at the rear of the building and it is to have an area of approximately one-half of that needed for the automobile.

    Requirements:

    The over-all width of the building is to be about 14 feet. The walls are to be constructed of concrete blocks and will be founded upon an in-place concrete slab. The roofing material is to be of corrugated aluminum or galvanized steel sheets. These sheets may be supported as seen fit, but at least one steel Fink truss should be involved.

    Materials:

    The following materials are readily available at the site:
    8" x 8" x 16" concrete blocks
    mortar
    3 1/2" x 2 1/2" x 5/16" angles
    3/8" steel plates
    1/2" dia. bolts
    nuts and washers
    roofing sheets
    ridge rolls
    nails, timber, concrete, roofing
    anchor bolts
    2" x 6" lumber
    standard doors and windows

14. Large construction projects are often sold to the appropriating authorities by means of three-dimensional models. This problem is to make the necessary working drawings for the fabrication of a table-size, working model of locks in a river dam. For a general concept of the function and layout of locks, refer to the topic, "Canals," in any encyclopedia.

    Data. The model must be simple and illustrate by the actual flow of water how the locks raise the water level to that of the reservoir and lower it to that of the downstream side. Therefore, the model will require a water-tight reservoir, a dam, locks, downstream channel, and adjoining downstream topography. To show the flow within the sluices, use transparent materials for the wall of the locks having the sluice opening, and for the sluices themselves contained in a cutaway section. Include a small, electrically-operated pump (purchase item) to recirculate the water from the downstream side to the reservoir.

    Requirements:
    a. The necessary detail drawings for fabrication of the individual parts of the system.
    b. Assembly drawings.
    c. Bill of Material.
    d. Pictorial illustration of the finished model. (See Unit 30)
    e. A carefully worded instruction sheet telling the layman how to build and operate the model.
    f. Preliminary sketches.

unit 24

Electrical and Electronic Diagrams

By Gary A. Granneman[*]

24.1 Introduction

The advancement of technology in all fields of engineering during the past ten years is truly astounding. Of special interest is the advancement of technology in the electronics field. Through the last decade, the progressive development of the transistor has contributed to the improvement in the design and application of the digital computer. These two devices, the transistor and the computer, have not only greatly affected the technology of all fields of engineering and science but also the everyday life of all people.

As future engineers, a basic knowledge of electronics will probably be required in all types of design. The objective of this unit is to introduce the basic language of electrical and electronic symbols and diagrams.

24.2 Symbols

In electrical and electronic diagrams graphical symbols are used to represent the function and the interconnection of the parts of the circuit. Graphical symbols, shorthand method of circuit description, consist of unique or characteristic shapes and are generally given an alpha numeric designation which provides for correlation between parts lists and written descriptions or instructions.

On page 349 of the Appendix is an abbreviated tabulation of the more commonly used graphical symbols. For a more complete listing see ANSI Y32.2-1967 (reference A). At times clarity and readability of a diagram can be enhanced by inverted or mirror-imaged symbols without altering their meaning. Symbol size and line width should be uniform and selected by considering the desired readability of the final drawing.

Part I

ELECTRICAL DIAGRAMS

24.3 Diagram Types

An electrical circuit may be described by many different diagrams. Each diagram has a specific but different purpose based on the type of information to be conveyed. The choice of diagram to be used is determined by the type of information to be communicated to the diagram's user.

The following definitions are abridged from ANSI Y14.15-1966 and ANSI Y32.14-1962 (References B and C respectively).

Block Diagram. A block diagram shows by means of single lines and labeled rectangular blocks, the interconnection of the major functions

---

[*]Assistant Professor of Engineering Graphics, Iowa State University; Masters degree in Electrical Engineering; Industrial experience; Lieutenant Commander in United States Naval Reserve.

ELECTRICAL AND ELECTRONIC DIAGRAMS

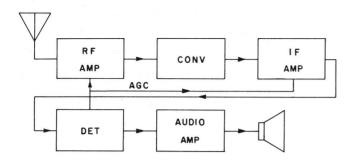

Fig. 24-1. *Block Diagram—Broadcast Band Receiver Circuit.*

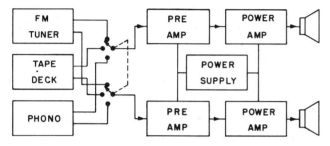

Fig. 24-2. *Block Diagram—Streo Audio System.*

of the component devices or parts of an electrical circuit or system. See Figs. 24-1 and 24-2.

Single-line or One-line Diagram. A single-line diagram shows by means of single lines and graphic symbols, the course of an electric circuit or system of circuits and component parts. Much of the detail shown in a schematic diagram is omitted. See Fig. 24-3.

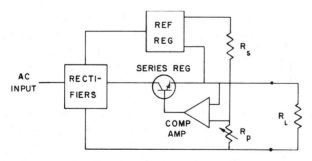

Fig. 24-3. *Single Line Diagram—Voltage Regulated D.C. Power Supply.*

Schematic or Elementary Diagram. A schematic diagram shows the electrical connections and functions of a specific circuit arrangement by means of graphical symbols. The schematic diagram facilitates tracing the circuit and its functions without regard to the actual physical size, shape, or location of the component device or parts. See Fig. 24-4.

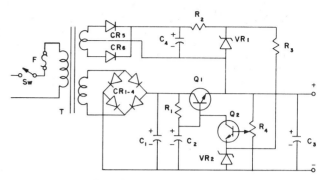

Fig. 24-4. *Schematic Diagram—Voltage Regulated D.C. Power Supply.*

Connection or Wiring Diagram. A wiring diagram shows the connections of an installation and its component devices or parts. It may cover internal or external connections, or both, and contains such detail as needed to make or trace the connections involved. The connection diagram usually shows the general physical arrangement of the component devices or parts. See Fig. 24-5.

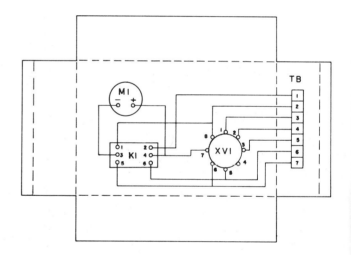

Fig. 24-5. *Point-to-Point Connection Diagram—Relay Control Box.*

Interconnection Diagram. An interconnection diagram is a form of connection or wiring diagram showing only external connections of the unit assemblies or equipment. Internal connections are usually omitted. See Fig. 24-6.

Terminal Diagram. A terminal diagram relates the functionally depicted internal circuit of an item or device to its terminal physical configuration, and locates the terminals with respect to the outline or orientation markings of the device.

ELECTRICAL AND ELECTRONIC DIAGRAMS

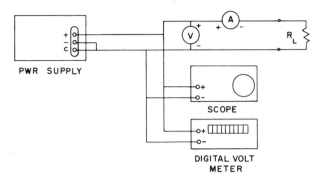

Fig. 24-6. *Interconnection Diagram—Power Supply Test Set Up.*

Logic Diagram. A logic diagram depicts by logic symbols and supplementary notations the details of signal flow and control, but not necessarily the point-to-point wiring in a system of two-state or binary logic devices. See Fig. 24-7.

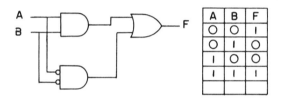

Fig. 24-7. *Logic Diagram with Truth Table. (ANSI Y32.14)*

24.4 General Standards for Diagrams

The following paragraphs extracted from ANSI Y14.15-1966 present basic material applicable to all types of diagrams.

Diagram Titles. When a diagram is in one of the forms defined in Art. 24.3, the name of that type should be included in the title. *Example*: SINGLE-LINE DIAGRAM PULSE AMPLIFIER.

Many diagrams become more useful if more than one type is combined on the same drawing. For example, wiring information such as terminal numbers, wire color or number designations may be included on a schematic diagram. In these combination diagrams the selection of the title should be based on the major purpose of the diagram as stated in Art. 24.3.

Line Conventions and Lettering. Fig. 24-8 shows recommended line weights and line applications for use on electrical diagrams. Both upper and lower case letters are used but with enlarged lower case letters where necessary for reproduction.

Line Application	Line Type and Thickness
General Use	Medium (.022)
Mechanical Connection: Shielding	Medium (.022)
Mechanical Grouping Boundary Line	Thin (.015)
For Emphasis	Thick (.030)

Fig. 24-8. *Line Conventions for Diagrams.*

Abbreviations. Abbreviations used on diagrams should be from ANSI Z32.13 (reference E) or other national standards. If no standard exists, a special abbreviation may be used if it is explained in the notes.

Representation of Electrical Contacts. Switch and relay symbols should be shown with contacts in the position of no applied operating force or in a functional position specified by a note.

Layout of Diagrams. The layout of diagrams should have the main features prominently shown. The parts of a well-designed diagram should be spaced to provide an even balance between blank spaces and lines. Sufficient blank area should be provided in the vicinity of symbols to avoid crowding the necessary notes or reference information.

In general all diagrams should be shown with signal *input on the left* and signal *output on the right* side of the diagram. For longer diagrams the signal flow should be from upper left to lower right with the circuit drawn in rows, like lines on a printed page.

Grouping of Parts. Auxiliary parts, resistors, capacitors, etc., should be grouped with their associated major parts to form a functional whole. If associated parts are physically grouped, a phantom line may be used to show their relationship. Typical groupings are circuit boards, mechanically shielded components, and hermetically sealed units.

Color Code. Wiring, component designations, and component values are often indicated by use of a color code. Fig. 24-9 lists a standard color code and abbreviations commonly used for this purpose.

Color	Designation	Numerical Code
Black	BK	0
Brown	BR	1
Red	R	2
Orange	O	3
Yellow	Y	4
Green	G	5
Blue	BL	6
Violet	V	7
Gray	GY	8
White	W	9

Fig. 24-9. *Color Code.*

Wiring Junctions and Crossovers. All junctions of connecting lines should be shown by one of two methods, the NO-DOT or the DOT method with only one method used on a given drawing.

The single junction is the preferred junction for both methods. For complex circuits the multiple junction may be used to simplify the diagram. This requires the use of the DOT method only to describe a junction versus a crossover. See Fig. 24-10.

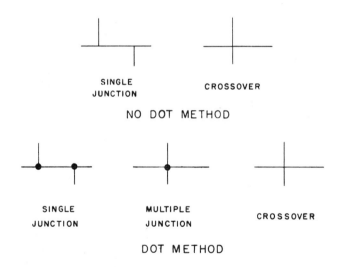

Fig. 24-10. *Junctions and Crossovers.*

Connecting Lines. Lines connecting the symbols on a diagram should be drawn as horizontal or vertical lines with as few bends and crossovers as possible. Exceptions are such circuitry as bridges, multi-vibrators, and other unique circuits where clarity of presentation is important.

24.5 Guide to Drawing Schematic Diagrams

The same procedure, slightly modified, should be used for drawing all diagrams. It should be realized that it is a basic guide and that only experience and practice will produce a high quality drawing. As in most types of drawings, clarity is one of the most important requirements.

1. *Planning.* Make a rough sketch as a guide to the final schematic:
   a. Subdivide the circuit into sections with regard to functions of the major components, i.e., tubes, transistors, transformers, filters, etc.
   b. Divide an appropriate size area on paper into the same number of subdivisions maintaining their relative size. Note that filament circuits are generally shown as a separate circuit on the diagram.
   c. Sketch estimated best position of major components and adding the rest of the circuit. Rearrange positions as necessary for clarity and simplicity, but do not crowd the parts. Use mirror or inverted images if needed.
2. *Execution.* The layout of the final drawing is as follows:
   a. Lay out subdivisions on the paper using very light lines. Leave room for title block, general notes, specific notes, etc.
   b. Place major components in subdivisions on a common horizontal line starting at the left, working towards the right.
   c. Add power supply lines, generally horizontal and spaced as follows: B+ line *above* major components with ground and bias line *below* major components.
   d. Add auxiliary components (R, L, C) along horizontal or vertical feeder lines (guide lines). Position components at same level if possible without crowding.
   e. Complete connections using horizontal and vertical lines at common levels or positions.
   f. Add auxiliary lines (shielding, mechanical linkage, etc.) sharpen and clean up. Label components using *unidirectional* lettering.

Part II

LOGIC DIAGRAMS

24.6 Logic for Design

Many electrical and electronic circuits are used to perform logic operations on information that is in digital form. Typical examples range in complexity from the modern high speed computer to the simple ON-OFF switch control of a light

ELECTRICAL AND ELECTRONIC DIAGRAMS

bulb. The design and analysis of digital circuits can be greatly simplified by the use of logic elements and their graphical symbols. The system designer is then free of the hardware details and may concentrate his efforts towards a solution in terms of simple logic elements or operations.

Electronic circuits used as logic elements are designed with two physical states possible at each signal terminal. These physical states may be two discrete voltage or current levels. Therefore, at a specific instant, a given terminal may have only one state, either H the highest or L the lowest level. Although not as common, pneumatic and hydraulic devices may also operate with two possible physical states and the design and analysis of these systems may be simplified by logic diagrams.

24.7 Binary Logic

A two-state-terminal device is characteristic of *binary logic*. The two physical states in binary logic are referred to as the 0-state (zero) and the 1-state (one.). The 0-state is called the reference or inactive state. The 1-state is called the significant or active state.

The 1 and 0 states of binary logic may be assigned to the H, high, and L, low, levels of an electronic device terminal in two possible ways described in the following article.

24.8 Positive and Negative Logic

Positive logic is said to result when the 1-state is assigned to the H-level and the 0-state to the L-level. In *Negative logic* the 1-state is assigned to the L-level and the 0-state to the H-level. For binary mechanical devices positive logic may be assigned to a position or physical state as desired. Negative logic is then defined by the opposite state.

24.9 Logic Elements

The basic building blocks of a complex digital system are called *gates*. Fig. 24-11 shows a simple electrical circuit, the distinctive logic symbol, and a truth table for the OR, AND, NOR, and NAND Gates. Positive logic is assigned with the 1-state being *switch closed* for inputs and the *highest voltage* for the output. The 0 on a terminal indicates negation or inversion of the normal terminal logic state. The student is referred to C and D in the reference list for a more complete discussion.

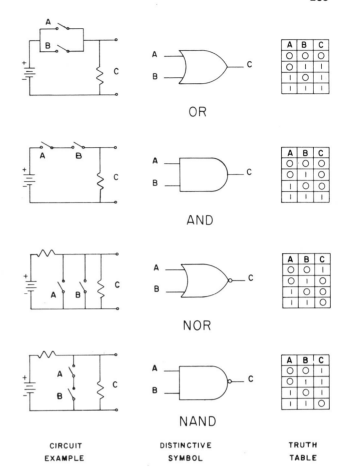

Fig. 24-11. *Logic Gates.*

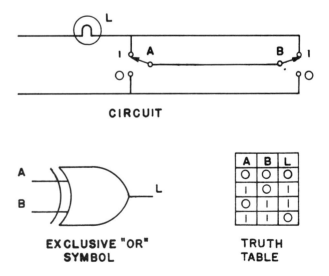

Fig. 24-12. *Three Way Light Switch.*

REFERENCES

A. ANSI Y32.2-1967 Graphics Symbols for Electrical and Electronics Diagrams.
B. ANSI Y14.15-1966 Electrical and Electronics Diagrams.
C. ANSI Y32.14-1962 Graphic Symbols for Logic Diagrams.
D. Mil Std 806 B Graphic Symbols for Logic Diagrams.
E. ANSI Z32.13-1960 Abbreviations for Use on Drawings.

PROBLEMS

1. (a) What is the difference between ordinary shielded wire and coaxial cable? (b) What is a wave guide? Give an application. (c) Name the major features found in most schematic diagrams. (d) What type of tabulated information is given on a schematic? (e) What is color coding? Give several examples.
2. Change Fig. 24-1 into a pictorial block diagram.
3. Sketch a circuit illustrating a rough layout in which at least 10 different symbols are used.
4. Make a wiring diagram for a stairway light having an on-off switch at both the top and bottom of the stairs.
5. Make a sectional drawing of a lamp socket showing internal construction.
6. Make a pictorial drawing of an octal socket.
7. Write a report with sketches and drawings that completely describes the making of a printed circuit.
8. (a) Make a schematic diagram for a simple radio receiver. (b) Make a pictorial assembly of the circuit showing point-to-point wiring. (c) Design a chassis and enclosure case for the radio. Include complete information for manufacture.
9. Design a *model* of a Triode Vacuum Tube showing the basic components and construction of the tube. Assume that this model will be used to explain construction, operation, and function of the triode.

    Data: The model should be at least four times actual size. It should be constructed of modeling materials that are readily available such as plastic, wood, screen, metal rods and plate and/or other materials that would be satisfactory to simulate the actual parts. Any triode vacuum tube may be used as a basis for the model.

    Requirements:
    a. Preliminary design sketches.
    b. Details of all model parts, unless standard items.
    c. Subassemblies, if used in construction of the model.
    d. Assemblies of the model.
    e. An exploded pictorial (See Fig. 26-8 and 26.9).
    f. A bill of materials.
    g. A concise written description or explanation of the function and operation of the tube.
10. Sketch block diagrams of the following:
    a. Television receiver.
    b. Sine square wave generator
    c. Oscilloscope
    d. Vacuum-tube voltmeter
11. Sketch the necessary orthographic views of a power transformer, one view to be a full section. Use appropriate shading symbols.
12. Make necessary enlarged orthographic views of:
    a. Toggle switch
    b. Fuse holder
    c. Octal tube socket
    d. Single-section wafer switch
13. Draw a schematic diagram for the following items:
    a. Volt-ohmeter
    b. Voltage regulated power supply
    c. Sine wave generator
    d. Simple radio receiver
14. Sketch a pictorial of a flat pack integrated circuit.
15. Draw logic and schematic diagrams of a single decade counter.
16. Draw a logic diagram for an electronic garage door opener.
17. Draw a logic diagram of the control system for a furnace.

unit 25

Computer Aided Design

by Gus Aronson*

25.1 Introduction

In recent years, the use of the digital computer to aid engineers in the design of machine tools, automobiles, and consumer products has become commonplace. Until a few years ago, the use of the computer was limited to record keeping and data processing areas; however, the ability to do repetitive calculations and comparisons has made it a tool more suitable for engineering analysis and design. The following articles will relate examples of a few of the many areas where computers are used in design and manufacture.

25.2 Computer Aids to Engineering

In the design of machines, certain engineering calculations occur many times. Examples of these calculations are the determinations of bearing load for gears, gear center locations, gear strength and wear life, moment of inertia of areas, feed system efficiencies, and others. Some calculations deal with measurements required for inspecting parts such as dimensions over pins for checking the pitch of gears. Another case involves calculations needed for finding compound angles accurately to set up part fixtures.

25.3 Design Data Processing

The modern company that utilizes computer facilities sets up standard forms so that data for these calculations can be readily converted by the programmers into computer language for a quick solution to the problem. An example of the generalized problem of calculating a compound angle is shown in Fig. 25-1. Fig. 25-2 shows the form which related the known information to the computer programmer.

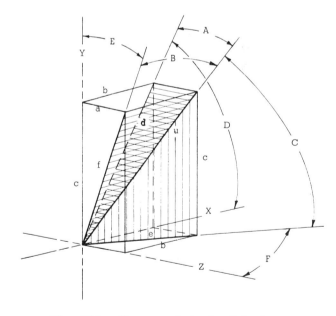

Fig. 25-1. *Compound Angle Calculation.*

---

*Advanced Technical Planning, Ingersoll Milling Machine Company, Rockford, Illinois. Formerly Assistant Professor of Engineering Graphics, Iowa State University.

SHOP ORDER								LAYOUT								NAME											DATE										TN	
1	2	3	4	5	6	7	8	9	10	11	12	13	14	15	16	17	18	19	20	21	22	23	24	25	26	27	28	29	30	31	32	33	34	35	36	37	38	39

1. TN indicates total number of problems.
2. See reverse side of this sheet for Nomenclature of the angles.
3. Give the known angles in alphabetical order.
4. Return to Technical Service Department

DRAWING NUMBER							FIRST KNOWN ANGLE																	SECOND KNOWN ANGLE															
							Angle (A → E)					Deg.				Min.				Sec.				Angle (B → F)					Deg.				Min.				Sec.		
1	2	3	4	5	6	7	8	9	10	11	12	13	14	15	16	17	18	19	20	21	22	23	24	25	26	27	28	29	30	31	32	33	34	35	36	37	38	39	40
1	2	3	4	5	6	7	8	9	10	11	12	13	14	15	16	17	18	19	20	21	22	23	24	25	26	27	28	29	30	31	32	33	34	35	36	37	38	39	40

Fig. 25-2. *Information Form for Compound Angle Calculation. (Courtesy Ingersoll Milling Machine Co.)*

SHOP ORDER	LAYOUT	NAME	DATE		TN
1–10	11–20	21–34	35–45	46–63	64–65
				PROGRAM NO C-0401	

1. TN indicates the total number of problems.
2. Fill in Shop Order, Layout, Name, Date, and TN.
3. Do not leave blanks after the decimal point.
4. Fill in Data Processing Instruction Sheet.
5. Return to Engineering Standards Department.

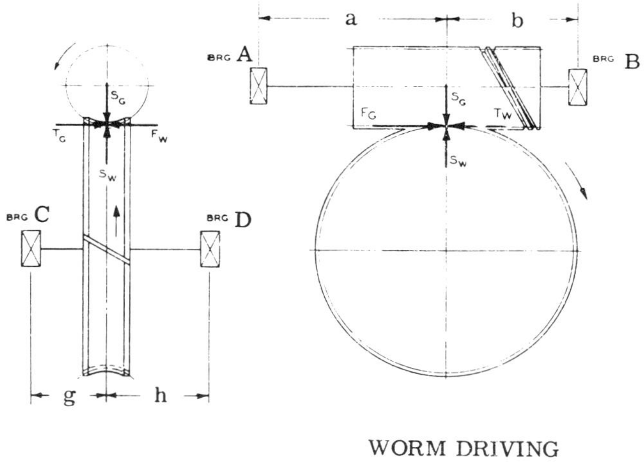

Drawing Or Part Number	Pitch Diameter Of Worm	Pitch Diameter Of Gear	No. Of Teeth Worm	No. Of Teeth Gear	RPM Of Worm	HP Of Worm	Helix Angle Deg.	Helix Angle Min.	Pressure Angle In Decimals	Distance a (Inches)	Distance b (Inches)	Distance h (Inches)	Distance g (Inches)	Co-Efficient Of Friction
1–7	8–12	13–17	18–20	21–23	24–27	28–33	34–35	36–37	38–42	43–47	48–52	53–57	58–62	63–65
	.	.				.			.	.	.	.	.	.
	.	.				.			.	.	.	.	.	.
	.	.				.			.	.	.	.	.	.
	.	.				.			.	.	.	.	.	.
	.	.				.			.	.	.	.	.	.
	.	.				.			.	.	.	.	.	.
	.	.				.			.	.	.	.	.	.
	.	.				.			.	.	.	.	.	.
	.	.				.			.	.	.	.	.	.
	.	.				.			.	.	.	.	.	.

Fig. 25-3. *Data for Bearing Load Calculations.*
(Courtesy Ingersoll Milling Machine Co.)

CALCULATION OF LOADS ON BEARINGS OF GEAR

Gear Forces	Calculations	Loads on Bearings	
		Bearing C	Bearing D
Tangential - F_G	$F_G = F_W \left(\dfrac{\cos\beta \cos\varphi - \mu \sin\beta}{\cos\varphi \sin\beta + \mu \cos\beta} \right) \otimes$	$F_G \times \dfrac{h}{j} \otimes$	$F_G \times \dfrac{g}{j} \otimes$
Thrust - T_G	$T_G = F_W \rightarrow$	—	$T_G \rightarrow$
Thrust Couple TC_G	$TC_G = T_G \times P.R._G \;\curvearrowright$	$\dfrac{TC_G}{j} \uparrow$	$\dfrac{TC_G}{j} \downarrow$
Separating - S_G	$S_G = S_W \downarrow$	$S_G \times \dfrac{h}{j} \downarrow$	$S_G \times \dfrac{g}{j} \downarrow$
Bearing Radial Load	$\sqrt{\Sigma \,(Loads)^2}$	$\sqrt{\left(F_G \times \dfrac{h}{j}\right)^2 + \left(\dfrac{TC_G}{j} - S_G \times \dfrac{h}{j}\right)^2}$	$\sqrt{\left(F_G \times \dfrac{g}{j}\right)^2 + \left(\dfrac{TC_G}{j} + S_G \times \dfrac{g}{j}\right)^2}$
Bearing Thrust Load	Σ Thrust Forces	—	T_G

$\updownarrow$ or $\leftrightarrow$ Indicates forces in the direction shown. $\otimes$ Indicates forces acting inward, normal to plane of paper.
$\odot$ Indicates forces acting outward, normal to plane of paper.

The sample output of the program is as follows:

SHOP ORDER 1234567   LAYOUT 1234567   ANEJA   PROG C-0401

DRAWING OR PART NO. 1234567

LOAD ON BRG. B DUE TO THRUST = 1637.49
LOAD ON BRG. D DUE TO THRUST = 1188.68

BEARING	LOAD DUE TO FW TANG. FORCE	THRUST COUPLE	LOAD DUE TO SEP-ARATING FORCE SW
A	489.46	204.20	480.12
B	699.22	204.20	685.89
C	948.02	860.85	675.06
D	689.47	860.85	490.95

BEARING	RESULTANT RADIAL LOAD
A	561.87
B	1131.89
C	966.05
D	1517.48

Fig. 25-4. *Computer Output for Bearing Load.*
(Courtesy Ingersoll Milling Machine Co.)

SHOP ORDER	LAYOUT	NAME	DATE		TN
1–10	11–20	21–34	35–45	46–63	64–65
				PROGRAM NO C-0801	

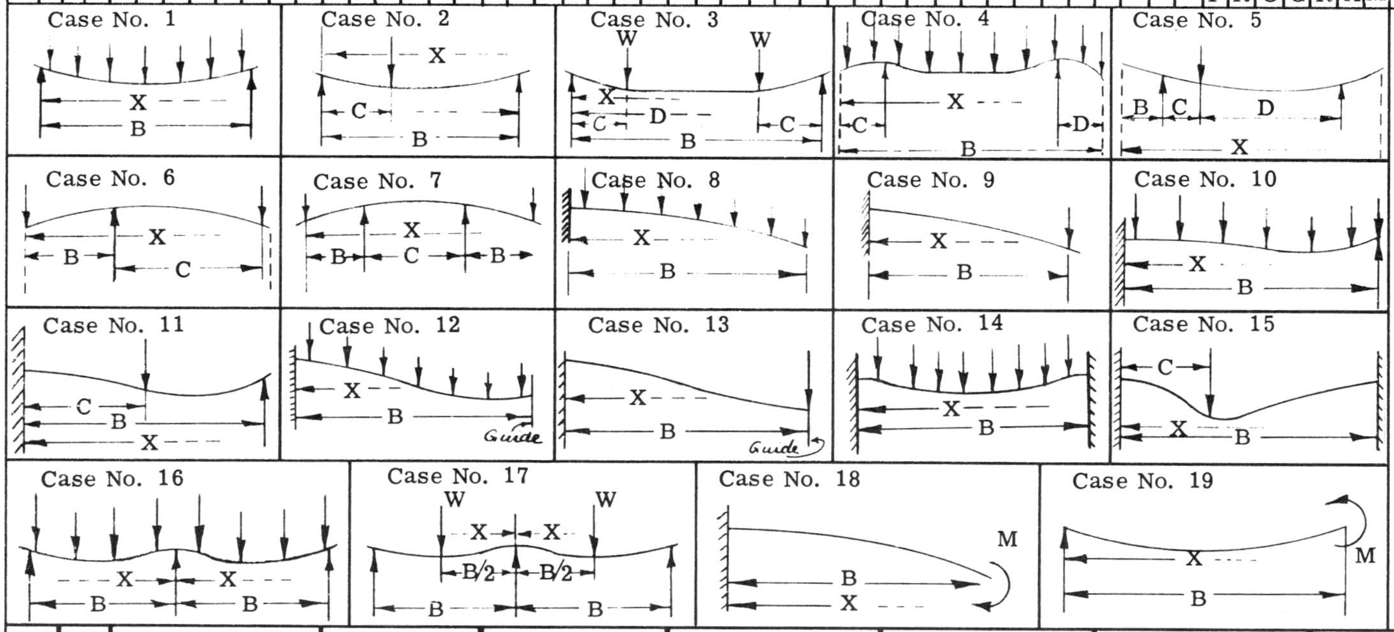

1. Fill in Shop Order, Layout, Name, Date & TN where TN indicates total number of blems.
2. Fill in Data Processing Instruction Sheet.
3. For Moment of Inertia, Program C-3101 is available from the Engineering Standards Department.
4. Return to the Standards Department.

Beam No.	Case No.	Modulus of Elasticity of the Material	Moment Of Inertia (Inch ^{4})	Total Weight W Or Moment M	Distance From Neutral Axis To Extreme Fiber	Distance X (In Inches)	Distance B (In Inches)	Distance C (In Inches)	Distance D (In Inches)
1–2	3–4	5–12	13–18	19–25	26–33	34–41	42–49	50–57	58–65
			•	•	•	•	•	•	•
			•	•	•	•	•	•	•
			•	•	•	•	•	•	•
			•	•	•	•	•	•	•
			•	•	•	•	•	•	•
			•	•	•	•	•	•	•
			•	•	•	•	•	•	•
			•	•	•	•	•	•	•
			•	•	•	•	•	•	•
			•	•	•	•	•	•	•
1–2	3–4	5–12	13–18	19–25	26–33	34–41	42–49	50–57	58–65

Fig. 25-5. *Input Data for Beam Stress and Deflection. (Courtesy Ingersoll Milling Machine Co.)*

A second example of a computer input form is shown in Fig. 25-3 for calculating bearing loads for a worm and worm gear combination. The computer output (solution to the problem) is shown in Fig. 25-4.

Another example of standardized input forms to the computer department for calculating the stress and deflection in a beam after the moment of inertia of the beam cross-section has been calculated is illustrated in Fig. 25-5.

25.4 Optimum Design Parameters

Some engineering design problems are not standard, but are complicated enough to require a computer solution. The computer plot of the cross-section of a welded machine member is a special problem of this type, Fig. 25-6. In this problem the end points of the cross-section center line of each part, the thickness of each part, the moment of inertia about the X and Y axis of each part, and the areas which make up the complete member are entered into the computer as shown in Fig. 25-7. The computer first solves the problem to find the moment of inertia about the X and Y axis, the torsional constant, the coordinates of the centroid, and the stiffness factors of the complete welded member, Fig. 25-8. The computer then plots the cross-section of the member, Fig. 25-6, to scale for the engineer to use in designing the part. Variations in the solution for better stiffness factors can be made by changing the thicknesses of the members which make up the weldment, thus determining the optimum cross-section.

25.5 Numerical Control for Production

Modern production facilities use Numerical Controlled (N/C) machines for manufacturing parts which require a high degree of accuracy. The N/C controller can be divided into two different classifications: First, the point-to-point positioning control; and second, the contouring control.

25.6 Point-to-Point Control

The point-to-point controller is used on machines which move a drilling or tapping spindle to various hole locations where machine motion during the move is not critical, but the locations must be accurate. By using this type of control, hole locations on mating parts can be matched. To assure that these parts will be machined correctly, the computer can use the N/C tape output from the computer to plot the locations of each hole to scale. This plot can be placed over the drawing of the part to visually check the program. Fig. 25-9 is an example of a plot of a point-to-point numerical control tape.

25.7 Contouring Control

The contouring numerical controller is used on milling machines and lathes where the surface generated by the cutting tool may be curved or straight. To produce the tape for this type of control, the computer is used to process input instructions using languages which describe the geometrical shapes of the part and change this information to instructions for moves, punched on tape, which are understood by the machine controller. The computer can also be used to draw plots of the path of the cutting tool as long as

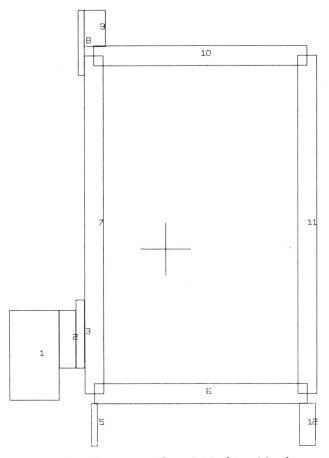

Fig. 25-6. *Computer Plot of Machine Member.*
(Courtesy Ingersoll Milling Machine Co.)

COMPUTER AIDED DESIGN

IDENTIFICATION EXAMPLE RUN

COORDINATE DATA

POINT	X	Y
1	1.964	3.500
2	1.964	10.428
3	4.589	6.000
4	4.589	10.428
5	5.594	6.000
6	5.594	11.250
9	6.690	0.000
10	6.690	3.250
11	6.690	4.000
12	23.680	4.000
14	6.690	30.000
15	5.685	28.500
16	5.685	33.500
17	6.797	30.700
18	6.797	33.500
20	23.680	30.000
23	23.680	3.250
24	23.680	0.000
25	7.440	10.750
26	13.440	4.750

MEMBER DATA

MEMBER	POINT	POINT	THICKNESS	IX	IY	AREA
1	1	2	3.928	1923.714	2990.049	27.213
2	3	4	1.322	289.594	356.598	5.853
3	5	6	0.691	161.861	167.428	3.627
5	9	10	0.500	297.825	52.730	1.625
6	11	12	1.500	3162.031	812.897	25.484
7	11	14	1.500	2333.314	1272.031	39.000
8	15	16	0.512	650.050	114.961	2.559
9	17	18	1.713	1384.332	150.923	4.796
10	14	20	1.500	5639.604	812.897	25.484
11	20	12	1.500	2333.314	4983.153	39.000
12	23	24	1.250	744.563	518.845	4.062

TORSION DATA

CELL	CONNECTING ELEMENTS			
1	10	11	6	7

Fig. 25-7. *Computer Output for Trial Section.*
(Courtesy Ingersoll Milling Machine Co.)

IDENTIFICATION EXAMPLE RUN

```
MOMENT OF INERTIA
    ABOUT X AXIS   18920.20
    ABOUT Y AXIS   12232.51

TORSIONAL CONSTANT   13617.18


COORDINATES OF CENTROID
    X = 12.384
    Y = 15.130


CROSS SECTIONAL AREA 178.708

    IX/A =105.8718
    IY/A = 68.4494
    K/A  = 76.1977
```

Fig. 25-8. *Computer Output of Critical Information. (Courtesy Ingersoll Milling Machine Co.)*

these moves are in a plane. Two and three-dimensional pictures of the part and the cutting tools can be drawn by the computer to check correctness of tool motion.

Fig. 25-10 shows a helical cutter being contoured on a N/C machine. Fig. 21-8 illustrates a contouring process which would be readily adapted to numerical control.

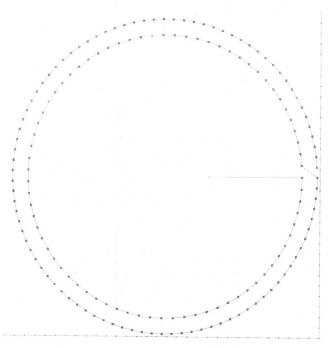

Fig. 25-9. *Point-to-Point Numerical Control Tape Plot. (Courtesy Ingersoll Milling Machine Co.)*

Fig. 25-10. *Contouring Helical Cutter on N/C Machine. (Courtesy Ingersoll Milling Machine Co.)*

COMPUTER AIDED DESIGN

25.8 Direct Computer Control to Machine

Industry has found that computers are not only necessary for use in the design and engineering of products, but can also be used in controlling an efficient flow of parts through the mass production lines of today's plants.

Numerical control is taking a further step in some industries by changing from a tape-fed controller at each machine to direct computer of many machines.

The information obtained at an inspection station can be directed to the tool adjusting mechanisms on a machine tool. Computers are also used to match parts during the assembly of machines such as automobiles so that body panels are all of the right color and wheels of the same size.

Finally and probably most useful, the computer can examine machinery along a production line and determine what is causing a delay. The computer can notify the repairman and the material handler so that operations of the other machines in the line will not be affected by a shutdown.

25.9 Design by Cathode Ray Tube

The two principal difficulties most people encounter in dealing with computers are 1), the computer language and 2), the batch mode of data processing. The latter is a term used for periodic "runs" where data for several problems are supplied to the computer which then solves them successively as a group with no interruptions. These problems have long been recognized by computer experts; and the cathode ray tube represents one solution which is extremely desirable for design applications.

The cathode ray tube console, Fig. 25-11, consists of three basic parts: A keyboard for commands and numerical input; the tube screen which projects graphical images of the computer's activity; and a light pen, to which the screen is sensitive. The tube face becomes a sketch pad on which the designer can communicate graphically with the computer by using the light pen, and verbally by using the keyboard. Highly sophisticated design techniques are being used in conjunction with the cathode ray tube. For a simple example, consider how the design of the weldment of Art. 25.4 might be accomplished using a graphic console.

The program of calculations, basic elements, etc., is supplied to the computer in the usual manner. The designer requests a work area display

Fig. 25-11. *Cathode Ray Tube in Operation.*
(*Courtesy IBM*)

on the screen. Symbols for design components available and perhaps instructions will also be shown at the screen's edge. A rectangular component is selected by pointing at its symbol with the light pen. The computer recognizes this request and will display a rectangle on the work area of the screen after the designer has specified its four corner coordinate locations or other complete description. This is repeated for each desired member until the assembly is completed on the screen.

Calculation of stiffness factors, etc., may be requested, with results displayed almost instantly on the screen. If values are unsatisfactory the designer may relocate, resize, add, or delete components by repeating the procedures discussed, thereby obtaining a different configuration. The designer is not bogged down with computational details; he is able to use his engineering judgment and intuition to observe the influence of design changes, an act presently beyond the capacity of the computer.

When an acceptable design is obtained, the designer may be able to rotate the image for more comprehensive examination. The finally accepted image, which is stored digitally in the computer's memory, can be recorded on a punched tape and an accurate drawing of the design made on a tape-controlled drafting machine such as the one shown in Fig. 25-12.

25.10 Conclusions

Before graduation, the engineering student will become familiar and somewhat proficient with computer operation. He will understand its language and have had practice in writing programs for the solution of problems. Most of his experience will be limited to programming mathematical equations and analyzing its numerical results.

It is not desired to belittle the importance of mathematical problems; however, the engineer should be aware of the graphical logic diagrams used to set up the successful program of any complicated problem. Also, if he becomes associated with design in any field, his company may use several graphical adjuncts to the digital computer. These include the X-Y plotter, the tape controlled drafting machine, and the Cathode Ray Tube. In addition, tape-controlled machine tool operation is basically a graphical output.

Bear in mind that the computer is more than a high-speed mathematician, it also serves as an important "graphician."

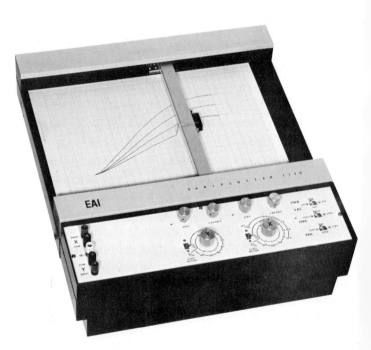

Fig. 25-12. *Tape Controlled Drafting Machine. (Courtesy Electronic Associates, Inc.)*

unit 26

Design Drawings

26.1 Introduction

The preceding units of this text have been largely devoted to visualizing and utilizing orthographic projection along with the standards and conventions used on engineering drawings.

In any creative design, however, the final product must be functional, and in order to assure proper operation of the device, its behavior as a system or assembly must be determined. Only then can the details of the design be finalized.

The purpose of this Unit is to discuss the drawings required to completely describe a working system. The remarks made in the following articles about mechanical systems apply generally to many other types of systems previously discussed.

26.2 Systems

A *system* may be defined in several ways. One engineering concept defines a system as a group of basic components, real or conceptual, arranged in such a way that they can accomplish some useful task. A list of systems would include such items as power transmissions, steering linkages, hydraulic and heating systems, electrical distribution systems, electronic circuits, pitch and yaw controls on space vehicles, logic diagrams, buildings or bridge structures, artificial limbs and organs, and orthographic projection, to name a few.

In some system drawings, components are illustrated realistically, while in others, they are shown symbolically. In the latter case, it is important to be able to read and understand the symbolic notation.

26.3 Working Drawings

In communicating a creative design, after all the ideas and concepts have been compared and decisions made, it is necessary to assemble all the graphical and written work that completely describes the design for production and assembly. These drawings are known as a *set of working drawings* and should include most or all of the following types:

1. Ideation and Conceptual Sketches
2. Layouts
3. Assemblies and Subassemblies, sectioned as necessary
4. Pictorials
5. Details
6. Parts List
7. Calculations and Supporting Data

Different types of system drawings have several common characteristics. These are discussed in general terms in the following paragraphs with specific references to mechanical systems. Appearance, clarity, and ease of duplication of all working drawings are improved by good line contrast. The alphabet of lines from ANSI Y14.2 is shown on page 339.

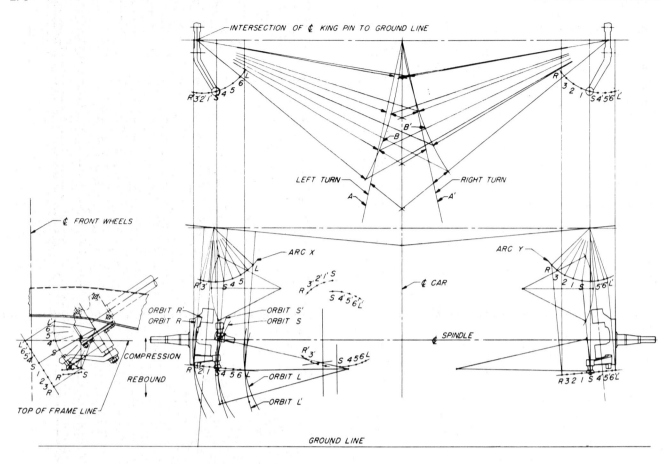

Fig. 26-1. *Geometric Layout—Automobile Steering.*
(Courtesy General Motors Corp.)

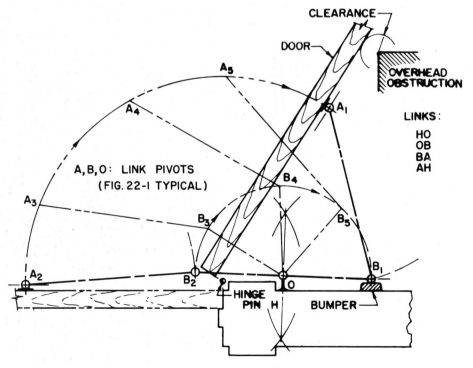

Fig. 26-2. *Doorstop Layout.*

DESIGN DRAWINGS

26.4 Layouts

The *layout* is probably the most important and often the least discussed drawing of the set. A layout is a very accurate drawing used to determine and verify the operation, circuit function and continuity, kinematic behavior, clearances, or other required characteristics of the system. It reflects the original conception of a design, and forms the basis for detail and assembly drawings.

For efficiency, the layout shows the minimum information required to determine the correct operation.

Fig. 26-1 is a geometric layout for a proposed automobile steering arrangement which, when properly interpreted, shows "wheel fight" during a turn.

Complicated layout drawings often make use of mathematics and physics beyond the freshman year. Fig. 26-2 illustrates a relatively simple example of a layout for a four-bar linkage doorstop.

Fig. 2-3, showing the spare tire lifter, is essentially a layout drawing. The geometrical requirements were first laid out after which the resulting tool configuration was superimposed with consideration for such items as strength, weight, appearance, and cost.

26.5 Assemblies and Subassemblies

As soon as a layout has been completed and the system operation verified, the hardware that supports the operation is added to complete the design. Standard components are used where possible along with specially designed features to hold or connect the system. In mechanical applications, section views are used extensively to clarify all aspects, as in Fig. 26-3. The designer then has sufficient information to develop each non-standard part. Notice that the hardware is not determined until the operation has been checked thoroughly. This is described in the discussion on Creative Design, Unit 2, where it was observed that the form of the design can only follow, not precede, its function.

The difference between an assembly, Fig. 26-4, and a subassembly, Fig. 26-3, is largely conceptual. A subassembly generally is a functioning component required for system operation but of little value by itself. Examples of subassemblies would be an airplane landing gear, automobile

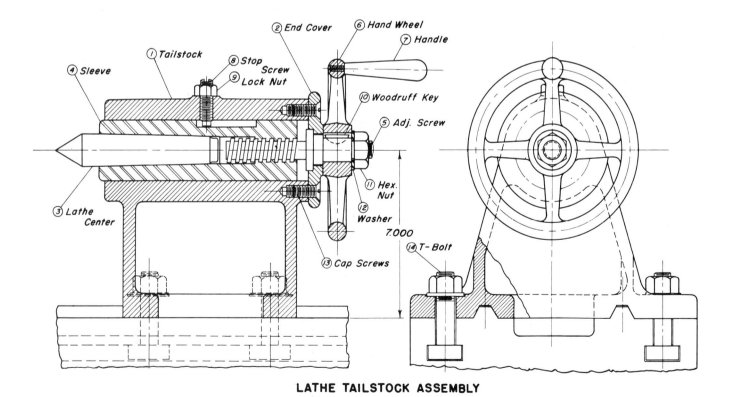

LATHE TAILSTOCK ASSEMBLY

Fig. 26-3. *Subassembly Drawing.*

carburetor, radio tuner circuit, prefabricated bar joist, or lathe tailstock as in Fig. 26-3.

In the assembly views, numbered circles called "balloons" are used to point out part numbers for details and parts list reference.

Only those dimensions which relate to the *use* of the product should be shown in the assembly drawing. This is also illustrated in Figs. 26-3 and 26-4.

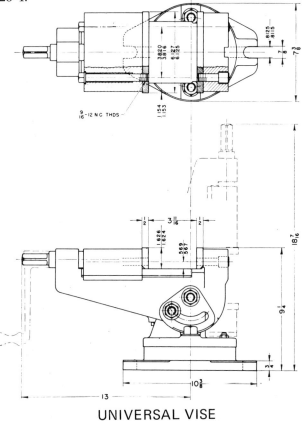

UNIVERSAL VISE

Fig. 26-4. *Assembly Drawing for Catalog. (Courtesy Cincinnati Milling Machine Co.)*

26.6 Details

A detail is a drawing of a single part completely dimensioned and specified for production. Purchased parts or standard components should not be detailed. The detail drawings of the parts of a system are made after sufficient information has been gained from the assemblies. Frequently it will be necessary to work back and forth between assemblies and details as new conflicts or better ideas are discovered. Although each item is shown by itself, detail drawings of more than one object may appear on a single sheet.

Certain information should be included on the detail drawing of *each* part:

1. Necessary orthographic views
2. Complete dimensions, with limit dimensions where necessary
3. Part number and name
4. Project name
5. Material and hardness if applicable
6. Finish notes
7. Number required

Study Figs. 12-24, 20-14 and 26-14 to find these elements.

26.7 Parts List

The *Parts List* is made up informally as work progresses and formally after all details of the design are confirmed. Every item in the system should appear on the Parts List, with references to its location on assembly and detail drawings. Fig. 26-5 is a parts list with typical columns. Note the progression is from the bottom up.

NO	NAME	REQ.	MAT'L. SIZE	NOTES
14	T-Bolt	4	$\frac{3}{8} \times 1\frac{1}{2}$	Std.
13	Flat Hd. Cap Sc.	4	$\frac{1}{4}$-20 NC-2	$\frac{5}{8}$ Long
12	Washer	1	$\frac{3}{8}$ Light	Std.
11	Hex. Nut	1	$\frac{3}{8}$-16 NC-2	Reg. Semi-fin.
10	Woodruff Key	1	#404	Std.
9	Lock Nut	1	$\frac{1}{4}$-20 NC-2	ASA-Light-Semi
8	Stop Screw	1	$\frac{1}{4} \times \frac{15}{16}$	Special
7	Handle	1		CI
6	Hand Wheel	1		CI
5	Adjust. Screw	1	Steel	SAE 1112
4	Sleeve	1	Steel	SAE 1020
3	Lathe Center	1	Tool Steel	#2 Morse Taper
2	End Cover	1		CI
1	Tailstock	1		CI

Fig. 26-5. *Bill of Material for Lathe Tailstock, Fig. 26-3.*

26.8 Title Blocks

Every sheet used in the set of working drawings must have a complete title block as in Fig. 26-6. This should include:

DESIGN DRAWINGS

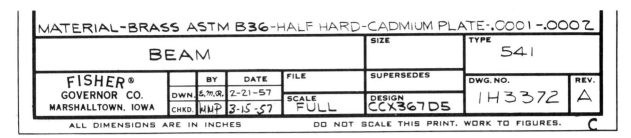

Fig. 26-6. *An Industrial Title Strip. (Courtesy Fisher Governor Co.)*

1. Company name
2. Project name
3. Sheet subject matter
4. Scale
5. Designer's name
6. Draftsman's name
7. Checker's name
8. Date drawn
9. Date checked
10. Sheet or drawing number
11. Revisions

Revisions are usually recorded by a numbered diamond or other symbol at both the title block and the revision on the drawing.

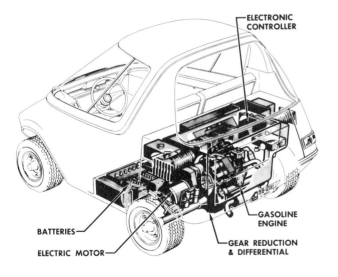

Fig. 26-7. *Phantom Pictorial—ES 512 Hybrid Commuting Vehicle. (Courtesy General Motors Corp.)*

26.9 Illustrations

Pictorials of mechanical devices are used to:

1. Clarify position of components in the assembly
2. Clarify method of operation
3. Educate assembly line and integrated job production workers
4. Advertise products
5. Provide maintenance information for customer service personnel
6. Provide a quick means for management to review the device

The engineer should be familiar with pictorials since as project leader he may be expected to select the best illustrations to satisfy these goals.

Fig. 26-7 shows a phantom section of an experimental automobile. The exterior appears to dissolve to show interior detail. Obviously an illustration like this represents a large investment and should be carefully planned before execution.

Fig. 26-8 employs sections and an exploded pictorial to clarify the assembly. In the exploded view the various parts are shown by moving them away from the main body along the axes of assembly.

Shading techniques are used extensively in pictorials to enhance appearance and depth. Several techniques readily adaptable to printing, microfilming, etc., are in common use. Line shading is used in Unit 10 and in Fig. 26-9. Block shading is used in Fig. 26-10 to give the illusion of a polished surface. The stippled areas in Fig. 26-11 look like rough cast surfaces. Many shading effects in half-tone dots as in Fig. 26-12, section ruling, etc., in black, white, and colors are commercially available preprinted on transparent gummed plastic or on waxed-back sheets. Either will adhere to an area on the drawing when rubbed with a blunt

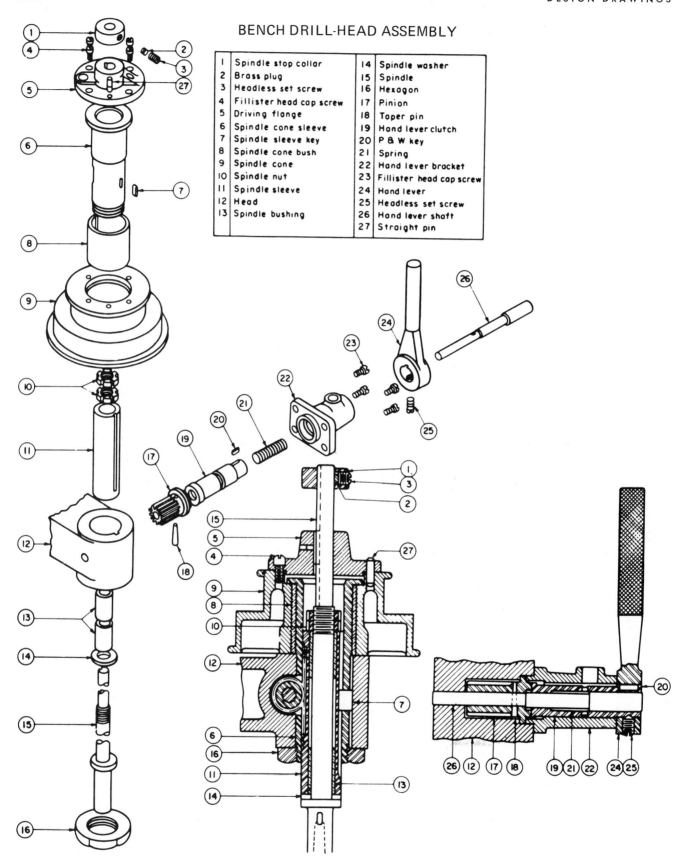

Fig. 26-8. *Orthographic and Pictorial Views.*

DESIGN DRAWINGS

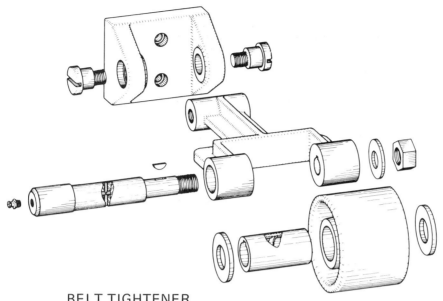

BELT TIGHTENER

Fig. 26-9. *Dimetric Illustration Using Line Shading.*

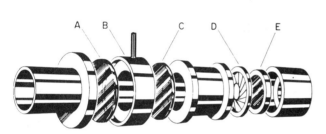

DIAGRAM OF PHOTOCELL HOLDER
(A & C)   GLASS PLATES
(B)   INFRARED ABSORPTION CELL
(D)   LIMITING DIAPHRAGM
(E)   PHOTOCELL

Fig. 26-10. *Block Shading. (Courtesy Ames Laboratory, USAEC)*

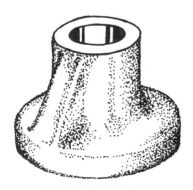

JOURNAL CASTING

Fig. 26-11. *Hand Stippling.*

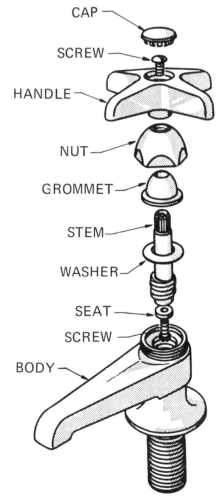

Fig. 26-12. *An Exploded View. (Courtesy Higgins Ink Co.)*

tool. No trimming is required with the waxed-type transfer, since only the area rubbed transfers to the drawing.

Other more difficult shading methods using continuous tone techniques with soft pencils, colored pencils or chalk, water colors, or air brush can be used with outstanding results but require considerable practice. An ink wash was used in shading the sketch in Fig. 26-13.

Fig. 26-13. *Conceptual Sketch of a Town Car.*

26.10 Simplified Drawing

Simplified drafting practices have not been widely adopted due to human resistance to change. Computer controlled machines use similar standards, however, and are becoming more common. This factor and rising labor costs may cause a large growth in the use of simplified drafting in the future.

Many of the recommendations of simplified drafting are a natural evolution of drawing standards. Each new set of standards shows an increased use of simplified symbols and short cut methods for representing standard specifications.

Methods of simplified drafting formulated by some of the larger industries vary somewhat since they are designed by the company to fit particular needs. The more common standards can be listed as follows:

1. Do not draw any standard parts.
2. Indicate standard hardware using only ballons and numbers on an assembly. Show outlines only if necessary to indicate position.
3. Provide written descriptions of parts when it is more efficient than detailing.
4. Use word descriptions to eliminate projected views where possible.
5. Omit elaborate, pictorial, or repetitive details.
6. Show hidden lines only where necessary for clarification.
7. Avoid section ruling unless necessary for clarification.
8. Indicate holes by centerline intersection and do not show outlines. Give hole diameters and processing specifications by word notation.
9. Use typewritten notes where possible. Avoid hand lettering.
10. Use freehand drawings where possible.
11. Omit all arrowheads and note leaders.
12. Use datum lines and reference numbers instead of dimension lines.

Rules should not interfere with the basic purpose of a drawing, which is to provide specifications for the particular part or assembly with sufficient clarity and detail to insure proper interpretation by anyone using the drawing.

An example of a simple part, comparing the standard representation at (a) and simplified drawing at (b), is shown in Fig. 26-14.

References for a more complete treatment on simplified drawing include: *Simplified Drafting Practice*, by W. L. Healy and A. H. Rau; *Simplified Drafting* by J. H. Bergen; *Functional Drafting* by Navy Department, Bureau of Ships.

26.11 Drawing Reproduction

Most design drawings are made directly on a transparent or translucent medium to provide an original that can be reproduced economically. Tracing paper, etc., is discussed in Art. 5.17.

Many different methods of reproduction are available, the choice depending on the number of copies needed, size desired, cost, and use.

Reproduction methods can be divided into three general classes with many subdivisions, a few of which are listed here.

1. *Mechanical*: Transfer of dye or ink in a semi-fluid state to the print paper.
    a. Duplicating
        1. Carbon Paper
        2. Hectograph
        3. Mimeograph

DESIGN DRAWINGS

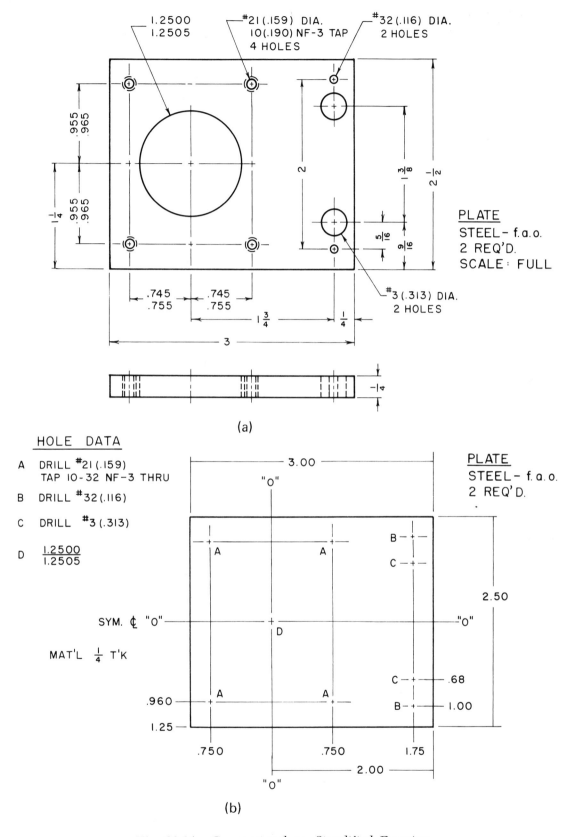

Fig. 26-14. *Conventional vs. Simplified Drawing.*

b. Printing
   1. Letterpress
   2. Photogravure
   3. Lithography
2. *Photochemical*: Formation of an image on a light-sensitive medium which is developed chemically.
   a. Transmitted Light (contact)
      1. Dry Diazo (ammonia vapor)
      2. Blueprint
      3. Moist Diazo (fluid developer)
      4. Photographic Contact Print
   b. Reflected Light (projection)
      1. Microfilm and Microfiche
      2. Photostat
      3. Photographic Reflex Print
3. *Special*: Fast image reproduction of single copies by special processes; generally convenient but expensive.
   a. Thermofax (heat)
   b. Xerox (electrostatic)
   c. Verifax (light)
   d. Other trade name processes

26.12 Reproduction Process Description

The *hectograph* principle, employed under several trade names, consists essentially of the transfer of the original, by pressure of the pencil or typewriter, to a master copy by means of a special "carbon" paper containing an analine dye. The copies are transferred from the master by contacting it, usually on a revolving drum, with paper moistened by a special duplicating fluid. Although the process is inexpensive and rapid, the number of copies that can be produced is limited.

The *mimeograph* is an inexpensive and widely used method for reproduction of small drawings in relatively large numbers. In this process a stencil is cut by drawing or typing. The prepared stencil is stretched over an inked pad on the drum of the mimeograph machine which allows ink to pass through the cut outlines to sheets of paper fed under the revolving drum. A photochemical method of cutting stencils for complicated or intricate details has been developed.

Letterpress printing using flat-bed or cylindrical presses is done commercially. The print is made from raised surfaces formed by electroplating a plastic, wax, or lead mold. Although the plates are expensive, the process is suitable for large quantity production.

Lithography or offset printing is economically used to produce fairly large numbers of drawings, maps, and other printed material. It is the process of making ink impressions of designs photographed on the surface of zinc or aluminum plates. The surface of the master plate is coated so that ink adheres only to the outlines of the image as it is transferred to a roller and then to the final copy. The cost of the lithography master sheets is less than the plates used in the letterpress process.

Dry Diazo is the most common method of reproducing drawings. Light sensitive paper or film is exposed through the tracing and developed immediately in hot ammonia vapor. Black, blue, or red lines can be formed on different media available, but commonly on a white background, hence the name *blueline* print. *Sepia* intermediates produced this way have brown lines on a translucent sheet. These are used like tracings to make more prints and for design modification work.

Blueprints were once widely used but are being displaced by the simpler dry diazo process. Blueprints are made by exposing and developing light-sensitive paper, but must be washed and dried. Intermediates produced by this process, called *Van Dyke* prints, have a brown background on translucent paper, and act like tracings to produce blue line prints indirectly on a blueprint machine.

Microfilm and Microfiche are becoming very popular methods of recording information. Storage space is negligible for the greatly reduced film print, and retrieval can be computerized. Prints for shop use are made from these films by an enlarging process.

Photostats are made with a camera which focuses the desired size of the image directly on a sensitized paper. The resulting print, developed and dried in the machine, is a negative with white lines on a dark background. A positive print with dark lines on white background is produced by photostating the negative print. Reproduction by this means is relatively expensive.

Thermofax employs heat reflected by the image of the original to cause discoloration of a special heat-sensitive reproduction paper.

Xerography employs a selenium-coated drum on which the optical image is electrostatically recorded. Powdered graphite particles are collected on the drum and then transferred to ordinary paper or film.

Verifax uses light to produce the image of the original on special paper.

More detailed explanations of all reproduction methods, especially the newly developed special processes, can be found in manufacturer's manuals.

PROBLEMS

Working Drawings

1. From knowledge, past experience, and reference material, compile a complete list of all working drawings necessary for the design and manufacture of the following mechanisms.
   a. Hand powered lawn mower.
   b. Bicycle
   c. Lawn sprinkler
   d. Bench vise
   e. Adjustable hacksaw
   f. Pop-up toaster

2. Fig. 26-15.

   a. Make detail drawings of each nonstandard part from the Rocker Arm assembly. Give complete specifications. Supply missing sizes for good design. Use "nominal" allowances and tolerances as necessary.
   b. Make an assembly drawing.
   c. Make a complete bill of material.

3. a. Compute the dimensions indicated on Fig. 26-16 using the basic shaft system.

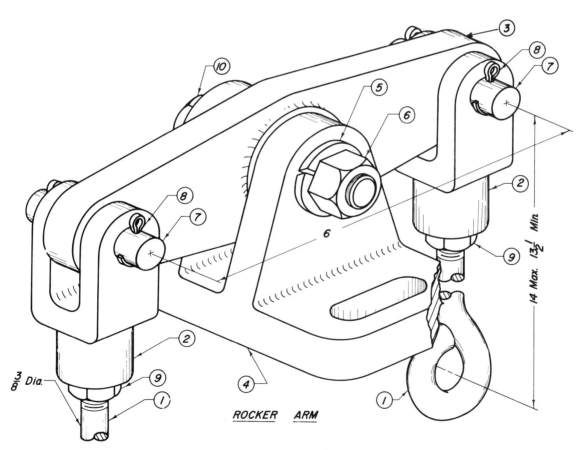

Fig. 26-15.

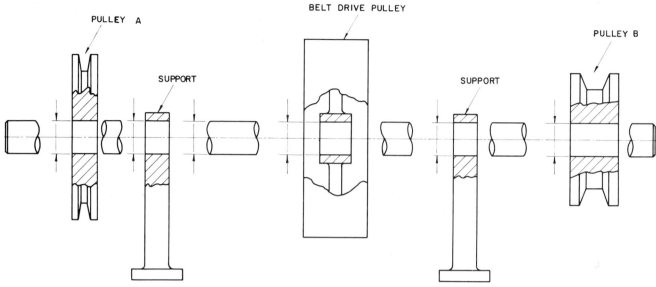

Fig. 26-16.

Specifications:
  The nominal size of the shaft is 1/2 inch.
  Pulley A is to have a shrink fit.
  Belt drive pulley has a wringing fit.
  Pulley B has a snug fit.
  Supports have a free fit.
  b. Make design and detail sketches of each pulley and one shaft support. Also, sketch and detail collars or spacers (not shown) to keep component parts in their proper position in assembly. Incorporate such features as keys, pins, setscrews, oilholes, bushings, counterbored or countersunk holes where appropriate.
  c. Sketch or draw an orthographic assembly with a bill of material.
4. Design an adjustable pipe hanger to support a 6 inch O.D. pipe whose centerline will vary from 12 to 15 inches below an I beam of a roof truss. The pipe lies in a vertical plane perpendicular to the beam and runs the entire length of the building.
5. Draw the necessary details of a mechanism that will change the direction of a reciprocating 1-inch diameter rod around a 45° bend. The rod has a throw of 6 inches. Make the attachment so it may be fastened to the floor 10 inches below the centerline of the rod.
6. Design a *Working Scale Model* of a hydraulic hoist to operate under a wagon box or truck box to facilitate unloading grain or similar material. The forward end of the box is to be raised until the bed has a 45° slope.

Requirements:
a. Exact reproduction of the structural details on the model is not essential, however, the same lifting action must be duplicated. Include schematic drawings to prove that the action will work.
b. Provide a complete set of working drawings for your design. This includes detail drawings of the individual parts and assembly drawings that are considered necessary for construction of the unit.
c. Assuring that a *model* wagon box or truck body with running gear is available, provide a drawing illustrating your model hoist mounted in working position on this equipment.
d. In considering your design any materials and building processes available to model builders may be considered.
e. Actual building of the model is not required.
7. Design a working model of a two cycle engine to be used as a visual aid for a class studying internal combustion engines.
  Data: The model should be designed to show the action of a piston in a cylinder and illustrate an arrangement of intake and exhaust ports. The materials to be used in the design should include the normal model building materials such as wood, plastic, wire, etc. as well as any other inexpensive shapes or forms that are readily available and could be used to advantage. It should be arranged by either cutaway portions or

transparent parts so that the relative position of the piston with respect to the intake and exhaust ports can be seen as the order of events in the cycle takes place.

Requirements. It is suggested that the model design be simple so that a small number of parts are involved. The following types of drawings are suggested to be included:
a. Preliminary sketches.
b. A layout to show the basic details of the design.
c. Detail drawings of the individual parts that could not be bought as a standard part.
d. A series of four drawings illustrating the position of the moving parts during the four events of the cycle which include compression, power (combustion), exhaust, and intake (with scavenging).
e. Assembly drawings. This might include section views or perhaps some type of pictorial.
f. Bill of material.
g. Written instructions for making and operating model.

8. Design a model of a pin tumbler lock showing the basic components and construction of the lock. The basic operating mechanisms of the lock should be stressed. Assume that this model will be used to explain the general construction, operation and function of the lock.

Data: The model is expected to be approximately 5 to 10 times actual size and should be constructed of materials readily available, such as plastic, wood, glue, nails, screws, dowel rods, wire, or other materials that could be used to simulate the actual parts.

Requirements:
a. Preliminary design (sketches).
b. Detail of all model parts, unless they are standard items.
c. Subassemblies, if useful in explaining the construction of the model.
d. Assemblies of the model.
e. Pictorial Drawing.
f. A bill of materials.
g. A concise written description or explanation of the basic operation of the model of the lock.

9. Supplies must be sent from earth to colonies established on the moon. The supply vehicle must soft land on the moon so a braking rocket is needed to slow the vehicle's descent. A pad will be placed on the end of each landing leg to help prevent sinking under load. The three legs will be constructed so they will cushion the landing. These legs must fold or retract in order to form the vehicle into a compact cylindrical package that can be carried aloft from the earth by booster rockets.

Design a model that could be used to help explain the general features of the spacecraft such as:
a. storage compartment
b. spherical fuel tanks
c. engine
d. folding legs

From your working drawings, a craftsman should be able to build this model. Any material, wood, wire, nails, etc., may be used in its construction.

Requirements:
a. Preliminary Sketches — assemblies, pictorials, etc.
b. Detail drawings of non-standard parts.
c. Assembly drawings.
d. Pictorial drawings.
e. Written instructions for making and operating model.

10. The International Sporting Goods Company has decided that their Clamp-on Boat Chair would sell better if it could be rotated and tilted backward. As an engineer for I.S.G.C. it is your responsibility to design a new bracket that will hold the chair to the boat seat and facilitate the rotation and tilt. Assume the chair seat has been manufactured of padded, one-inch plywood.

Specifications:
a. Full 360° rotation.
b. 30° tilt, backward only under pressure.
c. Maximum height of the bracket above the boat seat shall be no more than four inches.
d. Attachment and removal from the boat seat shall be simple and rapid.
e. Good resistance to moisture damage.

Requirements:
a. A design layout
b. Detail drawings
c. Assembly drawings
d. Parts list

Simplified Drawing Problem

11. Make details using simplified drawings of problems 2(a), 3(b), 4, 5 of this unit. Be consistent with the information in Art. 26.10 and Fig. 26-15.

Reproduction Problems

12. From reference material found in the library or other sources, write a report on the following reproduction processes:
    a. Blueprinting   d. Photolithography
    b. Ozalid   e. Black and White
    c. Van Dyke   f. Letterpress
13. Construct a comparative analysis chart to show classification, major similarities and differences, important uses, durability or permanence, cost, convenience, and other important factors of the reproduction processes named in Art. 26.11.
14. In problems A, B, C and D, make the following decisions, with reasons for each choice.

Problem A: You are the vice president in charge of service for a large automobile company. The latest model has developed a "bug" due to oil pressure building up in the crank case until it leaks past the retaining rings into the transmission. You wish to quickly produce an illustration which will show how and where 3 bleeder holes can be drilled to relieve the difficulty. The sheets are to be sent to 5,000 service managers to aid instruction of their mechanics.

Problem B: Assume that you, as a member of the troop committee of Boy Scouts, wish to assist on several projects relating to electricity. You find that the boys have difficulty reading schematic diagrams for wiring an electric bell, installing equipment to practice sending code between two rooms, and building a small radio receiving set. You decide to draw up some pictorials which the boys will be able to read. Remember, you have a limited budget and wish a copy for each of 12 boys.

Type of drawing:
1. Axonometric
2. Oblique
3. Perspective
4. Orthographic

Type of paper:
1. Heavy white
2. Tracing paper
3. Tracing cloth
4. Other

Method of drawing:
1. Pencil
2. Ink
3. Color
4. Other

Type of shading or rendering:
1. Straight line
2. Wash
3. Air brush
4. Stipple
5. Smudge
6. None

Method of reproduction:
1. Carbon Paper
2. Ditto
3. Blueprint
4. An original for each user
5. Ozalid, black line
6. Photolithograph
7. Letterpress
8. Etching

DESIGN DRAWINGS

Problem C: As a sales engineer with X company, you are charged to make the decisions listed. New catalog illustrations for an extensive advertising campaign are to be prepared in the drafting room of your company. The catalog is to be printed on slick paper representing a considerable expense to the company. The talent in the drafting room is able to produce the drawing in any way you direct.

Problem D: An assembly operation in your plant requires that the following items are to be installed on a small shaft in the order given: (a) Collar, attached by a set screw; (b) Spacer; (c) Coil spring; (d) Collar free to slide; (e) Two plain washers; (f) Hand wheel keyed to the shaft; (g) Nut. Because the 10 girls who are each doing this same assembly have been unable to read the assembly drawing, you decide to have the drafting room produce an exploded drawing.

Illustration Problems

1. Refer to Problems 14A, 14B, 14C, 14D. Prepare the appropriate illustration for each problem. Make assumptions concerning the designs and their sizes that are compatible with the problem statements.

2. Fig. 26-17. Make an exploded pictorial drawing of both assemblies. Combine the use of instruments and freehand drawing for circles, shading, etc., as you desire.

3. Make engineering illustrations of any mechanism in problem 1.

4. Refer to Fig. 26-15.
   a. Trace freehand and shade the pictoral assembly of the Rocker Arm.
   b. Make exploded-view pictorial, sketch, or instrument drawing of the Rocker Arm.

5. Refer to Fig. 26-3.
   a. Make a shaded pictorial, sketch or instrument drawing with appropriate broken-out section view of the Lathe Tailstock Assembly.
   b. Make exploded pictorial drawing with appropriate shading.

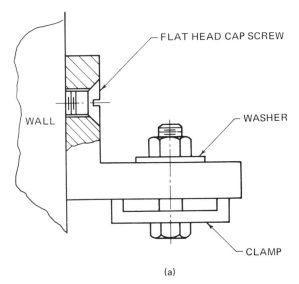
(a)

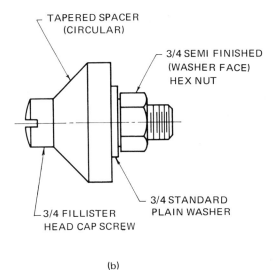
(b)

Fig. 26-17.

unit 27

Graphical Representation of Design Data

27.1 Why Graphical Representation

During the design process much data must be interpreted or evaluated. Since interpretation of numerical data is often a lengthy and difficult procedure, it may be expedient to convert the data to a more readable form, such as a chart, graph, or diagram.

Since the effectiveness of a chart depends to a great degree on its appearance, the details of construction must be carefully considered. The engineer who presents, evaluates, or uses graphical data must not only be familiar with the principles of correct design and construction but also understand the advantages and limitations of graphical representations.

27.2 Classification of Charts

The entire field of graphical representation can be separated into two general catagories: advertising or popular appeal, and scientific or technical.

These divisions include many different kinds of charts which can be used to depict specific types of data.

Part I
ADVERTISING OR POPULAR APPEAL CHARTS

27.3 Bar and Column Charts

The *bar chart*, one of the more common pictorial graphs appearing in many different forms or combinations, is effective for nontechnical use because it is particularly appropriate for comparing the magnitude of the parts of a whole. Quantities, values, and relations are illustrated by heavy bars whose lengths are proportional to the amounts represented, see Fig. 27-1. The bars, beginning at a base line, may be horizontal or vertical. When the bars are vertical, the diagram is called a *column chart*.

Lettering must be done carefully. If it is necessary to indicate the value of each individual bar, the figures should be placed along the bar and parallel to it. Values placed at the ends of the bar create an optical illusion of increased length. Names of items represented by individual bars are usually lettered outside the starting base line.

If the vertical bars are drawn touching each other, the resulting diagram is known as a *staircase chart*. More often only the profile of the tops of the bars are plotted on coordinate paper and the resulting stepped-line curve becomes a *staircase curve*.

The bar chart is frequently used to represent proportional composition as shown in Fig. 27-2.

27.4 Area Charts—Pie Diagrams

Comparisons of percentages of related quantities can also be represented by a pie chart or 100% circle. The pie chart, Fig. 27-3, has great popular appeal even though it is inferior to a

GRAPHICAL REPRESENTATION OF DESIGN DATA   293

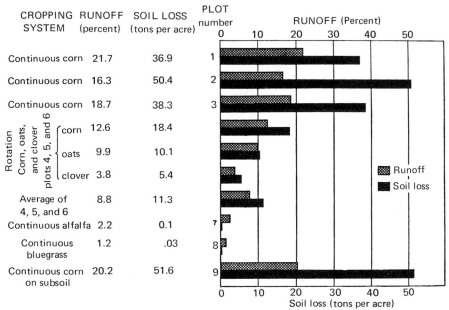

Fig. 27-1. *Bar Chart. (Courtesy Iowa State University)*

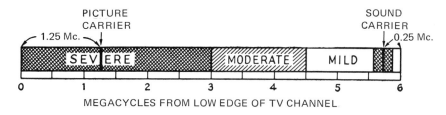

Fig. 27-2. *Proportional Bar Chart. (Courtesy American Radio Relay League)*

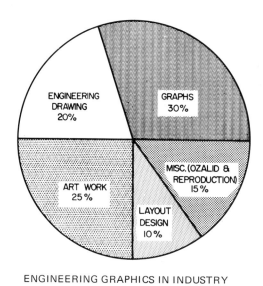

Fig. 27-3. *Pie Diagram. (Courtesy Ames Laboratory, USAEC)*

subdivided bar chart. The data is converted into corresponding degrees on the circumference of a circle whose diameter is any desired value. It is recommended that no more than five categories be portrayed since it becomes increasingly difficult to differentiate between the values shown especially if several of the sectors are approximately equal in size. All lettering and percentages should be on the sector or adjacent to the circumference of the sector so that the meaning of the chart will be clear. For contrast, the segments should be crosshatched, colored, or marked in some distinctive manner. The title, source and other pertinent data are placed to balance the chart.

27.5 General Information Charts

Flow charts are used to explain the steps in a series of operations, or to follow the progress of raw material through the stages of manufacture.

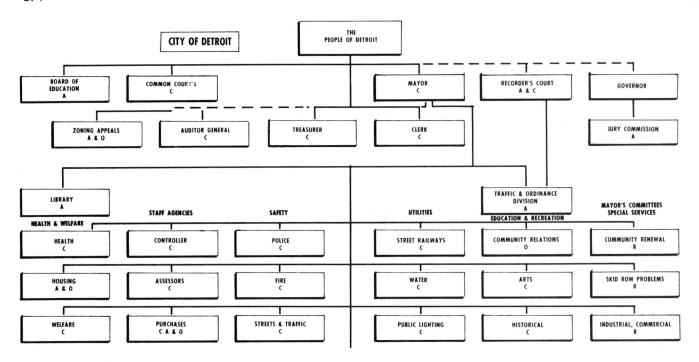

Fig. 27-4. *Organization Chart. (Courtesy City of Detroit)*

Organization charts show the relationship of all parts of an organization to each other, or the divisions of authority or responsibility existing within an organization, Fig. 27-4.

Ranking or *rating charts* show the position or rank of a series of items arranged usually in order of their frequency of occurrence.

Progress charts are schedule or production control charts used in planning and coordinating administrative, procurement, production, and distribution methods.

Distribution charts, usually in the form of maps, are used to show a wide variety of information such as density of population, accident frequency, and sales, Fig. 27-5.

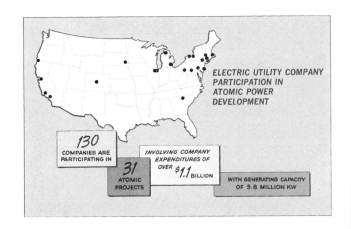

Fig. 27-5. *Distribution Chart. (Courtesy Edison Electric Institute)*

27.6 Pictorial Charts

Pictorial charts are effective in the presentation of data to people who have little interest in statistical tables. This simple form of chart, because of eye appeal, creates more interest than many conventional types of graphical communication.

Basically there are two forms of pictorial charts. One uses symbols whose proportions represent values, Fig. 27-6, and the other uses symbols, called counting units, to portray fixed quantities or values, Fig. 27-7.

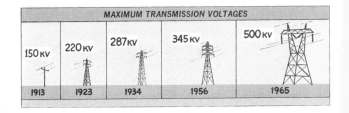

Fig. 27-6. *Proportional Pictorial Chart. (Courtesy Edison Electric Institute)*

GRAPHICAL REPRESENTATION OF DESIGN DATA

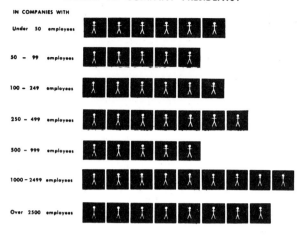

Fig. 27-7. *Counting Chart. (Courtesy Dun's Review & Modern Industry)*

Part II
SCIENTIFIC OR TECHNICAL CHARTS

27.7 Grid Paper

Technical or computation charts are usually plotted on commercially prepared grid paper, available in different spacings and sheet sizes. Some of the available grid rulings are rectangular, semilogarithmic, logarithmic, polar, trilinear and many specialized rulings.

27.8 Trilinear Charts

The trilinear chart plotted on a special grid, is used in the analysis of compounds, mixtures, or any relationship composed of three variables. The theory of the chart is based on the geometric principle that *the sum of the perpendiculars to the three sides from any interior point is equal to the altitude of the triangle.* The sides of the equilateral triangle are calibrated from 0% to 100%.

In Fig. 27-8, perpendiculars to the three sides from a point P' have been drawn and the three sides have been calibrated from 0 to 100%. The left side represents percentages of P'Z', the base shows percentages of P'Y' and the right side shows percentages of P'M'. Reading the scales shows that P'Z' = 40%, P'Y' = 30% and P'M' = 30%. Their sum is equal to 100%, and point P' represents a quantity composed of 40, 30 and 30 parts of the three variables.

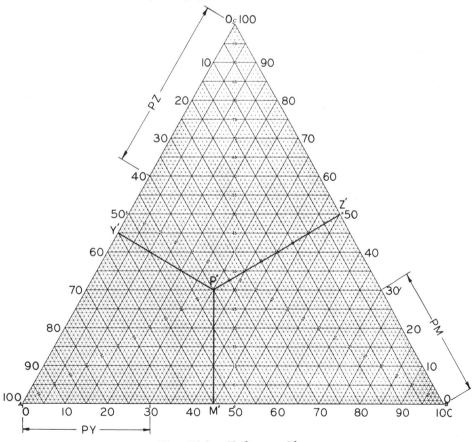

Fig. 27-8. *Trilinear Chart.*

27.9 Polar Charts

Polar coordinate paper consists of a series of equally spaced concentric circles with equally spaced radii emanating from the center called a pole. Two variables, one linear and the other angular, can be plotted to form a continuous curve. The angular measure is expressed in either degrees or radians and the linear variable in standard linear units.

The polar diagram of Fig. 27-9 illustrates horizontal patterns of radiation from an antenna two wave lengths long.

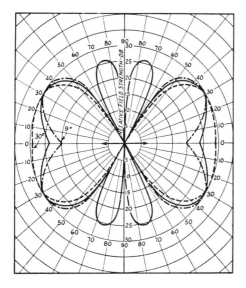

Fig. 27-9. *Polar Diagram. (Courtesy American Radio Relay League)*

27.10 Graphing Procedure

The construction of a graph can be divided into a sequence of eight steps applicable to graph construction on any type of grid.

1. Select the type of graph paper and grid spacing for best representation of the given data.
2. Choose the proper location of the horizontal and vertical axes.
3. Determine the scale units for each axis so as to appropriately display the data.
4. Identify, graduate, and calibrate the axes.
5. Plot the points representing the data.
6. Draw the curve or curves.
7. Identify the curve or curves. Add title and necessary notes.
8. Ink the graph for reproduction.

27.11 Selection of Suitable Grids

Commercially prepared rectilinear grid rulings are available in many different spacings such as 4, 5, 6, 8, 10, 12, 16, and 20 rulings to the inch. These grids can be obtained in different colors, such as green, blue, or orange, printed on opaque or translucent papers. The grids for graphs used in publications or for reading by the general public should be as coarse as possible with the minimum number of grid lines to insure the desired reading accuracy. See table below.

Technical graphs used for computation and analysis should be plotted on finely ruled grids so that necessary interpolations of the readings can be performed.

27.12 Location of the Axes

The axes of a graph consist of two intersecting straight lines. The horizontal line is the abscissa or X axis, the vertical line is the ordinate or Y axis and the point of intersection is called the origin. *Customarily the independent variable is plotted on the X axis and the dependent variable is placed on the Y axis.*

The origin point of the axes is placed in the lower left part of the grid paper when the data involves only positive values. If positive and negative values are to be graphed it will be necessary to shift the origin point accordingly.

Since many of the commercially prepared grids do not include sufficient border space for proper labeling, the axes should be placed approximately

Grid Paper	Plotted Curve	Equation	Form
Rectangular	Straight Line	$Y = mX + C$	Linear
Semilog	Straight Line	$Y = ae^{mX}$	Exponential
Log-log	Straight Line	$Y = aX^m$	Power

one inch inside the edge of the printed grid in order to allow ample room for graduation, calibration, and labeling of the axes. Note that the edge of the grid must be used on log paper.

27.13 Determination of Scale Units

Scale units for the axes must be selected so as to fit the grid subdivisions and also fit the available grid space to best advantage without distorting the picture of the data. A steep sloping curve on a graph indicates a significant change while a flat curve indicates an insignificant change. The scale units must be selected to ensure proper interpretation of the given data. The slope of a curve can be made steeper by increasing unit distances on the ordinate and decreasing unit distances on the abscissa. Refer to Unit 28 for a more detailed discussion of scale design.

27.14 Identification of Scales

The ordinate and abscissa should be labeled, graduated, and calibrated in such a manner that their readability is insured. Calibrations and captions are placed below the abscissa and to the left of the ordinate.

Calibration values are placed only on the accented lines of the printed grids. These identifying values are written completely when they contain *no more than three digits*. Identifying scale captions are placed parallel to the axes, and should read from the bottom and the right of the page. These scale captions should include word specification of the variable represented, symbol for the variable, and unit of measure.

27.15 Plotting the Data

The points on a graph prepared from experimental data are designated by symbols shown in Fig. 27-10. It is recommended that open-point symbols be used and that the curve shall not pass through these located points since it may become necessary to check the accuracy of the plotted points. If several curves are to be shown, each curve is designated by a different symbol. The filled-in symbols are used for three or more curves on one graph, on scatter diagrams or for accent on a graph. Plotted points are omitted on popular appeal graphs since in this case the purpose is to show only the significance of the curve. Fig. 27-11 illustrates the use of symbols and an identification table.

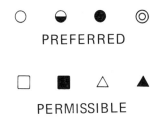

Fig. 27-10. *Symbols.*

27.16 Drawing of the Curve

The data involved determines whether the points are to be connected by a smooth curve or by a series of straight lines.

Continuous data is represented by a smooth curve which strikes an average path through the plotted points. This average curve can be located by eye, or if greater accuracy is required, by means of one of the methods mentioned in Art. 27.22.

Discrete or discontinuous data, not supported by theory or law, is connected by straight lines. The resulting graph is often called a broken-line curve.

27.17 Titles and Identification of Curves

All graphs should have a clear, concise title which identifies the data being represented, including source references. Titles for experimental data should include the date of the experiment and the name of the person who collected the data.

These descriptive titles are generally placed so as to balance the graph, but this placement should not interfere with the readability of the graph. If a special grid is being ruled, omit the grid lines in the area selected for the title.

When more than one curve is presented on a graph, a solid line should be used for the most important curve. The other curves can be identified by use of dash lines, dotted lines, etc. Each

Resistance Coefficient K, Equivalent Length L/D, And Flow Coefficient C_v — continued

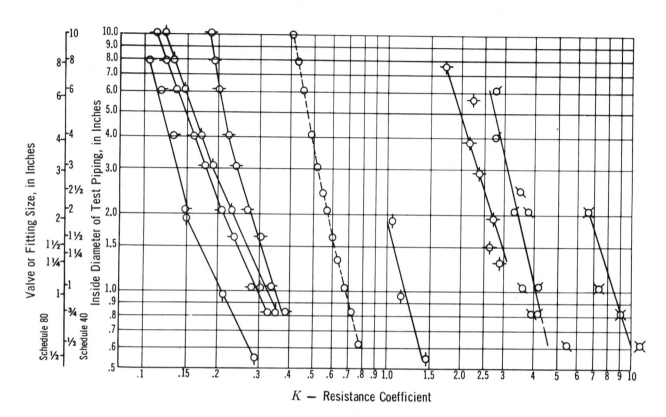

Variation of Resistance Coefficient K ($=f\, L/D$) with size

Symbol	Product Tested	Authority
○	Schedule 40 Pipe, 30 Diameters Long ($K = 30\,f$)	Moody A.S.M.E. Trans., Nov.-1944[1]
○-	125-Pound Iron Body Wedge Gate Valves	Univ. of Wisc. Exp. Sta. Bull., Vol. 9, No. 1, 1922[16]
⊖	600-Pound Steel Wedge Gate Valves	Crane Tests
-○	90 Degree Pipe Bends, $R/D = 2$	Pigott A.S.M.E. Trans., 1950[6]
⊙	90 Degree Pipe Bends, $R/D = 3$	Pigott A.S.M.E. Trans., 1950[6]
-○-	90 Degree Pipe Bends, $R/D = 1$	Pigott A.S.M.E. Trans., 1950[6]
⟡	600-Pound Steel Wedge Gate Valves, Seat Reduced	Crane Tests
-⟡-	300-Pound Steel Venturi Ball-Cage Gate Valves	Crane-Armour Tests
⌀	125-Pound Iron Body Y-Pattern Globe Valves	Crane-Armour Tests
⌀	125-Pound Brass Angle Valves, Composition Disc	Crane Tests
⌀	125-Pound Brass Globe Valves, Composition Disc	Crane Tests

Fig. 27-11. *Multiple Symbol Application.*
(*Courtesy Crane Co. Tech. Paper 410, Flow of Fluids*)

of the curves should be designated by brief labels placed close to the curves if possible.

27.18 Reproduction

Graphs which are to be published or are to be used in projectors should be inked. Since such graphs are rarely used to determine accurate readings, the grids should be as coarse as possible as long as the essential facts are presented clearly. Charts to be used on an overhead projector must be reproduced on film. They should be inked for satisfactory use in an opaque projector.

27.19 Semilogarithmic Charts

The semilogarithmic chart is extremely valuable for portraying proportional and percentage relationships. This type of chart not only represents correct relative changes but at the same time indicates absolute amounts.

On semilogarithmic paper, the vertical axis is ruled logarithmically and the horizontal axis, arithmetically. Since semilogarithmic grid paper combines the features of logarithmic and uniform rulings, this grid paper is very useful in statistical analysis because functions whose values are in geometric progression plot as straight lines. The slope of this straight line indicates *the rate of change* of the function. As a result, semilog paper is sometimes called rate of change paper. The rate of increase or decrease at any point on the curve may be found by determining its slope.

It is also possible to determine the percent increase or decrease since the vertical distance of a curve on semilogarithmic paper indicates the percentage change regardless of location of the curve on the grid. The distances from 10 to 20, 100 to 200 or 1000 to 2000 on a logarithmic scale represent increases of 100 percent, and the three curves representing these 100% increases have the same slope. A percentage scale can be constructed along one edge of the grid paper. Note that the distance from the beginning of a cycle, called the zero point, to the first major division, (1 to 2 on a log scale) is 100%; the distance 1 to 3 represents 200% and so on. The distances below the zero point represent percentage decreases. Percentage increases or decreases shown by the curve can be determined by transfer of the proper vertical distances to the percentage scale. A more practical method is to construct a percentage scale on a piece of durable, transparent material and to use this scale as a measure of the desired vertical distances as illustrated in Fig. 27-12.

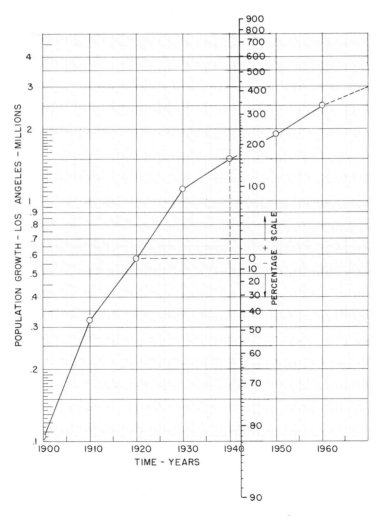

Fig. 27-12. *Use of Percentage Scale.*

27.20 Logarithmic Grids

Grid paper with parallel horizontal and parallel vertical rulings spaced in proportion to logarithms of numbers is available, not only in differing cycle lengths, but also in varying numbers of cycles. On this type of ruling, called logarithmic, graphs of products, quotients, or variables in the form of exponents plot as straight lines.

27.21 Empirical Equations

An equation derived from test data or experience is defined as an empirical equation. Sometimes it becomes necessary to supplement graphical analysis of test data by determining the equation which best fits the data. The usual procedure is to try to reduce the data to a straight line graph by plotting on rectangular, semilog or logarithmic grid paper. If the graph becomes *a straight line* on rectangular paper, the equation belongs to the family of curves whose equation is $Y = mX + C$. If the chart is *a curve* passing through the origin, the data may fit a power curve, and if the curve intersects one axis only, the data may fit an exponential curve. In either case it must be replotted. Then if the curve seems to fit a power curve, plot the data on logarithmic paper, or if exponential, plot on semilog paper. When the data plotted on either the semilog or log paper is a reasonably straight line, the form of the equation is known and the determination of the exact equation can begin. If it is not possible to reduce the resultant plot to a straight line on either the semilog or logarithmic paper, the empirical equation is probably of a more complicated form, or may contain some initial constant. The methods of rectifying this type of equation will not be considered.

27.22 Curve Fitting

A straight line may lie in many slightly different positions and yet fit the plotted points. The problem then resolves itself into one of finding the best possible fit.

There are three methods of finding the best fitting straight line. These are called:

1. Graphical Method or Selected Points
2. Method of Averages, Graphic Solution
3. Method of Least Squares

Of these three methods, the first is the simplest; the second, the most satisfactory since it is fairly simple with reasonable accuracy; the third is the most accurate but often too time-consuming. Only the first two methods will be discussed.

27.23 Graphical Method of Selected Points

In the graphical method of selected points, a line is drawn through a series of plotted points in such a manner that this line passes through a maximum number of the plotted points, with approximately an equal number of missed points on each side of the line. The missed points should *not* be bunched at one end of the line, but should be spread along its length.

27.24 Graphical Method of Averages

In the graphical method of averages, a plotted curve is reduced to two average points, through which a straight line is drawn. The reduction is accomplished in the following manner.

After the original data has been plotted, join each adjacent pair of points by a straight line. Determine the midpoints of each of these straight lines; these new points will be the average points. If, in turn, each adjacent pair of average points is joined by straight lines, new midpoints are determined. If this process is repeated a sufficient number of times, the original plotted points will be reduced to two average points, through which the straight line may be drawn.

27.25 Empirical Equations—Determination of Equations

Curves encountered in engineering are derived from test data plotted on graph paper with a smooth curve faired through the points. From this curve intermediate values of the variable not determined during the test can be found by interpolation. If the curve is a smooth line it may be possible to express the curve as an equation. However, because of the approximations made in reading of instruments, the calculation of the data, and the plotting of the graph, the resulting curve and its equation can represent the data only approximately. Such a curve is called *empirical*, and its equation is an empirical equation. This type of equation can be used only within the limits of the original data and predictions beyond this range should not be made.

Discussion of empirical curves will be limited to three fundamental types of basic equations; the straight line, $Y = mX + C$; the power curve, $Y = aX^m$; and the exponential curve, $Y = ae^{mX}$.

27.26 Empirical Equation—Straight Line

When test data plots as a straight line on rectangular grid paper, the equation of that line belongs to the family of curves whose basic equation is $Y = mX + C$. From analytical geometry we recognize this as the slope-intercept form of the straight line, where m is the slope and C is the intercept on the vertical axis. The value of the constant C may be determined directly from the graph. Slope m equals the tangent of the angle formed by the plotted line and a horizontal line, if the vertical and horizontal scale units are equal. However, if the axes have different scale units, then the slope must be determined by finding the coordinates of two points on the curve and using them to solve the following equation, $m = \dfrac{Y_2 - Y_1}{X_2 - X_1}$

For example, plot the curve on rectangular grid paper and derive the empirical equation of the surface conductance, F, of concrete at 20° F under varying air velocities in miles per hour, V, using the coordinates given in the table.

V	0	5	10	15	20	25
F	2.0	3.5	5.1	6.8	8.5	10

Plotting the data on rectangular graph paper, Fig. 27-13 indicates that the family curve is $F = mV + C$, where $m = slope$ and $C = intercept$ on the vertical axis

The value of C, read directly from the graph, is 2.

The value of slope m must be calculated by substituting the coordinate values of two points on the curve into the slope equation. Choosing $F_2 = 10$, $V_2 = 25$ and $F_1 = 2$, $V_1 = 0$ from the coordinate table, the slope becomes:

$$m = \frac{F_2 - F_1}{V_2 - V_1} = \frac{10 - 2}{25 - 0} = \frac{8}{25} = 0.32$$

Substituting $C = 2$ and $m = 0.32$ into the family equation, we derive $F = 0.32V + 2$, the empirical equation for this given data.

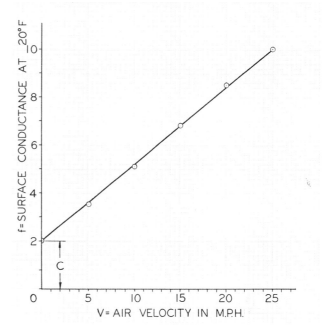

Fig. 27-13. *Surface Conductance of Concrete.*

27.27 Empirical Equations—Exponential Curves

If test data plots on semilog or ratio paper as a straight line, the curve belongs to the exponential curve family whose most common equation has the form $Y = Ae^{mX}$. By determining the values of the constants A and m, the equation for the given set of data can be expressed.

Reducing the family equation $Y = Ae^{mX}$ to an equation of the first degree by means of logarithms, we obtain:

$$\log Y = m(\log e) X + \log A$$

The value for m is a measure of the slope. The constant e, base of the Naperian logarithms, equals 2.718 and its logarithm is 0.434. The value of A is the vertical axis intercept when $X = 0$. The values of X and Y are coordinates of any point P lying on the curve. Solving the equation for the slope

$$m(\log e) = \frac{\log Y - \log a}{X}$$

$$\text{or } m = \frac{\log Y - \log a}{X(\log e)}$$

As an illustration, using semilog paper, plot the points whose coordinates are given in the following table and determine the equation which best fits the given data. See Fig. 27-14.

Y	9	5.2	3	1.74	1
X	0	2	4	6	8

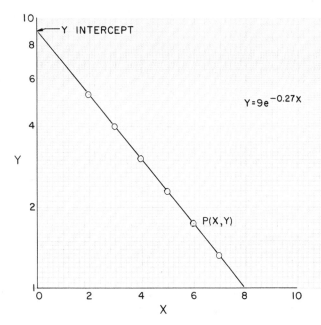

Fig. 27-14. *Exponential Curve and Equation.*

$$m = \frac{\log Y - \log A}{X(\log e)} = \frac{\log 1.74 - \log 9}{6(0.434)}$$

$$m = \frac{0.241 - 0.954}{6(0.434)} = \frac{-0.713}{2.6}$$

$$\therefore m = -0.271$$

The empirical equation for the given data is
$Y = 9e^{-0.27X}$

27.28 Empirical Equations—Power Curves

When test data plots as a straight line on logarithmic graph paper, the curve belongs to the family curves whose general equation is $Y = AX^m$. By determining the values of A and m, we may write the equation for a given set of data.

Problem. Plot the points on log paper, whose coordinates are given in the table below, and determine the equation which best represents the given conditions.

X	0.1	0.3	0.6	1.0	6.0
Y	4.0	1.5	0.8	0.5	0.1

The plot of this data on log paper, Fig. 27-15, shows that a straight line closely fits the points and hence, the equation will be of the form $Y = AX^m$.

Taking logarithms of both sides, we obtain $\log Y = \log 9 + m \log X$, a first degree equation which will plot as a straight line on rectangular grid paper.

Solving for the slope, $m = \dfrac{\log Y - \log A}{\log X}$, where X and Y are the coordinates of any point lying on the curve.

"Log A" will be the vertical axis intercept when the above first degree equation is plotted on rectangular grid paper. However, the constant A can be determined on log paper if we remember that this *vertical axis* is located where $\log X = 0$. In this problem, referring to Fig. 27-15, the value of A is 0.5.

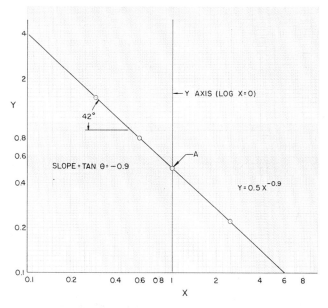

Fig. 27-15. *Power Curve and Equation.*

Using $A = 0.5$, $X - 0.1$, and $Y = 4.0$, solve for slope m.

$$m = \frac{\log Y - \log A}{\log X} = \frac{\log 4 - \log 0.5}{\log 0.1}$$

$$m = \frac{0.602 - (-0.301)}{-1.0} = \frac{0.903}{-1.0} = -0.903$$

The empirical equation for this set of data is $Y = 0.5X^{-0.9}$.

Since the cycle distances are equal, it is possible to determine the value of slope m by measuring the angle θ, formed by the plotted line and a horizontal axis. In Fig. 27-15 the angle θ, measured with a protractor, equals 42°.

Slope $m = \tan 42° = -0.9$, which checks the mathematical solution above.

PROBLEMS

1. From the latest census figures for United States construct a graph comparing the population of the five largest states.

2. By referring to manufacturers tables, plot on rectangular grid paper and on log-log paper the loss of head for water flowing in iron pipes whose diameters vary from 0 to 5 inches.

3. Construct a pie diagram illustrating your use of the 24 hours in an average day.

Rectilinear Charts

Prepare correctly drawn graphs of the following data. Use suitably subdivided rectangular coordinate paper.

4. *Calibration table* for thermocouples. Temperature, degree C, vs. EMF values, millivolts for Iron-Constantan Thermo-elements.

Deg. C	E.M.F.	Deg. C	E.M.F.	Deg. C	E.M.F.	Deg. C	E.M.F.
0	0	150	8.01	400	21.83	700	39.19
50	2.61	200	10.77	500	27.41	800	45.49
100	5.28	300	16.30	600	33.13	900	51.83

5. *Solubility table*: Grams of cane sugar in 100 grams of water, temperature in degrees Centigrade.

Temp.°C	0	10	20	30	40	50	60	70	80	90	100
Sugar	179.2	190.5	203.9	219.5	238.1	260.4	287.3	320.5	362.1	415.7	487.2

Exponential Curves

Draw correctly constructed, fully labeled graphs of the following data. Determine the equations.

6. Resistance per 1000 feet for various gages of Copper wire at 20°C.

Gage No.	Resistance, Ohms
0	.098
1	.124
2	.156
3	.197
6	.395

Gage No.	Resistance, Ohms
10	.999
12	1.59
16	4.02
18	6.39
22	16.1

7. Breaking strength of steel wire based on a tensile strength of 100,000 pounds per square inch. Weight is in pounds per 1000 feet.

Gage No.	Breaking Stress	Weight
0	7400	248.7
1	6290	211.4
2	5430	182.5
3	4680	157.1
4	3980	133.6
5	3365	113.1

Gage No.	Breaking Stress	Weight
6	2895	97.3
7	2460	82.7
8	2060	69.3
9	1720	57.8
10	1430	48.1
12	866	29.1

8. Obtain 1910 to 1970 census figures for several of the larger cities of the world.

   a. Plot graphs of this data.

   b. Determine from the graph, the percent increase or decrease for each 10-year period and for the 60 year period.

   c. Plot the percentage on semi-log paper using percent as the ordinate.

Power Curves

Prepare correctly constructed, fully labeled graphs of the following data. Determine the equations.

9. *Spherical Tanks*: Surface area in square feet. Volume in cubic feet.

Diameter Feet	Surface Area Sq. Ft.	Volume Cu. Ft.
6	113.10	113.10
8	201.06	268.08
8	314.16	523.60
10	314.16	523.60
12	452.39	904.78
14	615.75	1436.80
16	804.25	2144.70

Diameter Feet	Surface Area Sq. Ft.	Volume Cu. Ft.
18	1017.9	3053.6
20	1256.6	4188.8
22	1520.5	5575.3
24	1809.6	7238.2
26	2123.7	9202.8
28	2463.00	11494.0

10. *Heat Losses*: Horizontal bare steel hot-water pipes at 210°F to surrounding still air at 70°F. Heat losses in BTU/hr./lin.ft./°F.

Nom. Pipe Size Inches	Heat Loss
1/2	.470
1	.877
1 1/2	1.200
2	1.512

Nom. Pipe Size Inches	Heat Loss
2 1/2	1.896
3	2.173
3 1/2	2.500
4	2.787

Nom. Pipe Size Inches	Heat Loss
5	3.338
6	3.886
8	4.960
10	6.090

unit 28

Graphical Mathematics – Alignment Charts

28.1 Graphical Solution

Mathematical problems can be solved either algebraically or graphically. In many cases, a practical solution will require a combination of both methods.

Graphical methods are advantageous because they provide a visual explanation of the behavior of the variables and also assist in explaining the steps involved in the calculations. In many cases graphical methods can be used to quickly check the validity of a mathematical solution.

The accuracy of graphical solutions are sometimes questioned. Remember that measuring devices are graphical and the data read from these instruments is often recorded to three significant figures. This degree of accuracy is practical and can be substantiated by graphical computation.

Part I
MATHEMATICS

28.2 Network Charts

A network chart is the graphical representation of a family of curves plotted on an appropriate grid so that they appear as straight lines. The basic operations of addition, subtraction, multiplication, and division can be performed by use of network charts.

Fig. 28-1 illustrates a chart, constructed for the equation $TU = Z$, which can be used for either multiplication or division. The dash lines illustrate the method of solution for $10 \div 4$ and 3.5×4.

28.3 Powers and Roots of Numbers

A network chart for the determination of powers and roots of numbers can be constructed by plotting values determined from the equation $Y = A^n$.

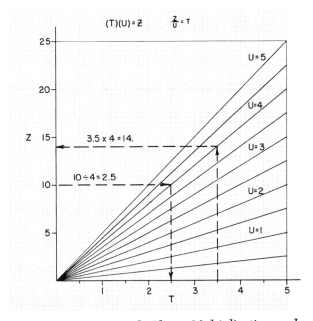

Fig. 28-1. *Network Chart—Multiplication and Division.*

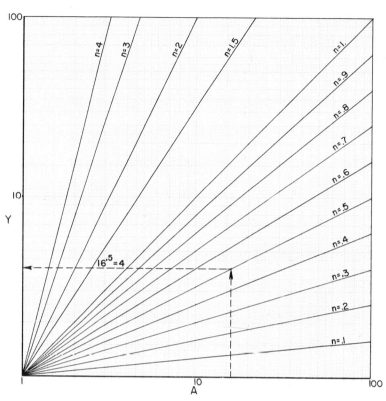

Fig. 28-2. *Network Chart—Powers and Roots.*

If the given equation is reduced to the form $\log y = n \log A$, the graph for each different value of n, plotted on logarithmic grid paper, will be a straight line as shown in Fig. 28-2. Any other required values of n can be plotted to increase its usefulness. Fig. 28-3 illustrates a chart showing plate-tank capacitance required for a quality factor of 10.

28.4 Simultaneous Equations

Simultaneous equations which consist of two or more equations expressed in terms of common variables, appear frequently in engineering calculations. These equations are usually solved by algebraic methods. In some instances the mathematical approach becomes too cumbersome or time consuming. In other cases where the data is empirical and the equations are unknown or impossible to derive, the mathematical approach cannot be used. For these reasons it may be advantageous to become familiar with the graphical method of finding values of unknowns to fit the specific conditions of an engineering problem.

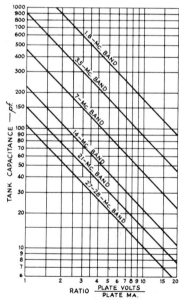

Fig. 28-3. *Network Chart Application.*
(*Courtesy American Radio Relay League*)

In the graphical solution, each equation or known condition is plotted on suitable grid paper. The X and Y coordinates of the point or points of intersection of the functions will satisfy the conditions and provide the required solution. The following problems illustrate the principles involved.

Problem 1. Determine graphically the values of X and Y which satisfy the simultaneous equations $10X - 5Y = 0$ and $8X - Y = 24$. Study Fig. 28-4. The graph shows that $X = 4$ and $Y = 8$ will satisfy the given conditions.

Problem 2. Determine the X and Y coordinates of the intersection point of two straight lines R and S whose coordinates are (6, 6) (-4, -1 1/2) and (5, 1 1/2) (-2, 5) respectively. Fig. 28-5.

Solution. Plot the given coordinates and draw the straight lines R and S. From the graph, the coordinates of the point of intersection are (2, 3).

The determination of the equations of the lines R and S is left for the student.

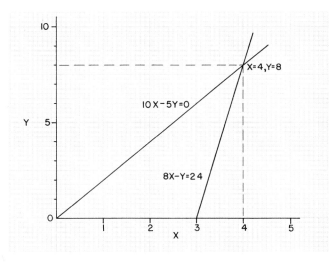

Fig. 28-4. *Prob. 1 Simultaneous Equations.*

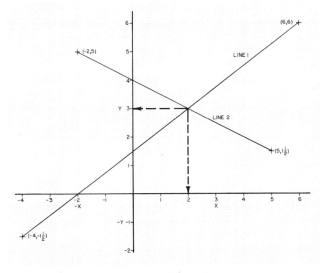

Fig. 28-5. *Prob. 2 Simultaneous Equations.*

28.5 Word Problems

Some difficulty may be encountered in determining the proper equations to fit specifications of a problem. In many instances, the graphical solution will aid the visualization of the specified conditions and thus help in the solution as well as showing the behavior of the variables involved in the specifications of a problem.

The following problem illustrates the principles of graphical addition and subtraction.

Problem. An empty 180-gallon tank is to be filled. Valve A is opened allowing water to enter at the uniform rate of 6 gallons per minute. Five minutes later, valve B is opened to allow water to enter at the uniform rate of 3 gallons per minute. Ten minutes after opening valve A, a drain valve C is opened allowing water to drain out at the uniform rate of 5 gallons per minute. All valves are open until the tank is full. Fig. 28-6.

1. Construct a graph showing the gallons of water in the tank at any instant.
2. Determine the time required to fill the tank.

Solution.

1. On suitable graph paper, let the ordinate represent gallons and the abscissa, time in minutes. Locate the 180 gallon line.
2. Plot the curves representing 6 gallons per minute, 3 gallons per minute, and -5 gallons per minute.
3. Add graphically the ordinates of the 3 and 6 gallon per minute curves.
4. Subtract the ordinates of the -5 gallon per minute curves.
5. Read time in minutes below the point where the combined curve intersects the 180 gallon line.
   Answer: 36.2 minutes to fill tank.

Part II

SCALES AND ALIGNMENT CHARTS

28.6 Computational Aids

Much of the computation in engineering requires repeated solution of certain formulas and equations. The engineer frequently uses the slide rule in solving these problems. Graphical aids, such as alignment charts, are often constructed for use by the less technically trained workers. These

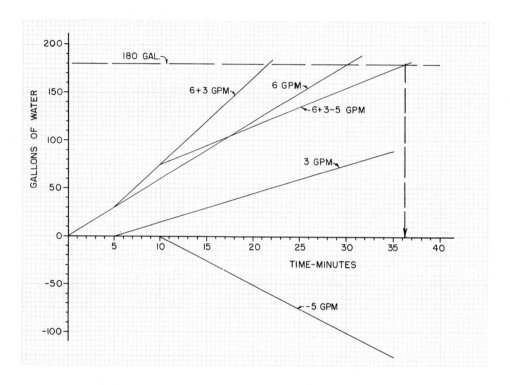

Fig. 28-6. *Graphical Addition and Subtraction.*

charts are advantageous because only a minimum of training instruction in their proper use is required.

28.7 Alignment Charts

The *Alignment Chart* is a fixed scale device used for the solution of one equation. The chart consists of two or more graduated and calibrated scales arranged in such a manner that solutions between given limits of an equation can be quickly determined. The arrangement of the scales will vary according to the type of equations being solved. The stems of the scales may be straight lines, curved lines, or combinations thereof.

Since alignment charts consist of scales properly arranged for the solution of equations, the chart designer must be familiar with the theory involved in scale design. He must also understand the meaning of mathematical terms such as linear and nonlinear equations, function, independent and dependent variable, limits, range and constant.

For example, in the equation $Y = X^2 - 3X + 10$, let X denote a quantity which varies from 0 to 8. This equation is nonlinear since it contains a variable whose exponent is greater than unity; Y is a function of X and is the dependent variable; X is the independent variable since it is to vary from 0 to 8; the values 0 and 8 are the lower and upper limits of the variable X; the algebraic difference between the upper and lower limits is the range of the variable; the fixed value 10 in the given equation is the constant.

GRAPHICAL MATHEMATICS—CHARTS

The discussion of charts will be limited to consideration of the simpler forms of charts. For a more complete explanation the student is advised to refer to the many excellent nomography texts available.

28.8 Scales

A *Scale* represents the relationship existing between two or more quantities varying according to some physical law. The scale consists of a series of marks, called *graduations*, laid down at predetermined distances along a line, called the *stem* or *axis*, for purposes of measurement or computation. Numerical values assigned to the major graduations are called *calibrations*. Scales are designed and constructed to represent *scale equations*.

Uniform Scale. The graphical representation of a first degree or linear scale equation will result in uniform spacing of the graduations along the stem of the scale. This type of scale is sometimes called arithmetic. Common examples, comparing magnitudes or amount of change, are the architects scale, engineers scale, yardstick, thermometer, and others.

Nonuniform Scale. If the scale equation contains a variable whose exponent is greater than one, or contains trigonometric or logarithmic functions of the variable, the graphical representation will show nonuniform or uneven spacing between graduations. This type of scale is called nonuniform, geometric, logarithmic, or functional.

Functional Scale. On a functional scale, the graduations are calibrated using the values of the variable. The distances from an initial point are laid off in proportion to the values of the *function* of the variable.

28.9 Scale Design

Before constructing a scale, the designer must consider such factors as:

Suitable scale lengths
Maximum and minimum spacing of graduations
Length of graduations and subdivisions
Proper calibrations

These factors are interrelated and depend to a great degree upon the type of problem and the desired reading accuracy. On all scales the subdivisions between major graduations should be such that the value of any desired point can be easily interpolated. See Fig. 28-7.

All scales must be completely labeled. A descriptive title and scale equation must be placed in a position which gives a balanced appearance to the finished drawing.

Scale Length. The scale length necessary for proper representation of the given maximum and minimum range of values of a variable may be controlled by the introduction of a constant, called a multiplier or scale modulus M, into the given scale equation.

The actual value of the scale modulus M can be determined by dividing the desired scale length S by the range of limits to be represented. The equations are:

For a uniform scale,

$$M = \frac{\text{scale length}}{\text{range of limits}} = \frac{S}{\text{max. limit} - \text{min. limit}}$$

For a logarithmic scale,

$$M = \frac{S}{\log(\text{max. limit}) - \log(\text{min. limit})}$$

For a functional scale,

$$M = \frac{S}{F(X_U) - F(X_L)}$$

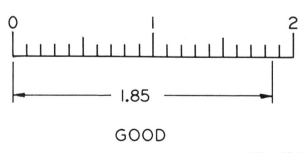
GOOD

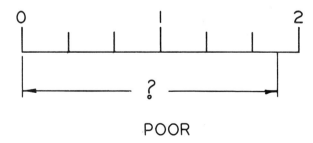
POOR

Fig. 28-7. *Scale Patterns.*

Scale Distance, Uniform Scale

Problem. To find the distance between any two points on a given uniform scale of length equal to 3 inches between limits of 2 and 8, Fig. 28-8.

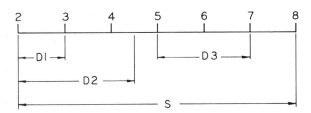

Fig. 28-8. *Distance on a Uniform Scale.*

Procedure:

1. Find the modulus, $M = \dfrac{S}{\text{range}} = \dfrac{3}{8-2} = 0.5$

2. On uniform scales the distance D between any two points on the scale can be found by multiplying the range of limits involved by the scale modulus M.

Using this equation, solve for any desired lengths, say from 2 to 3, 2 to 4.5, 5 to 7.
$D_1 = M(\text{range}) = 0.5\,(3-2) = 0.5$ inch
$D_2 = 0.5\,(4.5 - 2) = 1.25$ inches
$D_3 = 0.5\,(7 - 5) = 1$ inch

Scale Distance, Logarithmic Scale

Problem. To find the distance between any two points in one cycle of a logarithmic scale of length S, Fig. 28-9.

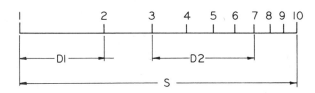

Fig. 28-9. *Distance on a Logarithmic Scale.*

Procedure:

1. Determine the modulus M. Let $S = 3$ inches.
$M = \dfrac{S}{\text{range}} = \dfrac{3}{\log 10 - \log 1} = 3$

2. $D_1 = M(\text{range}) = 3(\log 2 - \log 1)$
$D_1 = 0.903$ inch
$D_2 = 3(\log 7 - \log 3) = 3(.845 - .477)$
$D_2 = 1.10$ inches.

Scale Distance, Functional Scale

Problem 1. Construct a functional scale, 4.5 inches long, which represents $F(U) = U^2$ between the limits of 0 and 6. See Fig. 28-10.

Procedure:

1. Compute distances D for all values of the variable U necessary for desired graduations. Tabulate:

Variable U	0	1	2	3	4	5	6
$F(U) = U^2$	0	1	4	9	16	25	36
$D = M_u\,F(U)$	0	.125	.500	1.125	2.00	3.125	4.50

2. Plot distances D.
3. Letter calibrations, scale equation, and title.

The origin point of the scale or zero value of the function does not always appear on the actual scale. The following problem illustrates this case.

Problem 2. Construct a functional scale, 5 inches long, representing $F(W) = W + 2$. Let W vary from 0 to 20. See Fig. 28-11.

Procedure:

1. Solve for the modulus,
$M_w = \dfrac{S}{\text{range of function}} = \dfrac{5}{(20+2)-(0+2)} = \dfrac{5}{20} = 0.250$

2. Compute and tabulate distances D for all values of the variable W, necessary for the desired graduations.

W	0	1	2	4	5	10	12	15	18	20
$F(W)=W+2$	2	3	4	6	7	12	14	17	20	22
$D=M_w(W+2)$	.5	.75	1.0	1.5	1.75	3.0	3.5	4.25	5.0	5.5

3. Plot distances D.
4. Letter calibrations, scale equation and title.

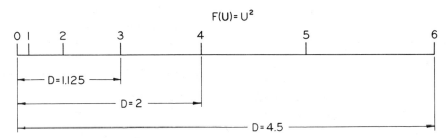

Fig. 28-10. *Distance on a Functional Scale.* $F(U) = U^2$

GRAPHICAL MATHEMATICS—CHARTS

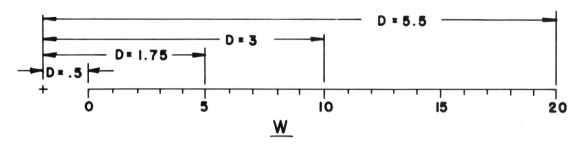

Fig. 28-11. *Functional Scale.* $F(W) = W + 2$

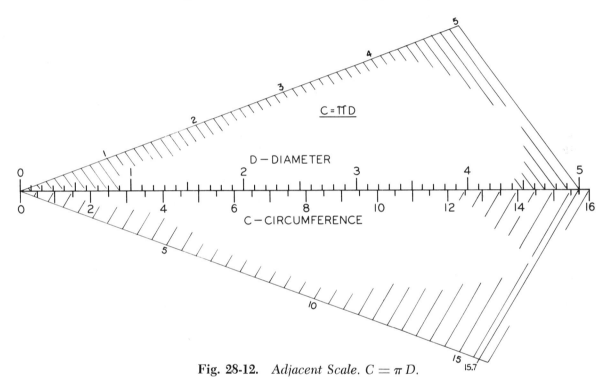

Fig. 28-12. *Adjacent Scale.* $C = \pi D$.

28.10 Adjacent or Conversion Chart

An adjacent scale, as the name implies, consists of two scales lying on opposite sides of a common stem. This chart is used to convert data from one unit to another, such as Inches to Centimeters, degrees Fahrenheit to degrees Centigade, and so on. The scale equation is of the form $F(X) = F(Y)$.

28.11 Adjacent Scale Construction

Problem. To construct a 6-inch adjacent scale chart for the equation $C = \pi D$, where D varies from 0 to 5 inches, and C is the circumference in inches. Graduate the D scale to tenths of inches and C to half inch divisions. Fig. 28-12.

Procedure:

1. Inspection of the given equation $C = \pi D$ shows it to be linear and therefore the C and D scales are uniform. The range of the independent variable D is from 0 to 5, therefore C must vary from 0 to 15.7.
2. Graduate each side of the stem, Fig. 28-12, by the method shown in Fig. A-1 of the Appendix.
3. Letter the calibrations, title and scale equation.

28.12 Parallel Scale Theory

Equations of the form $U \pm V = W$ and $\log U \pm \log V = \log W$ can be graphically represented by three graduated parallel scales placed so that a straight line connecting two of the scales cuts the third scale at a point which satisfies the given scale equation. In this construction the two outside scales, located some convenient distance apart, represent the independent variables and the third scale, representing the dependent variable, is placed so that the conditions of the scale equation are satisfied. Scale moduli for the independent variable scales can be determined by the method shown in Art. 28.9. The modulus for the dependent variable scale can be found by an equation expressed in terms of the scale moduli of the independent variable scales.

After the two parallel scales for the independent variables are constructed, graduated, and calibrated, it is necessary to find the proper location of the dependent scale. The equations needed for the mathematical solution show that the location of the dependent scale is related to the modulus of, and the distance from an independent scale, and also to the distance between the two independent scales.

$$M = \frac{\text{Scale Length}}{\text{Range of Variable}} \quad \text{Eq. 1}$$

$$\frac{M_X}{A} = \frac{M_Y}{B} = M_Z \left[\frac{A + B}{AB}\right] \quad \text{Eq. 2}$$

$$A + B = \text{Width of Chart} \quad \text{Eq. 3}$$

$$M_Z = \frac{M_X M_Y}{M_X + M_Y} \quad \text{Eq. 4}$$

Where M_X, M_Y, M_Z are the moduli of the three scales and A, B are the distances between the dependent and independent scales. For the derivation of these equations, the student should refer to available texts on Nomography.

28.13 Construction of Parallel Scale Charts — Addition

Problem. Construct a 4 x 6 parallel scale chart for the equation $P = C + 2M$, where P equals selling price, C is wholesale cost and M is the

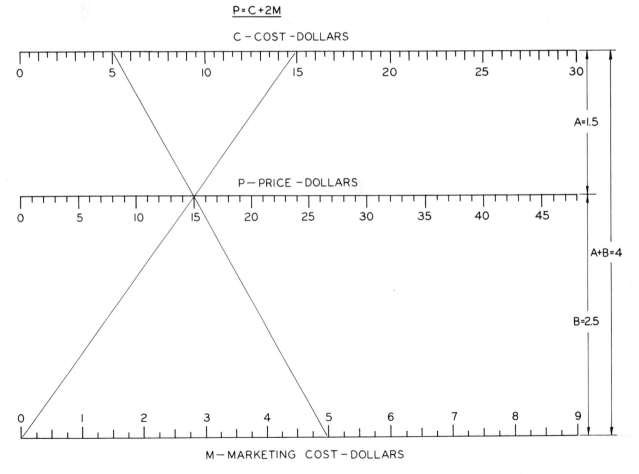

Fig. 28-13. *Parallel Scale—Addition.*

marketing cost. Let C vary from 0 to 30 dollars, and M vary from 0 to 9 dollars. Graduate C to 1/2 dollars, M to 1/4 dollars and P to dollars. Fig. 28-13.

Procedure: Semimathematical Method

1. Construct 2 parallel 6-inch stems, 4 inches apart, for the independent variables C and M. Graduate and calibrate these scales.
2. Determine value of M_C:

$$M_C = \frac{\text{scale length}}{\text{range}} = \frac{6}{30} = .20$$

3. Determine value of M_M:

$$M_M = \frac{6}{2(9)} = \frac{6}{18} = .33$$

4. Locate dependent scale P:

$$\frac{M_C}{M_M} = \frac{A}{B} = \frac{.20}{.33} = .6$$

$$A = .6B$$

$$A + B = 4$$

$$.6B + B = 4$$

$$\therefore B = 2.5 \text{ inches and } A = 1.5 \text{ inches}$$

5. Solve the given equation for the range of P, and graduate and calibrate the P scale for these limits.
6. Add title, scale equation and any necessary explanations for using the chart.

28.14 Construction of Parallel Scale Charts — Subtraction

Problem. Construct a parallel scale chart for the equation $X = X_L - X_C$, where X is combined reactance, X_L is inductive reactance, and X_C is capacitive reactance. Let X_L vary from 0 to 300 ohms and X_C from 20 to 200 ohms. Graduate scales to units of 10. Diagram 4 inches wide and scales 6 inches long. Refer to Fig. 28-14.

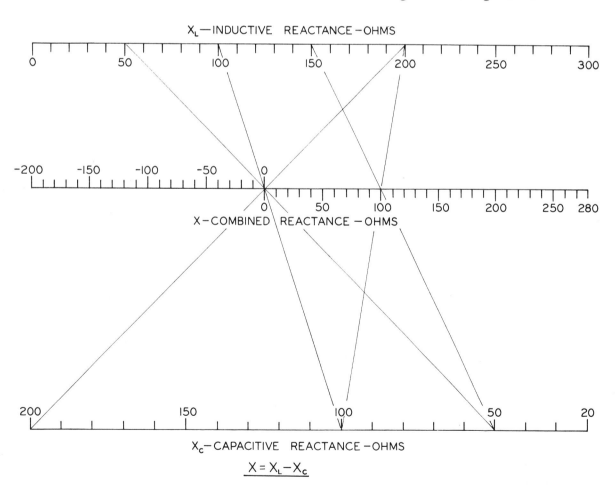

Fig. 28-14. *Parallel Scale—Subtraction.*

Procedure: Graphical Method
1. Construct two 6-inch stems 4 inches apart, for the independent variable scales X_L and X_C.
2. Graduate the X_L scale from 0 to 300.
3. Graduate the X_C scale from 20 to 200.

   Since X is determined by *subtracting* X_C from X_L, *the origin points of these two scales will be at opposite ends of the stems.*

4. Locate the dependent scale X.
   Graphically locate the zero point of the X scale by drawing straight lines between any quantities lying respectively on X_L and X_C whose difference is equal to zero. Establish the 100 mark on the X scale by subtracting quantities on X_L and X_C whose difference is equal to 100. A straight line through the zero and 100, parallel to X_L and X_C, establishes the stem location for the X scale.
5. Graduate and calibrate the X scale from −200 to +280.
6. Add lengths, title, scale equation and any necessary instructions for using the chart.

28.15 Construction of Parallel Scale Charts —Functional Scales

Equations of the form $XY = Z$ and $\frac{X}{Y} = Z$ may be plotted on parallel scale charts by reduction to a linear form. This is accomplished by taking the logarithms of both sides of the equation, thus: $XY = Z$ becomes $\log X + \log Y = \log Z$ and $\frac{X}{Y} = Z$ becomes $\log X - \log Y = \log Z$. Resulting plots will be functional scales with logarithmic spacing. The mathematical or graphical approach to the solution of these equations parallels that of the equations $X + Y = Z$ and $X - Y = Z$.

28.16 Construction of Parallel Scale Charts —Multiplication

Problem. Construct a 4 x 5 parallel scale chart for the equation $P = EI$, where P = power in watts, E = volts and I = amperes. Let E vary from 1 to 10 volts and let I vary from 1 to 100 amperes. Refer to Fig. 28-15.

Procedure: Semimathematical Method
After the given equation has been expressed in linear form as $\log P = \log E + \log I$, the procedure is the same as explained in Art. 28.15.

The computed moduli for I, E, and P scales are respectively 2.5, 5, and 1.67. Fig. 28-15 shows the computed cycle length distances on the I and P scales. The actual computations are left for the student.

28.17 Construction of Parallel Scale Charts — Division

Problem. Construct a 4 x 5 parallel scale chart for the equation $P = \frac{E^2}{R}$, where P is power in watts, E in volts and R in ohms. Let E vary from 10 to 250 volts and R vary from 10 to 100 ohms. See Fig. 28-16.

Procedure: Graphical
1. Locate 2 parallel stems 5 inches long and 4 inches apart.
2. Graphically graduate one stem from 10 to 100 and the other from 10 to 250. Since the equation is in the subtractive form, $\log X - \log Y = \log Z$, the 10 calibration must be on opposite ends of the 2 scales.
3. Locate the dependent scale by check points, such as 1, 10, 100 and 1000.
4. Graduate the cycle distances, calibrate selected graduations, label scales, and write the equation of the chart in the form $P = \frac{E^2}{R}$.

Cautions. The graphical method of layout in some cases reduces the time and effort needed in the preparation of a chart.

However, the graphical method of layout is recommended only when the chart equation is a readily solved simple form, and the arrangement and graduation of the scales are evident.

28.18 Construction of Parallel Scale Charts —Four or More Variables

Equations of the form:
$U+V+W=Z$ or $\log U + \log V + \log W = \log Z$
may be represented by a chart composed of two three-scale nomographs, one of the scales being common to both charts. The equation
$$U + V + W = Z$$
is rewritten in the form
$$U + V = S \text{ and } S + W = Z.$$
Scales are plotted for U, V, and S with no graduations on the S scale, since this scale serves only as a pivot line. Scales are then plotted

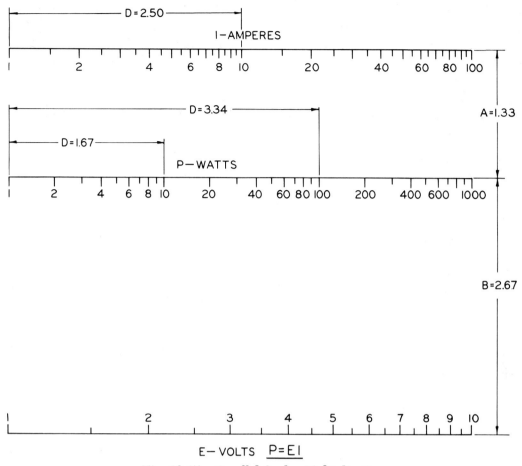

Fig. 28-15. *Parallel Scale—Multiplication.*

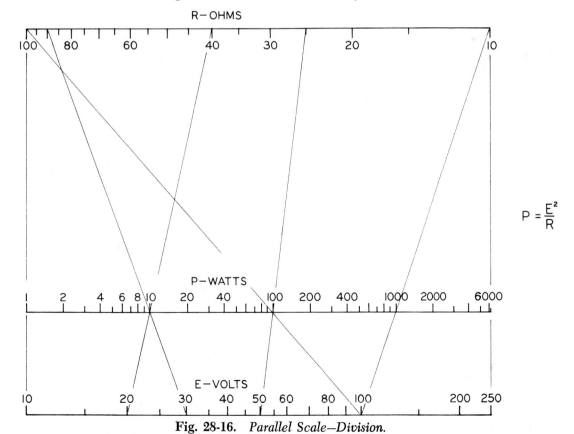

Fig. 28-16. *Parallel Scale—Division.*

for S, W, and Z, using the previously plotted S scale. Even though the S scale need not be graduated, it does represent a function and does have a scale modulus which must be computed and used in the plotting of the W and Z scales. The chart will be of a form similar to the illustration in Fig. 28-17. If values U, V, and W are given, the value of Z may be found by setting index line 1 across the given values of U and V, noting the point of intersection on scale S. Through this point of intersection and through the given value of W, a second index line 2 is drawn. The value of Z is indicated at the point of intersection of line 2 and the Z scale. Careful consideration should be given to the choice of scale arrangement in order to minimize magnification of errors. A combination Network and Alignment Chart is shown in Fig. 28-19.

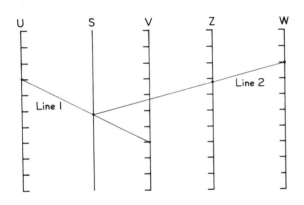

Fig. 28-17. *Combination Scales—4 Variables.*

Part III

N CHARTS

28.19 N or Z Charts

Many equations of the form $F(X) \cdot F(Y) = F(Z)$, after reduction to the form
$$\log F(X) + \log F(Y) = \log F(Z),$$
are best represented by a chart consisting of three parallel straight lines. However, in some cases, it may be advantageous to represent this form of equation by means of an N or Z chart, particularly if one or more of the functions are linear and if the limits of the variables are such that reading accuracy can be attained.

The N chart consists of two parallel scales, at some convenient distance apart, connected by a diagonal scale, Fig. 28-18. The parallel scales are graduated and calibrated to represent the dependent variable and one of the independent variables respectively. The diagonal, graduated and calibrated to represent the remaining independent variable, passes through the *zero points* of both parallel scales. The zero point of the diagonal coincides with the zero point of the scale representing the dependent variable.

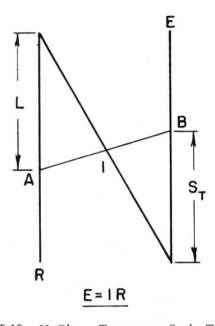

Fig. 28-18. *N Chart—Temporary Scale Theory.*

28.20 Graduation of Diagonal—Use of Temporary Scale

The diagonal scale of an N chart can often be graduated rapidly by using a temporary scale and a related focal point. The geometry involved is illustrated by the following problem.

Problem. Given an N chart for the equation $E = IR$. Determine the distance to a focal point for a temporary scale of random length.

Referring to Fig. 28-20, a temporary scale is located along the stem of the independent variable scale E. The zero point of the temporary scale coincides with the zero point of the scale E and the distance L is measured from the zero point of scale R. The distance L to the focus A is related to the temporary scale modulus M_T by the relationship

$$\frac{L}{M_T} = \frac{M_R}{M_E} \qquad \text{Eq. 1}$$

GRAPHICAL MATHEMATICS—CHARTS 317

*Equivalent Lengths L and L/D and Resistance Coefficient K

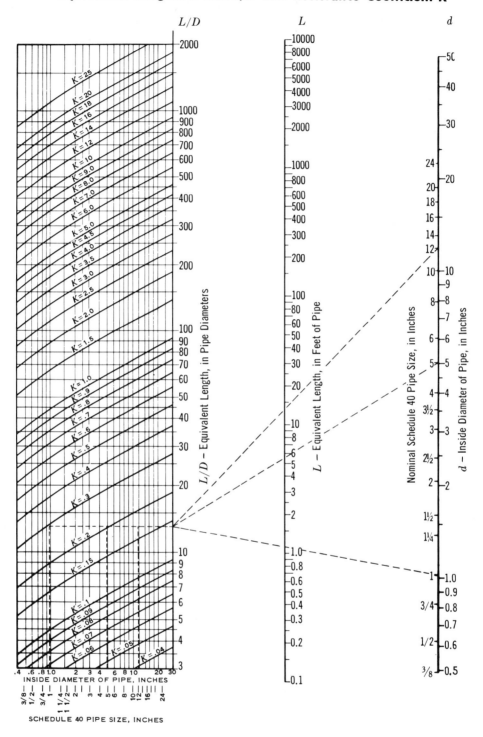

Problem: Find the equivalent length in pipe diameters and feet of Schedule 40 pipe, and the resistance factor K for 1, 5, and 12-inch fully-opened gate valves.

Solution				
Valve Size	1″	5″	12″	Refer to
Equivalent length, pipe diameters	13	13	13	
Equivalent length, feet of Sched. 40 pipe	1.1	5.5	13	Dotted lines on chart.
Resist. factor K, based on Sched. 40 pipe	0.30	0.20	0.17	

Fig. 28-19. *Combined Network Chart and Parallel Scale. (Courtesy Crane Co. Tech. Paper 410, Flow of Fluids)*

The relationship between the length of the temporary scale and the distance L can be found by combining the temporary scale equation and equation 1, as follows:

From equation 1, $M_T = L \left(\dfrac{M_E}{M_R}\right)$

Substituting for M_T in temporary scale equation,

$$S_T = M_T [F(I)] \qquad S_T = L \dfrac{M_E}{M_R} [F(I)]$$

Solving for L

$$L = \dfrac{S_T (M)_R}{M_E [F(I)]} \qquad \text{Eq. 2}$$

By substituting the proper values into equation 2, the focal distance L can be determined for any convenient length S_T of the temporary scale. After the diagonal has been graduated and calibrated, the temporary scale is erased. The length of the diagonal need not be known when using this temporary scale method.

28.21 Graduation of Diagonal —Mathematical

An N chart for the equation $E = IR$ is shown in Fig. 28-21. There are two equations, either one of which can be used to locate graduation marks along the diagonal. If it is desired to measure the graduation distances X from the zero point of the R scale, the equation becomes

$$X = \dfrac{K}{\left(\dfrac{M_E}{M_R}\right)[F(I)] + 1} \qquad \text{Eq. 1}$$

If the graduation distances are measured from the zero point of the E scale, the equation is

$$Y = \dfrac{K M_E [F(I)]}{M_E [F(I)] + M_R} \qquad \text{Eq. 2}$$

Where M_E, M_R = The moduli of the E and R scales

X = distance along diagonal from zero of R scale

Y = distance along diagonal from zero of E scale

K = length of the diagonal

$F(I)$ = some value of the current I.

The distance to any desired graduation can be found by substituting proper values of the

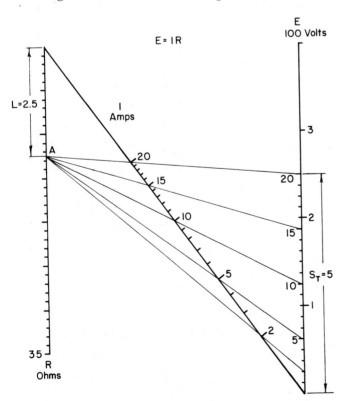

Fig. 28-20. N Chart—Graduation of Diagonal.

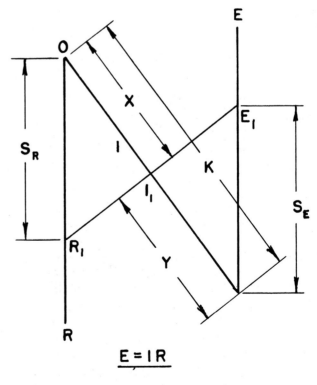

Fig. 28-21. N Chart—Mathematical Graduation of Diagonal.

GRAPHICAL MATHEMATICS—CHARTS

variables into these equations. The derivation for the equations can be found in texts pertaining to Nomography.

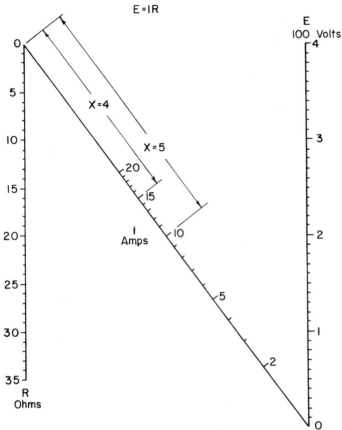

Fig. 28-22. N Chart—Ohms Law.

28.22 N or Z Chart Construction

Problem. Construct an N chart from Ohm's Law, $E = IR$. Let E vary from 0 to 400 volts; I, from 2 to 20 amperes; R, from 0 to 35 ohms. Fig. 28-22.

Considerations. The chart should be so designed that it can be constructed rapidly. Since the dependent variable E, located on one of the parallel lines, varies from 0 to 400 volts, a line 8 inches long will give the scale a good multiplier value. Since R has greater range than the variable I, let R be located on the remaining parallel line. For R, a scale length of 7 inches will provide a good multiplier value. The variable I is located on the diagonal whose length K should have a numerical value which will fit conveniently into the equation for graduation distances. In this problem a 10 inch diagonal is ideal. Note that Fig. 28-22 is half size.

Procedure:
1. Construct two parallel lines such that the diagonal joining them is 10 inches long.
2. On one parallel line graduate and calibrate the scale for the independent variable E.
3. On the remaining parallel line, graduate and calibrate the scale for the independent variable R.
4. Graduate mathematically and calibrate the scale for I on the diagonal from 2 to 20 amperes.
   $X =$ distance along diagonal from the zero point of scale R.

$$M_E = \frac{8}{400} = \frac{1}{50} = .02$$

$$M_R = \frac{7}{35} = \frac{1}{5} = .2;$$

$$K = \text{diagonal} = 10$$

Solving for X, $\quad X = \dfrac{K}{\left(\dfrac{M_E}{M_R}\right)[F(I)] + 1}$

F(I)	2	3	5	10	15	20
$\dfrac{M_E}{M_R}[F(I)]$	.2	.3	.5	1.0	1.5	2.0
$\dfrac{M_E}{M_R}[F(I)] + 1$	1.2	1.3	1.5	2.0	2.5	3.0
$X = K \div \dfrac{M_E}{M_R}[F(I)] + 1$	8.33	7.70	6.66	5.00	4.00	3.33

The complete table is not shown. Random values of F(I) have been chosen for illustration purposes.

5. Graduate and calibrate the diagonal. The distance Y could have been computed by using equation 2 of Art. 28.20.

28.23 Types of Alignment Chart Equations

The majority of the charts used in industry are solutions of the following types of equations:
   a. $F(X) = F(Y) \pm F(Z)$
   b. $F(X) = F(Y) \cdot F(Z)$

c. $F(X) = F(Y)/F(Z)$

d. $F(X) = F(X) \cdot F(Y) \pm F(Z)$

e. $F(Z) = (A + B) \cdot F(Y)$

The equations of the form given in (a) are very common and have been discussed in this text. The charts for the type of equations given in (b) and (c), when reduced to logarithmic form, are parallel scales with logarithmic graduations. These equations can also be represented by a Z or N chart. Multiplication and division can be performed on this style of chart without use of logarithmic scales.

The equation in (d) results in a chart with parallel Y and Z axes, the X axis usually curved. The equation (e) produces a chart which is similar to the chart of (d).

Thorough treatment of these charts, and many others, can be found in the many excellent texts devoted to Nomography.

PROBLEMS

Arithmetic and Algebra

1. Construct network charts for the graphical addition and subtraction of the following equations:
   a. $S + T = U$
      S varies from 0 to 50; T varies from 0 to 70.
   b. $A + B + C = D$
      A varies from 0 to 20; B varies from 0 to 30; C varies from 0 to 20.
   c. $J + K - L = M$
      J varies from 0 to 30; K varies from 0 to 25; L varies from 0 to 30.

2. Construct network charts for the graphical multiplication and division of the following equations:
   a. $E = RI$ R varies from 0 to 6
      I varies from 0 to 5 in 1/2 unit steps
   b. $T = \dfrac{S}{V}$ S varies from 0 to 10
      V varies from 2 to 8

3. Construct a network chart for the equation $W = Z^n$ where Z varies from 0 to 50 and n varies in tenths from 0.1 to 1.2.

4. Solve graphically the following equations:
   a. $4x - 2y = 2$ b. $2x + 3y = 13$
      $2x + y = 7$ $5x - 2y = 4$

5. The coordinates of the extremities of two straight lines are respectively $(-3, 0)$ $(6, 6)$ and $(0, 5)$ $(6, 3)$. Plot the lines and write the equations. Determine the x and y coordinates which satisfy the equations. What is the slope of each line?

6. Plot the graphs and determine the x and y coordinates which satisfy the equations:
   a. $x - 2y = 2$
      $3x - 2y = 6$
   b. $x^2 + y^2 = 25$
      $-x + y = 1$
   c. $x^2 + y^2 = 25$
      $4x^2 + 9y^2 = 144$

7. Three men, Joe, Pete and Stan, can do a piece of work in 8 days. Joe and Pete can do the same in 12 days, while Pete and Stan do the work in 10 days. How long would it take each man to do the work alone?

8. Jim requires 6 days to complete a job assignment while it takes 8 days for Henry to do the same assignment. Jim does 0.3 of the job on the first day, 0.1 the second day, 0.3 the third day, nothing on the fourth and fifth days, completing the job on the sixth day. Henry does 0.1 of the job the first day, is sick on the second and third days, does 0.3 on the fourth day, is sick again on the fifth and sixth days, does 0.4 on the seventh day and completes the job on the eighth day. How many days would be required if Jim and Henry worked together on a similar job if the above work history of each were repeated? If Jim and Henry worked together at their best rates, how many days would be required?

9. One pipe can fill a tank in 6 hours and a second can empty the tank in 4 hours. How much time is required to empty the tank, if it is full when both valves are opened?

10. One of two grades of ore contains 75% zinc and the other contains 25% zinc. How many pounds of each should be used to make a ton of mixture containing 50% zinc?

11. Mr. Smith walked from A to B at the rate of 2 mph. Returning, he ran part way at the rate of 6 mph and finished by walking 6

GRAPHICAL MATHEMATICS—CHARTS

minutes at his original rate. His roundtrip required 40 minutes.

Mr. Jones, leaving at the same time as Mr. Smith, walked at a uniform rate from A to B and back to A in 40 minutes.

a. At what times will the two men meet or pass?
b. What is the distance traveled by the men?
c. At what rate does Mr. Jones walk?

12. At what times of the day are the minute and hour hands of a clock together?
13. Bill and John began to swim the length of a 180-foot pool from opposite ends. Bill swims at 3 feet per second and John at 2 feet per second. If they swim back and forth for 24 minutes, find the number of times they meet in the pool assuming no loss of time for turning around.
14. An empty 200-gallon tank is to be filled. Valve A is opened, allowing water to enter at the rate of 5 gallons per minute. Ten minutes later valve B is opened, allowing water to enter at 3 gallons per minute. Five minutes after opening of valve B, a drain valve C is opened, allowing water to run out at the rate of 10 gallons per minute for 10 minutes.

a. Assuming uniform flow, construct a graph showing the gallons of water in the tank at any instant.
b. How many gallons are in the tank when valve C is closed?
c. Determine the time required to fill the tank.
d. If valve B had not been opened, how soon would the tank have been filled?

15. Three motorists, A, B, and C pass a known marker at speeds of 25, 40 and 50 miles per hour respectively. Driver B passes the marker 1 1/2 hours after A. How much later does Driver C pass the marker in order to overtake A at the same time that B overtakes A?
16. One of two grades of iron ore contains 60% iron and the other contains 20% iron. How many pounds each should be used to make a 2000-pound mixture containing 40% iron?
17. A man and a boy can row a boat 2 mph upstream and downstream at 4 1/2 mph. If they left at 2 o'clock, how far upstream can they travel if they must be back at the starting point at 5 o'clock? How much time is required for the return trip?

Alignment Charts

18. *Functional Scales*: Design functional scales for the following equations. Show any ecestions in a neat and orderly manner.

a. $f(T) = T^2$ T varies from 0 to 20
   Scale length, 8 inches
b. $f(W) = \log W$ W varies from 1 to 40
   Scale length, 6 inches
c. $f(V) = \sin V$ V varies from 10° to 180°
   Scale length, 6 inches
d. $f(S) = S^{1/2}$ S varies from 0 to 36
   Scale length, 6 inches

19. *Conversion Scales*: Design conversion scales for the following equation. Show any necessary calculations in a neat and orderly manner. If done graphically, show necessary construction.

a. $S = \sin S$ S, degrees, varies from 0 to 90.
   Scale length, 9 in.
b. $GMT = CST + 6$ GMT, Greenwich Mean Time, for one day or 24 hours.
   CST, Central Standard Time
   Indicate AM and PM on both scales. Scale length, 8 inches.
c. $P = .43H$; P, pressure, lbs. per sq. in.
   H, head of water, varies from 0 to 40 feet.
   Choose appropriate scale length.
d. $W = 550 P$; W, work, foot pounds per second
   P, power, horsepower varies from 0 to 10
   Choose appropriate scale length

20. Construct graphically a parallel scale chart for the equation $T + .6R = P$. T varies from 0 to 60; R varies from 20 to 140. Use 6 inch scales and chart width of 5.5 inches. Show graphical construction. Use good judgment in graduating and calibrating the scales. Solve mathematically for location of the dependent scale. Determine the multipliers of the 3 scales.
21. Construct graphically a parallel scale chart for the equation $V2 - V1 = AT$. V2, final velocity, varies from 0 to 24; V1, initial velocity, varies from 0 to 16; A is acceleration; let T, time, be a constant of one. Show graphical construction. Use good judgment in graduating and calibrating the scales. Use 8 inch scales and a chart width of 4 inches. Solve

mathematically for the dependent scale location.

22. Construct graphically a parallel scale chart for the equation $A = .26DL$, the surface area of a culvert. A, area in square feet; D, diameter, varies from 10 to 60 inches; L, length in feet, varies from 5 to 20 feet. Choose suitable scale lengths and chart width. Use good judgment in graduating and calibrating the scales.

23. Construct graphically a parallel scale chart for the equation $N = PD$. N, number of teeth; P, diametral pitch, varies from 2 to 24 teeth per inch; D, pitch diameter, varies from 3 to 30 inches. Choose suitable scale lengths and chart width. Use good judgment in graduating and calibrating the scales.

24. Construct parallel scale charts for the surface area and volume of any geometrical shapes. Refer to handbooks for the equations.

25. Construct a parallel scale chart for a problem assigned by your instructor.

26. *N Charts*: Construct N charts for the following equations. Show construction. Use judgment in graduating and calibrating the scales.

    a. $S = 1/2 AT^2$ S, distance, varies from 0 to 32,000; A, acceleration, varies from 10 to 40; T, time, varies from 15 to 40. Use 8″ parallel scales and a 10″ diagonal scale.

    b. $A = \dfrac{BH}{2}$ A, area, varies from 0 to 200 square inches; B, base, varies from 10 to 40 inches; H, altitude, varies from 5 to 40 inches. Use 5 inch parallel scales and a 6 inch diagonal.

    c. $N = PD'$ N, number of teeth, varies from 10 to 240; P, diametral pitch, varies from 3 to 30 teeth per inch; D′, pitch diameter, varies from 2 to 24 inches. Use 8 inch parallel scales and a 10 inch diagonal.

27. Construct an N chart for the following equations. Select reasonable values for the variables.

    a. $E = RI$; E = Voltage, R = Resistance, I = Current

    b. $P = EI$; P = Power, E = Voltage, I = Current

    c. $E = \dfrac{WV^2}{64.4}$; E = Kinetic Energy, W = Weight, V = Velocity

    d. $P = I^2R$; P = Power, I = Current, R = Resistance

    e. $A = \dfrac{BH}{2}$; A = Area, B = Base, H = Altitude

    f. $A = \dfrac{PN}{365}$ A = Amount of premium returnable
                   P = Annual insurance premium
                   N = Number of days prior to expiration

unit 29

Graphical Mathematics – the Calculus

Part I
DIFFERENTIATION

29.1 The Calculus

The calculus is defined as a highly systematic method of calculation by a special system of algebraic notation. The calculus provides methods of obtaining answers to many types of problems that otherwise would have to be solved by trial and error or other laborious methods. The calculus is invaluable in the solution of problems involving rate variation, growth, or change.

Differentiation and Integration are the two complementary operations of the calculus. Differentiation is used to find the *rate of change* of a variable, while Integration is used to find the *total change*.

If the relationship of the variables can be written in equation form to permit algebraic analysis, the calculus should be used for a mathematical solution. However, the solution of many problems becomes complicated because a great deal of the data used in engineering computations is either experimental or obtained by performance measurements. Under these conditions it becomes necessary to determine the basic equation of the data before any mathematical solution can be applied. Otherwise a solution by graphical, semi-graphical, or mechanical means may be more appropriate.

29.2 Function

If, for every change in one variable there is a corresponding change in a second variable, the second variable is considered to be a function of the first. This is true because the variation of the second variable depends upon the behavior of the first. For instance, a long rod is heavier than a shorter rod of the same diameter and material. Therefore, the weight is a function of the length, a relationship which can be expressed formally as $Y = F(X)$, where Y represents the weight and X represents the length of the rod.

29.3 Uniform Rate of Change

Rate of change may be uniform or nonuniform. Uniform rate of change is said to be linear and will plot as a straight line on rectangular coordinate paper, see Fig. 29-1, where the rate of change is represented by the slope of the line. The general equation for this line is $S = mT + C$, where T is the independent variable, S is the dependent variable, m is the slope of the line and C is the S intercept when $T = 0$. If there is no

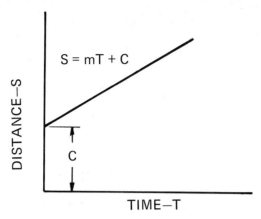

Fig. 29-1. *Uniform Rate of Change, Distance vs. Time.*

change of distance with respect to time, Fig. 29-2, the data will plot as a straight line parallel to the Time axis. The equation therefore becomes simply $S = C$.

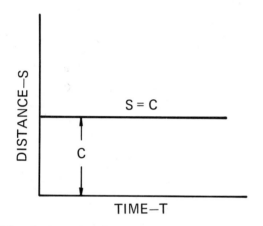

Fig. 29-2. *No Change, Distance vs. Time.*

29.4 Nonuniform Rate of Change

If the rate of change is nonuniform, the graph of distance as a function of time will be a curved line as shown in Fig. 29-3. For this type of data the average rate of change for a short interval, such as P to Q, can be found by drawing a chord from P to Q. The slope $\frac{S}{T}$ of this chord represents the change for this interval. As Q approaches point P as a limit, ΔT will approach zero as a limit. The chord approaches a tangent line through point P, while the angle θ approaches the value of angle ϕ as its limit.

$$\therefore \lim_{\Delta T \to 0} \frac{\Delta S}{\Delta T} = \frac{ds}{dt} \quad \text{or}$$

limit of tangent θ = tangent ϕ = $\frac{ds}{dt}$ = slope of tangent line at P. From this derivation comes the very important theorem: *The value of the derivative at any point on a curve is equal to the slope of the line tangent to the curve at that point.*

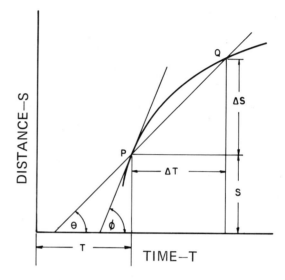

Fig. 29-3. *Tangent Line Definition.*

29.5 Derivative Curves

If the slopes of lines tangent to a curve are known, the derivative curve can be drawn by plotting the magnitudes of the slopes as ordinates at the respective abscissa values of the tangent points.

The slope at any point on a given curve is equal to the ordinate of the corresponding point on the next lower curve, called the *derived* or *derivative curve*.

Major drawbacks to this method of solution lie in the difficulty of determining accurate tangents and the necessity of computing the slope values of these tangents. These difficulties can be minimized by using a graphical approach called the *chord method*.

29.6 Tangent Line Construction

If a portion of a given curve is approximately circular, the location of a tangent point is the intersection with the curve of a line through the midpoint of, and perpendicular to, a chord. The tangent line will be parallel to the chord. See Fig. 29-4a.

If a portion of the arc approximates a parabola, the location of the tangent point and direction of tangent line may be found as shown in Fig. 29-4b.

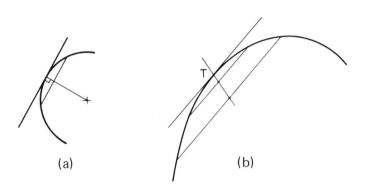

Fig. 29-4. *Tangents.*

29.7 Differentiation—Chord Method

Problem. Graphically differentiate the given volume curve Q. Construct the resulting area curve R. Fig. 29-5.

Procedure:

1. For reasons of clarity and convenience of projection, construct a set of axes for the derived curve directly below the given axes of the volume curve, both having the same X axes. The divisions on the Y and Y' axes are the same in length only. The calibration of this Y' axis will be discussed later.

2. Locate the tangent points T by the method of Fig. 29-4a.

3. Calibrate selected unit distances on the Y' scale as follows: The chord with the steepest slope, OH, will give the greatest ordinate distance. The slope of OH, as read from the graph, is $\Delta Y \div \Delta X = 20 \div 1$ or 20. This value of 20 could be used as the calibration for one unit distance $\Delta Y'$. However, the calibrations of the ordinate scales of both curves are then equal in length and value. This is not desirable in most problems since the resulting curve would be too flat. To overcome the flatness, a multiplier of 1/2 was used in the given example. Thus every unit distance $\Delta Y'$ on the Y' scale becomes equal to 10 square feet.

4. The ordinates of the derivative curve are located graphically by using a pole point P.

   The pole point is located to the left or right of the zero point of the derivative curve scale depending on whether the slopes of the derivative curve are positive or negative. In Fig. 29-5 the slopes are positive and therefore the pole point was located to the left at a multiple of ΔX distance from the origin point O' on the X' axis extended.

5. Through the pole point P draw rays parallel to the chords OH, HA, AB, etc., of the given curve until they intersect the Y' axis at points M, N, S, etc.

6. A line from M is drawn parallel to the X' axis until it intersects a straight line projected from point T on the arc OH, locating a point M' on the derivative curve. Points N', S', U', and W' are located in a similar manner. The desired curve R may now be constructed by drawing a smooth curve through the points.

29.8 Derivative Curve Scale

The overall length of the Y' scale for the derivative curve should be such that it can be readily graduated and calibrated to fit the existing grid on the graph. If the pole distance (P.D.) is selected at random, the resulting Y' scale often cannot be easily graduated and calibrated.

The most satisfactory method is to select a scale length, as long as possible, such that the derivative curve will fit the available space to best advantage and can be interpreted easily. The maximum reading of the Y' scale is estimated from the steepest slope of a tangent line to the given curve. The scale is graduated and calibrated to provide accurate reading and interpolation. In using this method it becomes necessary to compute the proper pole distance for the selected scale. An equation for computing P.D. can be expressed as:

$$\text{P.D.} = \frac{\text{Value of } \Delta Y}{(\text{Value of } \Delta X)(\text{Value of } \Delta Y')}$$

where: P.D. is in multiples of ΔX length, not value; ΔY and $\Delta Y'$ are of equal grid length.

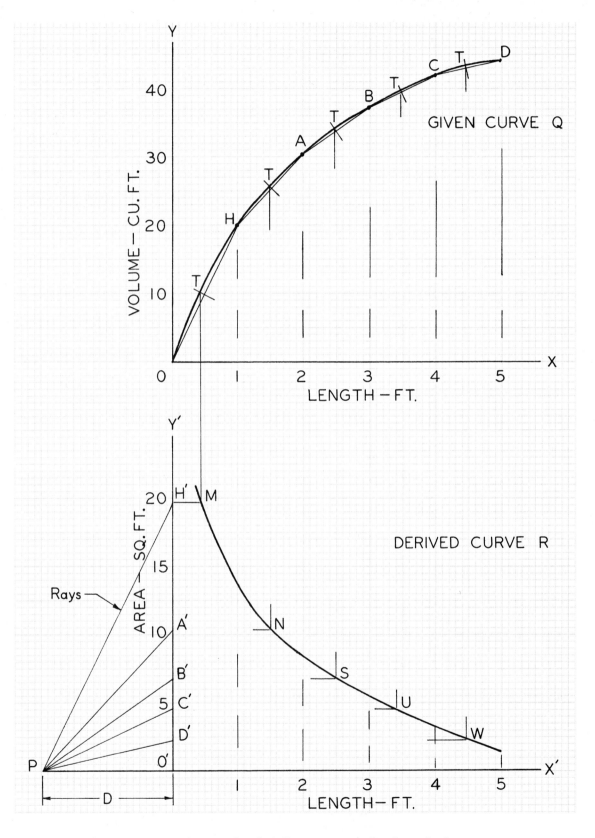

Fig. 29-5. *Graphical Differentiation—Chord Method.*

In Fig. 29-5, let ΔY = 20 cu. ft.; ΔX = 2 feet, then ΔY' = 10 sq. ft.

∴ P.D. in ΔX increments = $\dfrac{20 \text{ cu. ft.}}{(2 \text{ ft.})(10 \text{ sq. ft.})}$

P.D. = 1 ΔX increments.

The selected ΔX increment can be of any length since the pole distance is expressed in multiples of this increment.

For instance, in computing the above pole distance other values could have been used, such as:

P.D. = $\dfrac{30 \text{ cu. ft.}}{(1 \text{ ft.})(15 \text{ sq.ft.})}$

P.D. = 2 ΔX increments.

29.9 Differentiation of Broken Line Curve

The slopes of OP and PQ, Fig. 29-6, are constant and positive as shown by the rising straight lines from O to P to Q, etc. The slope of QR is equal to zero since there is no change in velocity and the slope of RS is constant but negative since the line has a downward slope. Since all the given slopes are constant, the derivative curves for the four segments will be horizontal lines.

The construction of the derivative curve, Fig. 29-6, is similar to the construction explained in Arts. 29.7 and 29.8.. Note that, since the velocities are constant throughout a given time segment, the acceleration remains constant for the entire corresponding time segment on the derivative curve.

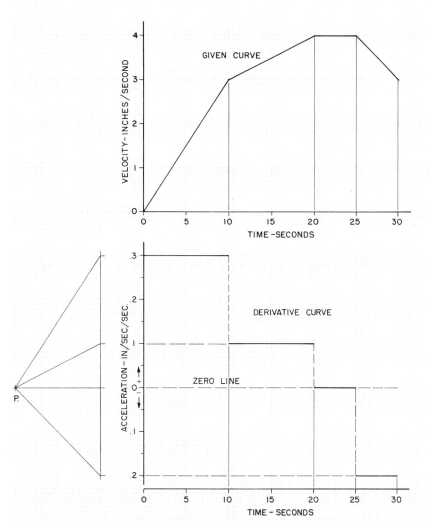

Fig. 29-6. *Graphical Differentiation—Broken Line Curve.*

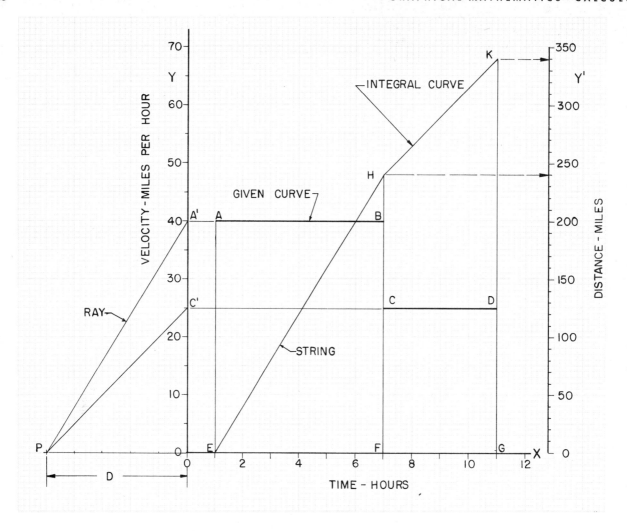

Fig. 29-7. *Graphical Integration Theory.*

Part II

INTEGRATION

29.10 The Integral Curve

Many of the most important applications of the calculus concern the determination of a function from a given derivative. This process is called Integration and the resulting function is called the Integral.

The graphical approach often offers a quicker means of solution for experimental data or functional relationships that do not lend themselves readily to mathematical equations; for example, the cut and fill areas on a dam or highway project, or the area of a hysteresis loop to determine circuit losses.

29.11 Graphical Method—String Polygon

A purely graphical method of solution for the area under a curve is possible by application of the Funicular or String Polygon method. This method is of value because the area can be determined accurately.

Problem. Determine the area under the broken line curve, ABCD. Fig. 29-7.

Procedure:
1. Project lines AB and CD horizontally until they intersect the Y axis at A' and C' respectively.
2. Locate a pole point P at a selected distance to the left of the origin 0 on the X axis extended.
3. Draw rays from pole point P to points A' and C'.

GRAPHICAL MATHEMATICS—CALCULUS

4. Draw a string from point E, under the given curve, parallel to ray PA′ until it intersects the ordinate FB extended at H.

5. From H draw a string, parallel to ray PC′ until it intersects the ordinate GD extended at K.

The resulting curve EHK is called the upper or integral curve and the ordinate FH represents the area under the line AB, while the ordinate GK represents the total area under the given line ABCD.

If the Y axis is calibrated in miles per hour, the X axis in hours, the area under the curve represents distance in miles as shown on the Integral ordinate Y′. These numerals can be used to prove that the values on the ordinate Y′ are equal to the area under the given curve between any desired limits of time. The given construction is a graphical verification of the Area Law which states: *The difference in the length of the ordinates in any continuous curve equals the total net area between the corresponding ordinates in the next lower curve.*

29.12 Graphical Integration—Ray Polygon

Let it be required to find the area under the curve, $Y = F(X)$, between the limits (S) and (T) as shown in Fig. 29-8.

Vertical strips are formed by drawing the ordinates Y_0, Y_1, Y_2, etc. The spacing between the ordinates need not be uniform but will depend on the slope of the given curve. Closer spacing should be used for a curve with a steep slope.

Determine the mean ordinate for each vertical strip by drawing horizontal lines AB, CD, and so on, through the arc between two consecutive ordinates so as to form two "triangles" of equal area, as shown by line AB between ordinates Y_0 and Y_1, Fig. 29-8. If the triangular areas are small, the mean ordinate can be approximated with fair accuracy since the eye is responsive to changes in the small area.

The horizontal lines AB, CD, EF, GH transform the given curve into a broken line or step curve. From this point on, the procedure of construction of the broken line upper curve is the same as described in Art. 29.11.

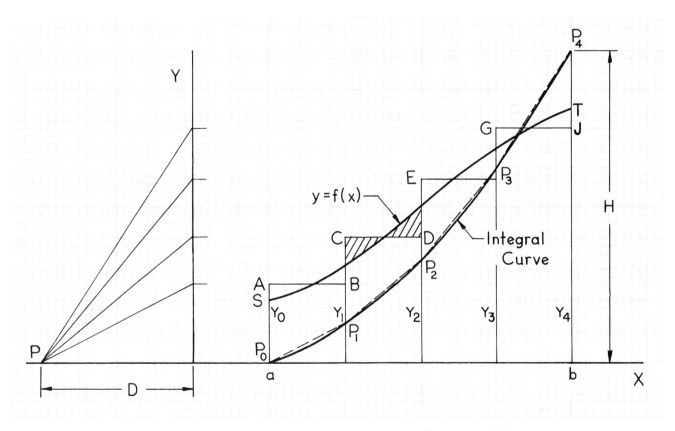

Fig. 29-8. *Graphical Integration—String Polygon Method.*

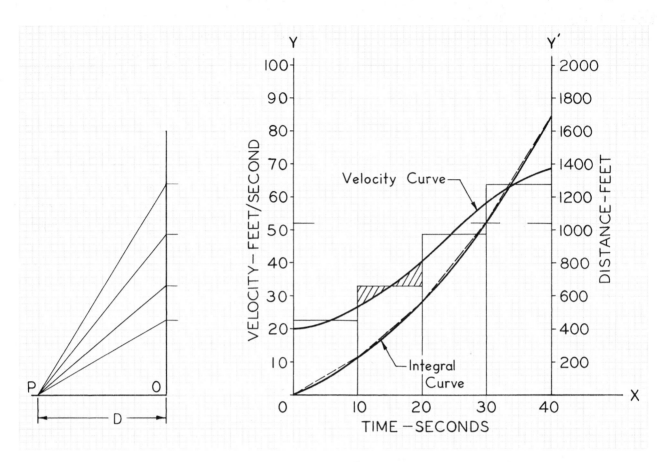

Fig. 29-9. *Calibration of Integral Scale.*

A smooth curve drawn through the ends of the straight line strings forms the integral curve. The final area indicated by the integral curve, between the limits S and T, is the product of the pole distance P.D. and the length of ordinate H. In order to determine the actual numerical value of the area at any point on the integral curve, values must be assigned to the X and Y axes. The principles involved will be discussed in the following article.

29.13 Calibration of Integral Scale

A curve representing the velocity in feet per second of a body during a 40 second interval is shown in Fig. 29-9. Construct an integral curve showing the total distance traveled and determine the distance traveled in 30 seconds.

The construction of the integral curve follows the procedure outlined in Art. 29.11.

In order to conserve space the integral curve is superimposed on the given velocity curve. The ray diagram has been moved away from the Y axis in order to leave space for the graduations and title of the velocity scale.

The time calibration on the X axis is common to both curves. The calibration for any $\Delta Y'$ distance on the scale for the integral curve can be determined as follows:

A suitable scale for an integral curve can be determined since the maximum value to be plotted is equal to the total area under the given curve. The approximate area under the curve will equal its base length X multiplied by the *mean ordinate* of the curve. After determining the approximate maximum value to be plotted, a scale unit can be selected to make the range of the vertical scale fit the available space and also permit logical graduation and calibration of the existing grid.

Applying this principle to the problem, Fig. 29-9, the product of the estimated mean ordinate, 35 f.p.s., and the total time interval, 40 seconds, is 1400 feet. This area can be located reasonably by making each major grid division 200 feet.

29.14 Integral Scale vs. Pole Distance

The required pole distance can be computed, as follows

$$\text{P.D.} = \frac{\Delta Y'}{(\Delta Y)(\Delta X)}$$

where $\Delta Y'$ = value of an increment of the ordinate integral scale

ΔY = value of an equal increment of the ordinate scale of the given curve. Note that the $\Delta Y'$ and ΔY increments are of *equal length*.

ΔX = value of an increment on the X axis of the given curve.

P.D. = number of ΔX length increments.

For example, referring to Fig. 29-9, the computation for P.D. is

$\Delta Y' = 400$ feet, $\Delta Y = (20$ ft./sec.$)$ and $\Delta X = 10$ seconds

$$\text{P.D.} = \frac{400 \text{ feet}}{(20 \text{ ft./sec.})(10 \text{ seconds})}$$

P.D. = 2 increments of ΔX

With the pole distance determined, the pole point can be located and the procedure of Art. 29.11 used to complete the integral curve.

To determine the distance traveled during 30 seconds, draw a horizontal line to the distance scale from the point on the integral curve directly above T = 30 seconds. This indicates a value lying between 1000 and 1200 feet. Graduating this distance more finely will give an answer of 1040 feet of travel for the elapsed time of 30 seconds.

29.15 Constant of Integration

All integral curves illustrated in this unit pass through the origin point of the X and Y axes. This is not true for all curves since the position of the curve with respect to the X axis depends on either known or assumed initial conditions.

If the equation $Y = 3X$ is plotted, the resulting straight line will pass through the origin. Not so, however, for the equation $Y = 3X + 2$ since it will cross the Y axis at 2 when $X = 0$. The shape of the plotted curves will not change and the two straight lines will be parallel. These equations are of the form $Y = mX + C$, the general equation for a straight line. The constant C controls the location of the line with respect to the X axis and is called the constant of integration.

When a curve is integrated by calculus and the limits are not specified, a constant of integration C must be included so that its position is fixed with respect to the X axis.

29.16 Semigraphical Methods

Problems which require finding the value of an area, but do not require an integral curve, can be solved by semigraphical methods. Values of the areas can be determined by applying one of the following several rules of approximation.

Trapezoidal Rule:

$$A_T = \Delta X \left[\frac{Y_0 + Y_n}{2} + Y_1 + Y_2 + Y_3 + \cdots Y_{n-1} \right]$$

Durand's Rule:

$$A_D = \Delta X \, [0.4(Y_0 + Y_n) + 1.1 \, (Y_1 + Y_{n-1}) + Y_2 + Y_3 + \cdots + Y_{n-2}]$$

Simpson's Rule:

$$A_S = \frac{\Delta X}{3} \, [(Y_0 + Y_n) + 4(Y_1 + Y_3 + Y_5 + \cdots + Y_{n-1}) + 2(Y_2 + Y_4 + Y_6 + \cdots + Y_{n-2})]$$

Weddle's Rule:

$$A_W = \frac{3}{10} \Delta X (Y_0 + 5Y_1 + Y_2 + 6Y_3 + Y_4 + 5Y_5 + Y_6)$$

These rules are semi-graphical because data scaled from a graph is substituted into the equation. Your instructor can supply further information concerning the use and relative accuracy of these equations.

29.17 Mechanical Methods

The area under a curve may be found by using a mechanical integrator called a *Planimeter*. If the curve is traced with the planimeter, the area is automatically recorded on dials. The most common planimeter, Fig. 29-10, has a pole arm 4 inches long and a measuring wheel whose circumference is 2.5 inches. One revolution of the wheel will equal 4 x 2.5 or 10 square inches. The dial, the wheel and vernier are so graduated that a reading to the nearest one hundredth of a square inch is possible. For accurate results, it is recommended that an average of several measurements be used.

Other types of instruments are the Amsler Integrator, the Integraph, and Linear Planimeters.

29.18 Applied Problem

An automobile is accelerated from a standstill. The resulting velocities in feet per second for each second of elapsed time are given below. See Fig. 29-11 for problem solution.

Time in Seconds	0	1	2	3	4	5	6	7	8	9	10
Velocity in ft./sec.	0	21	38	51	61	69	75	80	83	86	88

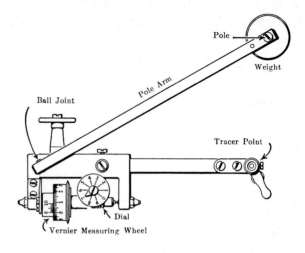

Fig. 29-10. *Planimeter. (Courtesy Keuffel and Esser Co.)*

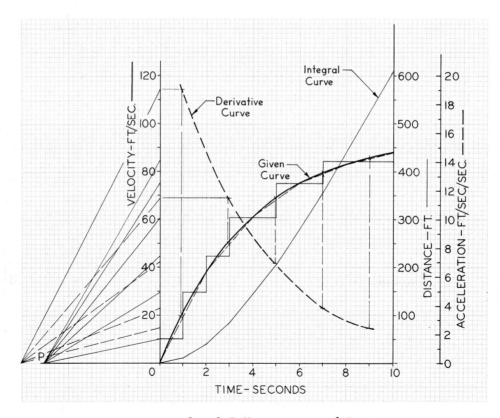

Fig. 29-11. *Combined Differentiation and Integration.*

1. Construct an acceleration curve and a distance curve.
2. What distance is traveled in 10 seconds?
3. What is the acceleration at the end of 3.5 seconds?

29.19 Summary

Problems, which involve methods of the calculus, may be roughly divided into one of several classifications:

1. If the algebraic equations involved in the problem are fully known, the underlying laws are also known. Problems of this type can be solved accurately by mathematical methods of the calculus.
2. If the exact equation is not known but data to plot the function is available, semigraphical methods may be applied to good advantage.
3. If the problem involves experimental data for which no law or equation is known, graphical methods or approximate analytical methods are the best approach to the solution. Sometimes the graphical and analytical methods can be advantageously combined.
4. If the problem involves a known equation, whose analytical solution is too time-consuming or too complicated, graphical methods are invaluable.

PROBLEMS

Differentiation

1. a. Plot the given information and draw a smooth curve.
    b. Construct a derivative curve.

X	0	4	8	12	16	20	24	28	32	36
Y	10	17	22	26	29	31	32	31	29	22

2. An automobile accelerated from a standstill attained the following velocities (in feet per second) at the end of various time intervals in seconds.

Time	0	1	2	3	4	5	6	7	8	9	10
Velocity	0	21	38	51	61	69	75	80	83	86	88

    By graphical differentiation determine the acceleration at any instant.

3. The temperature drop in a rapid cooling process following pasteurization of milk, is dependent on refrigerant effectiveness and the distance the milk flows at a uniform rate over cooling coils. Using the given data, plot a smooth curve showing the temperature change with respect to distance traveled. Construct a derivative curve showing the rate of temperature drop per foot of travel.

X axis, Distance in Ft.								
0	2	4	6	8	10	12	14	16
Y axis, Temperature in degrees F								
106	105	85	67	56	48	43	39	37

4. Work done during a nine-minute interval is shown by the tabular data. Plot the curve representing the given data. From this curve, derive the power curve. Express power, the rate of doing work, in terms of horsepower. (1 hp = 33,000 ft.lbs./min.)

Time, min.	1	2	3	4	5	6	7	8	9	10
Work, ft.lb × 1,000	94	90	84	76	68	58	48	40	34	30

5. The relationship of the velocity of a decelerating body to elapsed time T is shown by the tabular data.
    a. Plot the given information.
    b. Derive the curve showing the rate of deceleration at any time T.

Time Min.	Vel. Ft./Min.	Time Min.	Vel. Ft./Min.
0	90	10	24
1	80	12	20
2	70	14	17
3	61	16	14
4	53	18	12
5	46	20	11
6	40	22	10
8	30		

6. A cam follower on a rotating cam is at the tabulated distances from center, in inches, at intervals of one second during one complete revolution of the cam. Plot the given data. Construct the velocity and the acceleration curves.

Time Sec.	Dist. Inches	Time Sec.	Dist. Inches
0	1.0	9	3.0
1	1.1	10	3.0
2	1.6	11	3.0
3	1.9	12	2.7
4	1.9	13	1.8
5	1.9	14	1.2
6	2.0	15	1.0
7	2.3	16	1.0
8	2.8		

7. Creative Problems.
   a. Design a problem showing the relationship between the volume, area, and height or depth of a cylindrical container whose axis is horizontal. Supply required data and apply graphical differentiation.
   b. Design a problem showing distance, velocity, and acceleration realtionships for a uniformly falling body.
   c. Design a problem illustrating distance, velocity, and acceleration relationships for a complete rocket trajectory.

Integration

8. Water flows at an average velocity of 2 m.p.h. past a certain point in a river. The depth of the water is tabulated at various stations measured from one bank across the river.

   a. Construct an integral curve to determine the cross-sectional area of the river.
   b. Show a calibrated scale to give the amount of flow in cubic feet per second.

Station (ft.)	Depth (ft.)	Station (ft.)	Depth (ft.)
0	0	60	12.8
10	4.5	70	13
20	8.0	80	12
30	10	90	8.5
40	11	100	0
50	12		

9. Refer to Problem 2. Plot the given relationship and determine the distances traveled during each elapsed second.

10. Plot an acceleration curve from the equation for a freely falling body, $g = 32.2$ ft./sec.^{2}. Plot the relationship for a period of 10 sec. Using graphical integration determine the distance-time relationship at each second for the falling body.

11. Fig. 29-12. A plot of land is bordered by a creek, two highways, a street, a filling station and a railroad right of way (R.O.W.). Using graphical calculus, divide the land into four equal subdivisions whose east and west boundaries are parallel to Highway 60. Scale of map: $1'' = 100'$.

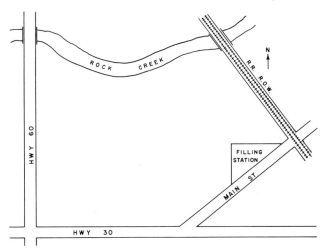

Fig. 29-12.

12. The velocity, V, of a moving object at various times, T, is given in the table. Determine graphically the total distance, S, traveled. Assume S = 0 when T = 0.

T, Sec.	0	3	5	10	15	18	20	25
V, ft./sec.	25	19	16.5	12	8.5	7	6	4

13. A plot of a given curve is a semicircle having positive Y ordinate values and a 6-inch diameter coincident with the X axis. The X axis units range from 0 to 60 feet; the Y axis units range from 0 to 90 feet. Find the area under the given curve by:

    a. Trapezoidal Rule
    b. Simpson's Rule
    c. Planimeter
    d. Graphical Integration

14. A plot of ground is a right triangle whose base is 100 feet and altitude is 60 feet. Using graphical calculus, subdivide the given plot of ground into three equal parts whose boundaries are perpendicular to the base.

15. By graphical integration, determine the area of a 3-inch radius semicircular surface. The graduated and calibrated area scale, 6 inches long, ranges from 0 to 30 square inches.

Appendix

CONTENTS

Table No.
1. Instrument Line Alphabet 339
2. Geometrical Constructions 340
   - Fig. A-1 Divide a given line into a desired number of equal parts.
   - Fig. A-2 Construct a circle through three points.
   - Fig. A-3 Transfer of an irregular figure from one position to another.
   - Fig. A-4 Construct a regular hexagon—3 methods.
   - Fig. A-5 Construct a regular pentagon.
   - Fig. A-6 Construct a circle tangent to two given lines.
   - Fig. A-7 Construct an arc tangent to a given circle and straight line.
   - Fig. A-8 Construct ellipses—3 methods.
   - Fig. A-9 Find the major and minor axes of an ellipse, given a pair of conjugate axes.
3. Table of Decimal Equivalents 342
4. Logarithms of Numbers 343
5. Trigonometric Functions 344
6. Areas of Plane Figures 345
7. Volumes of Solids 346
8. Equivalents 348
9. Electrical and Electronic Symbols 349
10. Twist Drill Sizes 350
11. Unified and American Screw Threads 351
12. American Standard Bolts 352
13. American Standard Cap Screws 354
14. American Standard Machine Screws 355
15. American Standard Washers 356
16. American Standard Lock Washers 357
17. American Standard Hex Nuts 358
18. American Standard Hex Slotted Nuts 359
19. Cotter Pins 360
20. Taper Pins 360

21	Setscrews, Heads and Points	361
22	Acme Threads	362
23	Square Threads	362
24	Running and Sliding Fits	363
25	Locational Clearance Fits	364
26	Locational Transition Fits	365
27	Locational Interference Fits	365
28	Force and Shrink Fits	366
29	Graphical Symbols	367
30	American Standard Wrought Iron and Steel Pipe and Taper Pipe Threads	368
31	Screwed and Flanged Fittings and Valve Sizes	369
32	Malleable Iron Screwed Fittings	370
33	Class 125 Cast Iron Flanged Fittings	371
34	Typical Drawn Thinwall Tubing	372
35	Typical Square and Rectangular Tubing	372
36	American Standard Square and Flat, Plain Taper, and Gib Head Keys	373
37	Pratt and Whitney Keys	373
38	Woodruff Keys	374
39	Bibliography	375

APPENDIX

1. INSTRUMENT LINE ALPHABET ANSI Y14.2

LINE   A: .030 to .038
WIDTH B: .015 to .022

	Width	
VISIBLE LINE	A	
HIDDEN LINE	B	
SECTION LINE	B	
CENTER LINE	B	
DIMENSION LINE EXTENSION LINE AND LEADERS	B / B	
CUTTING-PLANE LINE OR VIEWING-PLANE LINE	A / A	
BREAK LINES	A / B	
PHANTOM LINE	B	

2. GEOMETRICAL CONSTRUCTIONS

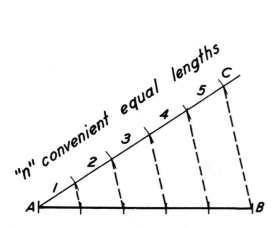

Fig. A-1. *Given line AB, to be divided into "n" equal lengths, n = 5.*

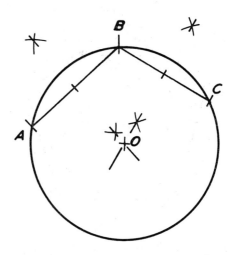

Fig. A-2. *Given points ABC, to locate the circle through the three points.*

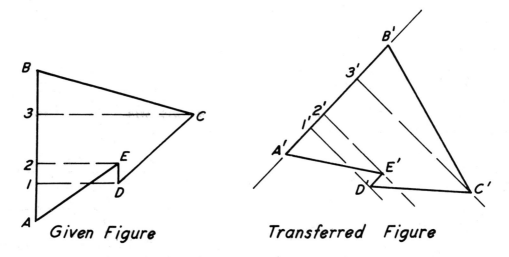

Fig. A-3. *To transfer position of a given figure by normal offsets.*

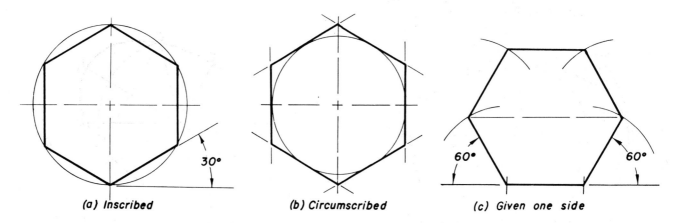

Fig. A-4. *To construct hexagons, given: a) cross diagonal; b) across flats; and c) one side.*

3. GEOMETRICAL CONSTRUCTIONS (Cont'd)

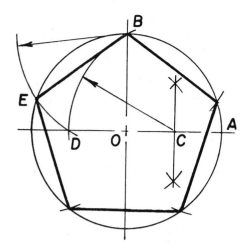

Fig. A-5. To inscribe a pentagon in a given circle.

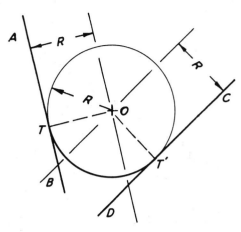

Fig. A-6. Given the straight lines AB and CD to locate tangent arc TT' of given radius, R.

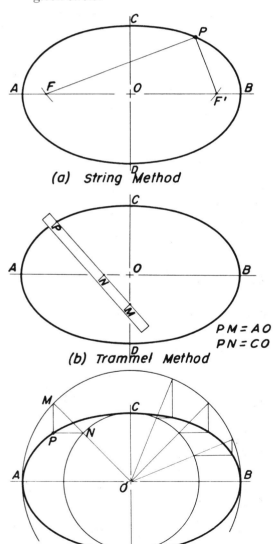

(a) String Method

(b) Trammel Method

PM = AO
PN = CO

(c) Concentric Circle Method

Fig. A-8. To construct true ellipses by three methods.

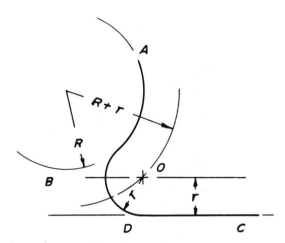

Fig. A-7. Given circle AB and straight line CD, to locate the tangent arc of radius r.

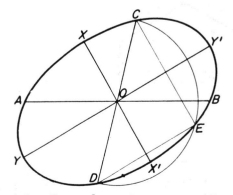

Fig. A-9. Given the conjugate axes AB and CD, to locate the major and minor axes, YY' and XX' of the ellipse.

3. TABLE OF DECIMAL EQUIVALENTS

Fraction	Decimal		Fraction	Decimal
1/64	.015625		33/64	.515625
1/32	.03125		17/32	.53125
3/64	.046875		35/64	.546875
1/16	.0625		9/16	.5625
5/64	.078125		37/64	.578125
3/32	.09375		19/32	.59375
7/64	.109375		39/64	.609375
1/8	.125		5/8	.625
9/64	.140625		41/64	.640625
5/32	.15625		21/32	.65625
11/64	.171875		43/64	.671875
3/16	.1875		11/16	.6875
13/64	.203125		45/64	.703125
7/32	.21875		23/32	.71875
15/64	.234375		47/64	.734375
1/4	.25		3/4	.75
17/64	.265625		49/64	.765625
9/32	.28125		25/32	.78125
19/64	.296875		51/64	.796875
5/16	.3125		13/16	.8125
21/64	.328125		53/64	.828125
11/32	.34375		27/32	.84375
23/64	.359375		55/64	.859375
3/8	.375		7/8	.875
25/64	.390625		57/64	.890625
13/32	.40625		29/32	.90625
27/64	.421875		59/64	.921875
7/16	.4375		15/16	.9375
29/64	.453125		61/64	.953125
15/32	.46875		31/32	.96875
31/64	.484375		63/64	.984375
1/2	.5		1	1.

APPENDIX

4. LOGARITHMS OF NUMBERS

No.	0	1	2	3	4	5	6	7	8	9	No.	0	1	2	3	4	5	6	7	8	9
10	0000	0043	0086	0128	0170	0212	0253	0294	0334	0374	55	7404	7412	7419	7427	7435	7443	7451	7459	7466	7474
11	0414	0453	0492	0531	0569	0607	0645	0682	0719	0755	56	7482	7490	7497	7505	7513	7520	7528	7536	7543	7551
12	0792	0828	0864	0899	0934	0969	1004	1038	1072	1106	57	7559	7566	7574	7582	7589	7597	7604	7612	7619	7627
13	1139	1173	1206	1239	1271	1303	1335	1367	1399	1430	58	7634	7642	7649	7657	7664	7672	7679	7686	7694	7701
14	1461	1492	1523	1553	1584	1614	1644	1673	1703	1732	59	7709	7716	7723	7731	7738	7745	7752	7760	7767	7774
15	1761	1790	1818	1847	1875	1903	1931	1959	1987	2014	60	7782	7789	7796	7803	7810	7818	7825	7832	7839	7846
16	2041	2068	2095	2122	2148	2175	2201	2227	2253	2279	61	7853	7860	7868	7875	7882	7889	7896	7903	7910	7917
17	2304	2330	2355	2380	2405	2430	2455	2480	2504	2529	62	7924	7931	7938	7945	7952	7959	7966	7973	7980	7987
18	2553	2577	2601	2625	2648	2672	2695	2718	2742	2765	63	7993	8000	8007	8014	8021	8028	8035	8041	8048	8055
19	2788	2810	2833	2856	2878	2900	2923	2945	2967	2989	64	8062	8069	8075	8082	8089	8096	8102	8109	8116	8122
20	3010	3032	3054	3075	3096	3118	3139	3160	3181	3201	65	8129	8136	8142	8149	8156	8162	8169	8176	8182	8189
21	3222	3243	3263	3284	3304	3324	3345	3365	3385	3404	66	8195	8202	8209	8215	8222	8228	8235	8241	8248	8254
22	3424	3444	3464	3483	3502	3522	3541	3560	3579	3598	67	8261	8267	8274	8280	8287	8293	8299	8306	8312	8319
23	3617	3636	3655	3674	3692	3711	3729	3747	3766	3784	68	8325	8331	8338	8344	8351	8357	8363	8370	8376	8382
24	3802	3820	3838	3856	3874	3892	3909	3927	3945	3962	69	8388	8395	8401	8407	8414	8420	8426	8432	8439	8445
25	3979	3997	4014	4031	4048	4065	4082	4099	4116	4133	70	8451	8457	8463	8470	8476	8482	8488	8494	8500	8506
26	4150	4166	4183	4200	4216	4232	4249	4265	4281	4298	71	8513	8519	8525	8513	8537	8543	8549	9555	9561	8567
27	4314	4330	4346	4362	4378	4393	4409	4425	4440	4456	72	8573	8579	8585	8591	8597	8603	8609	8615	8621	8627
28	4472	4487	4502	4518	4533	4548	4564	4579	4594	4609	73	8633	8639	8645	8651	8657	8663	8669	8675	8681	8686
29	4624	4639	4654	4669	4683	4698	4713	4728	4742	4757	74	8692	8698	8704	8710	8716	8722	8727	8733	8739	8745
30	4771	4786	4800	4814	4829	4843	4857	4871	4886	4900	75	8751	8756	8762	8768	8774	8779	8785	8791	8797	8802
31	4914	4928	4942	4955	4969	4983	4997	5011	5024	5038	76	8808	8814	8820	8825	8831	8837	8842	8848	8854	8859
32	5051	5065	5079	5092	5105	5119	5132	5145	5159	5172	77	8865	8871	8876	8882	8887	8893	8899	8904	8910	8915
33	5185	5198	5211	5224	5237	5250	5263	5276	5289	5302	78	8921	8927	8932	8938	8943	8949	8954	8960	8965	8971
34	5315	5328	5340	5353	5366	5378	5391	5403	5416	5428	79	8976	8982	8987	8993	8998	9004	9009	9015	9020	9025
35	5441	5453	5465	5478	5490	5502	5514	5527	5539	5551	80	9031	9036	9042	9047	9053	9058	9063	9069	9074	9079
36	5563	5575	5587	5599	5611	5623	5635	5647	5658	5670	81	9085	9090	9096	9101	9106	9112	9117	9122	9128	9133
37	5682	5694	5705	5717	5729	5740	5752	5763	5775	5786	82	9138	9143	9149	9154	9165	9165	9170	9175	9180	9186
38	5798	5809	5821	5832	5843	5855	5866	5877	5888	5899	83	9191	9196	9201	9206	9212	9217	9222	9227	9232	9238
39	5911	5922	5933	5944	5955	5966	5977	5988	5999	6010	84	9243	9248	9253	9258	9263	9269	9274	9279	9284	9289
40	6021	6031	6042	6053	6064	6075	6085	6096	6107	6117	85	9294	9299	9304	9309	9315	9320	9325	9330	9335	9340
41	6128	6138	6149	6160	6170	6180	6191	6201	6212	6222	86	9345	9350	9355	9360	9365	9370	9375	9380	9385	9390
42	6232	6243	6253	6263	6274	6284	6294	6304	6314	6325	87	9395	9400	9405	9410	9415	9420	9425	9430	9435	9440
43	6335	6345	6355	6365	6375	6385	6395	6405	6415	6425	88	9445	9450	9455	9460	9465	9469	9474	9479	9484	9489
44	6435	6444	6454	6464	6474	6484	6493	6503	6513	6522	89	9494	9499	9504	9509	9513	9518	9523	9528	9533	9538
45	6532	6542	6551	6561	6571	6580	6590	6599	6609	6618	90	9542	9547	9552	9557	9562	9566	9571	9576	9581	9586
46	6628	6637	6646	6656	6665	6675	6684	6693	6702	6712	91	9590	9595	9600	9605	9609	9614	9619	9624	9628	9633
47	6721	6730	6739	6749	6758	6767	6776	6785	6794	6803	92	9638	9643	9647	9652	9657	9661	9666	9671	9675	9680
48	6812	6821	6830	6839	6848	6857	6866	6875	6884	6893	93	9685	9689	9694	9699	9703	9708	9713	9717	9722	9727
49	6902	6911	6920	6928	6937	6946	6955	6964	6972	6981	94	9731	9736	9741	9745	9750	9754	9759	9763	9768	9773
50	6990	6998	7007	7016	7024	7033	7042	7050	7059	7067	95	9777	9782	9786	9791	9795	9800	9805	9809	9814	9818
51	7067	7084	7093	7101	7110	7118	7126	7135	7143	7152	96	9823	9827	9832	9836	9841	9845	9850	9854	9859	9863
52	7160	7168	7177	7185	7193	7202	7210	7218	7226	7235	97	9868	9872	9877	9881	9886	9890	9894	9899	9903	9908
53	7243	7251	7259	7267	7275	7284	7292	7300	7308	7316	98	9912	9917	9921	9926	9930	9934	9939	9943	9948	9952
54	7324	7332	7340	7348	7356	7364	7372	7380	7388	7396	99	9956	9961	9965	9969	9974	9978	9983	9987	9991	9996
No	0	1	2	3	4	5	6	7	8	9	No	0	1	2	3	4	5	6	7	8	9

5. TRIGONOMETRIC FUNCTIONS

ANGLE	SINE	COSINE	TAN	COTAN	ANGLE
0°	0.0000	1.0000	0.0000	Infin.	90°
1°	0.0175	0.9998	0.0175	57.290	89°
2°	.0349	.9994	.0349	28.636	88°
3°	.0523	.9986	.0524	19.081	87°
4°	.0698	.9976	.0699	14.301	86°
5°	.0872	.9962	.0875	11.430	85°
6°	.1045	.9945	.1051	9.5144	84°
7°	.1219	.9925	.1228	8.1443	83°
8°	.1392	.9903	.1405	7.1154	82°
9°	.1564	.9877	.1584	6.3138	81°
10°	.1736	.9848	.1763	5.6713	80°
11°	.1908	.9816	.1944	5.1446	79°
12°	.2079	.9781	.2126	4.7046	78°
13°	.2250	.9744	.2309	4.3315	77°
14°	.2419	.9703	.2493	4.0108	76°
15°	.2588	.9659	.2679	3.7321	75°
16°	.2756	.9613	.2867	3.4874	74°
17°	.2924	.9563	.3057	3.2709	73°
18°	.3090	.9511	.3249	3.0777	72°
19°	.3256	.9455	.3443	2.9042	71°
20°	.3420	.9397	.3640	2.7475	70°
21°	.3584	.9336	.3839	2.6051	69°
22°	.3746	.9272	.4040	2.4751	68°
23°	.3907	.9205	.4245	2.3559	67°
24°	.4067	.9135	.4452	2.2460	66°
25°	.4226	.9063	.4663	2.1445	65°
26°	.4384	.8988	.4877	2.0503	64°
27°	.4540	.8910	.5095	1.9626	63°
28°	.4695	.8829	.5317	1.8807	62°
29°	.4848	.8746	.5543	1.8040	61°
30°	.5000	.8660	.5774	1.7321	60°
31°	.5150	.8572	.6009	1.6643	59°
32°	.5299	.8480	.6249	1.6003	58°
33°	.5446	.8387	.6494	1.5399	57°
34°	.5592	.8290	.6745	1.4826	56°
35°	.5736	.8192	.7002	1.4281	55°
36°	.5878	.8098	.7265	1.3764	54°
37°	.6018	.7986	.7536	1.3270	53°
38°	.6157	.7880	.7813	1.2799	52°
39°	.6293	.7771	.8098	1.2349	51°
40°	.6428	.7660	.8391	1.1918	50°
41°	.6561	.7547	.8693	1.1504	49°
42°	.6691	.7431	.9004	1.1106	48°
43°	.6820	.7314	.9325	1.0724	47°
44°	.6947	.7193	.9657	1.0355	46°
45°	.7071	.7071	1.0000	1.0000	45°
ANGLE	COSINE	SINE	COTAN	TAN	ANGLE

6. AREAS OF PLANE FIGURES

Nomenclature
a, b, c, d — Lengths of Sides
A — Area
d, d_1, d_2 — Diameters
e, f — Lengths of Diagonals
h — Vertical Height or Altitude
l, l_1, l_2 — Length of Arc
L — Lateral Length or Slant Height
n — Number of Sides
θ — Number of Degrees of Arc
p — Perimeter
r, r_1, r_2, R — Radii

Right Triangle
$p = a + b + c$
$c^2 = a^2 + b^2$
$b = \sqrt{c^2 - a^2}$
$A = \dfrac{ab}{2}$

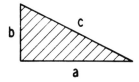

Equilateral Triangle
$p = 3a$
$h = \dfrac{a}{2}\sqrt{3} = .866\,a$
$A = a^2\,\dfrac{\sqrt{3}}{4} = .433\,a^2$

General Triangle
$s = \dfrac{a+b+c}{2}$
$p = a+b+c$
$h = \dfrac{2}{a}\sqrt{s(s-a)(s-b)(s-c)}$
$A = \dfrac{ah}{2}$
$A = \sqrt{s(s-a)(s-b)(s-c)}$

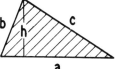

Square
$a = b$
$p = 4a$
$A = a^2 = .5e^2$
$e = a\sqrt{2} = 1.414\,a$

Rectangle
$p = 2(a+b)$
$e = \sqrt{a^2 + b^2}$
$b = \sqrt{e^2 - a^2}$
$A = ab$

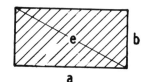

General Parallelogram or Rhomboid; and Rhombus
Rhomboid—opposite sides parallel
$p = 2(a+b)$
$e^2 + f^2 = 2(a^2 + b^2)$
$A = ah$

Rhombus—opposite sides parallel and all sides equal
$a = b$
$p = 4a = 4b$
$e^2 + f^2 = 4a^2$
$A = ah = \dfrac{ef}{2}$

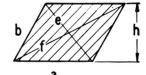

Trapezoid
$p = a+b+c+d$
$A = \dfrac{(a+b)}{2}\,h$

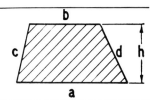

Trapezium
$p = a+b+c+d$
A = Sum of Areas of two major triangles
$A = \dfrac{(h_1 + h_2)\,g + f h_1 + j h_2}{2}$

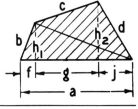

Regular Polygon
n = number of sides
$p = na$
$a = 2\sqrt{R^2 - r^2}$
$A = \dfrac{nar}{2} = \dfrac{na}{2}\sqrt{R^2 - \dfrac{a^2}{4}}$
$= n \times$ Area of each triangle

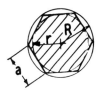

Circle
$p = 2\pi r = \pi d = 3.1416\,d$
$A = \pi r^2 = \dfrac{\pi d^2}{4} = .7854\,d^2$
$= \dfrac{p^2}{4\pi} = .07958\,p^2$

Hollow Circle or Annulus
$A = \dfrac{\pi}{4}(d_2^2 - d_1^2) = .7854\,(d_2^2 - d_1^2)$
$= \pi(r_2^2 - r_1^2)$
$= \pi\,\dfrac{d_1 + d_2}{2}(r_2 - r_1)$
$= \pi(r_1 + r_2)(r_2 - r_1)$

6. AREAS OF PLANE FIGURES (Cont'd)

Sector of Circle

$$l = \frac{\pi r \theta}{180} = \frac{r\theta}{57.3} = .01745 r\theta$$

$$= \frac{2A}{r}$$

$$A = \frac{\pi \theta r^2}{360} = .008727 \theta r^2$$

$$= \frac{lr}{2}$$

Segment of Circle

for $\theta < 90°$

$$A = \frac{r^2}{2}\left(\frac{\pi\theta}{180} - \sin \theta\right)$$

for $\theta > 90°$

$$A = \frac{r^2}{2}\left(\frac{\pi\theta}{180} - \sin(180 - \theta)\right)$$

for chord rise, etc., see "Properties of Circle"

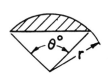

Sector of Hollow Circle

$$A = \frac{\pi\theta (r_2^2 - r_1^2)}{360}$$

$$A = \frac{r_1 - r_2}{2}(l_1 + l_2)$$

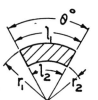

Fillet

$$A = .215 r^2$$

or approximately

$$A = \frac{r^2}{5}$$

Ellipse

$p = \pi(a + b)$ approximately
$\quad = \pi[1.5(a+b) - \sqrt{ab}]$ more nearly

$$A = \pi ab$$

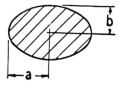

Parabola

$$A = \frac{2}{3} ab$$

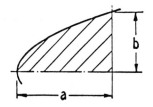

7. VOLUMES OF SOLIDS

Nomenclature

a, b, c, d	—	Lengths of Sides
C	—	Length of Chord
A	—	Total Area
A_B	—	Area of B
A_L	—	Area of Lateral or Convex Surfaces
A_R	—	Area of Right Section
A_T	—	Area of Top Section
h, h_1, h_2	—	Vertical Height or Altitude
h_G	—	Vertical Distance between Centers of Gravity of Areas
L, L_1, L_2	—	Lateral Length or Slant Height
L_G	—	Slant Height between Centers of Gravity of Areas
p	—	Perimeter
p_B	—	Perimeter of Base
p_R	—	Perimeter of Right Section
r, r_1	—	Radii
V	—	Volume

Cube

$A = 6a^2$

$V = a^3$

Parallelopiped

$A = 2(ab + bc + ac)$

$V = abc$

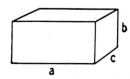

General Prism and Right Regular Prism

$A_L = p_R L = p_B h$

$A = A_L + 2A_B$

$V = A_R \times L = A_B h$

Frustum of Prism

$V = A_B h_G$

$V = A_R L_G$

Right Regular Pyramid or Cone

$A_L = \frac{1}{2} p_B L$

$V = \frac{1}{3} A_B h$

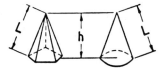

7. VOLUMES OF SOLIDS (Cont'd)

General Pyramid or Cone

$V = \dfrac{1}{3} A_B h$

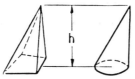

Frustum of Right Regular Pyramid or Cone

$A_L = \dfrac{1}{2} L (p_B + p_T)$

$A = A_L + A_B + A_T$

$V = \dfrac{1}{3} h (A_B + A_T + \sqrt{A_B A_T})$

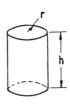

Frustum of General Pyramid or Cone (Parallel Ends)

$V = \dfrac{1}{3} h (A_B + A_T + \sqrt{A_B A_T})$

Right Circular Cylinder

$A_L = 2\pi r h$

$A = 2\pi r (r + h)$

$V = \pi r^2 h$

General Cylinder (Any Cross Section)

$A_L = p_B h = p_R L$

$A = A_L + 2 A_B$

$V = A_B h = A_R L$

Frustum of General Cylinder

$V = \dfrac{1}{2} A_R (L_1 + L_2)$

$V = A_B h_G$

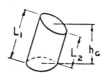

Frustum of Right Circular Cylinder

$A_L = \pi r (h_1 + h_2)$

$A_T = \pi r \sqrt{r^2 + \left(\dfrac{h_1 - h_2}{2}\right)^2}$

$A_B = \pi r^2$

$A = A_L + A_T + A_B$

$V = \dfrac{\pi r^2}{2} (h_1 + h_2)$

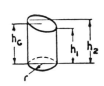

Sphere

$A = 4\pi r^2 = 12.566 r^2$

$V = \dfrac{4}{3} \pi r^3 = 4.189 r^3$

Spherical Sector

$A = \dfrac{\pi r}{2} (4h + C)$

$V = \dfrac{2}{3} \pi r^2 h = 2.0944 r^2 h$

Spherical Segment

$A_T = 2\pi r h = \dfrac{\pi}{4} (4h^2 + C^2)$

$V = \dfrac{\pi}{3} h^2 (3r - h) = \dfrac{\pi}{24} h (3C^2 + 4h^2)$

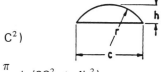

Spherical Zone

$A_L = 2\pi r h$

$A = \dfrac{\pi}{4} (8rh + a^2 + b^2)$

$V = \dfrac{\pi h}{24} (3C^2 + 3b^2 + 4h^2)$

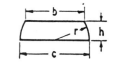

Torus

$A = 4\pi^2 r r_1$

$V = 2\pi^2 r^2 r_1$

8. EQUIVALENTS

Measure

1 in.	=	25.4 mm
1 in.	=	2.54 cm
1 mm	=	0.03937 in.
1 mm	=	0.00328 ft
1 micron	=	0.000001 meter
1 torr	=	1 mm mercury
10^{-8} torr	=	1 atom mercury
1 ft	=	304.8 mm
1 ft	=	30.48 cm
1 sq. in.	=	6.4516 sq cm
1 sq cm	=	0.155 sq in.
1 sq cm	=	0.00108 sq ft
1 sq ft	=	929.03 sq cm

Circumference of a circle $= 2\pi r = \pi d$

Area of a circle $= \pi r^2 = \dfrac{\pi d^2}{4}$

Weight

1 kg	=	2.205 lb
1 cu in. of water (60 F)	=	0.073551 cu in. of mercury (32 F)
1 cu in. of mercury (32 F)	=	13.596 cu in. of water (60 F)
1 cu in. of mercury (32 F)	=	0.4905 lb

Velocity

1 ft per sec	=	0.3048 m per sec
1 m per sec	=	3.2808 ft per sec

Density

1 lb per cu in.	=	27.68 gram per cu cm
1 gr per cu cm	=	0.03613 lb per cu in.
1 lb per cu ft	=	16.0184 kg per cu m
1 kg per cu m	=	0.06243 lb per cu ft

Equivalents of Temperature

To convert degrees Centigrade to degrees Fahrenheit:

$$t = 1.8\, t_c + 32$$

To convert degrees Fahrenheit to degrees Centigrade:

$$t_c = \frac{t - 32}{1.8}$$

Where: t_c = temperature, in degrees Centigrade

Prefixes

Prefix	Name	Value	Power
Atto	one-quintillionth	0.000 000 000 000 000 001	10^{-18}
Femto	one-quadrillionth	0.000 000 000 000 001	10^{-15}
Pico	one-trillionth	0.000 000 000 001	10^{-12}
Nano	one-billionth	0.000 000 001	10^{-9}
Micro	one-millionth	0.000 001	10^{-6}
Milli	one-thousandth	0.001	10^{-3}
Centi	one-hundredth	0.01	10^{-2}
Deci	one-tenth	0.1	10^{-1}
Uni	one	1.0	10^0
Deka	ten	10.0	10^1
Hecto	one hundred	100.0	10^2
Kilo	one thousand	1 000.0	10^3
Mega	one million	1 000 000.0	10^6
Giga	one billion	1 000 000 000.0	10^9
Tera	one trillion	1 000 000 000 000.0	10^{12}
	one quintillion	1 000 000 000 000 000.0	10^{15}
	one quadrillion	1 000 000 000 000 000 000.0	10^{18}

APPENDIX

9. ELECTRICAL AND ELECTRONIC SYMBOLS
Compiled from ANSI 32.2—1962 and Catalogs.

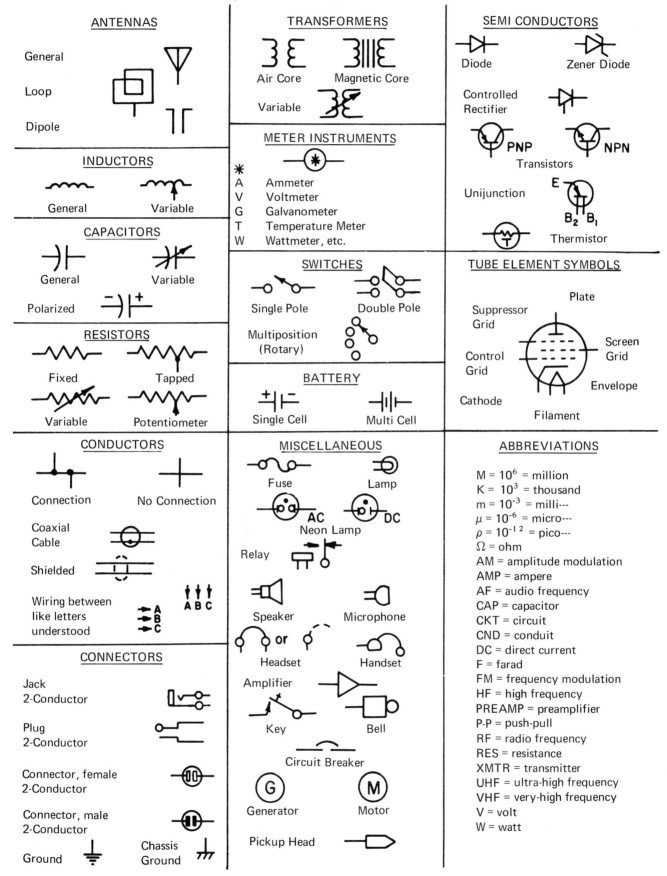

10. TWIST DRILL SIZES

Number Sizes						Letter Sizes	
No.	Diam.	No.	Diam.	No.	Diam.	Letter	Diam.
1	.2280	28	.1405	55	.0520	A	.234
2	.2210	29	.1360	56	.0465	B	.238
3	.2130	30	.1285	57	.0430	C	.242
4	.2090	31	.1200	58	.0420	D	.246
5	.2055	32	.1160	59	.0410	E	.250
6	.2040	33	.1130	60	.0400	F	.257
7	.2010	34	.1110	61	.0390	G	.261
8	.1990	35	.1100	62	.0380	H	.266
9	.1960	36	.1065	63	.0370	I	.272
10	.1935	37	.1040	64	.0360	J	.277
11	.1910	38	.1015	65	.0350	K	.281
12	.1890	39	.0995	66	.0330	L	.290
13	.1850	40	.0980	67	.0320	M	.295
14	.1820	41	.0960	68	.0310	N	.302
15	.1800	42	.0935	69	.0292	O	.316
16	.1770	43	.0890	70	.0280	P	.323
17	.1730	44	.0860	71	.0260	Q	.332
18	.1695	45	.0820	72	.0250	R	.339
19	.1660	46	.0810	73	.0240	S	.348
20	.1610	47	.0785	74	.0225	T	.358
21	.1590	48	.0760	75	.0210	U	.368
22	.1570	49	.0730	76	.0200	V	.377
23	.1540	50	.0700	77	.0180	W	.386
24	.1520	51	.0670	78	.0160	X	.397
25	.1495	52	.0635	79	.0145	Y	.404
26	.1470	53	.0595	80	.0135	Z	.413
27	.1440	54	.0550				

Other whole and fractional sizes in increments of 1/64 inch.

APPENDIX

11. UNIFIED AND AMERICAN SCREW THREADS

Basic Dimensions for Coarse, Fine, Extra-Fine, 8, 12, and 16 Thread Series.
Compiled from ANSI B1.1—1960, and SAE.

Nominal Diam.	Basic Major Diam.	Coarse UNC, NC Classes 1A, 1B, 2A, 2B, 3A, 3B, 2, 3		Fine UNF, NF Classes 1A, 1B, 2A, 2B, 3A, 3B, 2, 3		Extra-Fine UNEF, NEF Classes 2A, 2B, 2, 3		8 Thd. 8N Classes 2A, 2B, 3A, 3B, 2, 3		12 Thd. 12UN, 12N Classes 2A, 2B, 3A, 3B, 2, 3		16 Thd. 16UN, 16N Classes 2A, 2B, 3A, 3B, 2, 3	
		Thds. per in.	Tap Drill	Thds. per in.	Tap Drill	Thds. per in.	Tap Drill	Thds. per in.	Tap Drill	Thds. per in.	Tap Drill	Thds. per in.	Tap Drill
#0	0.0600			80	3/64								
#1	0.0730	64	53	72	53								
#2	0.0860	56	50	64	50								
#3	0.0990	48	47	56	45								
#4	0.1120	40	43	48	42								
#5	0.1250	40	38	44	37								
#6	0.1380	32	36	40	33								
#8	0.1640	32	29	36	29								
#10	0.1900	24	25	32	21								
#12	0.2160	24	16	28	14	32	13						
1/4	0.2500	20	7	28	3	32	7/32						
5/16	0.3125	18	F	24	I	32	9/32						
3/8	0.3750	16	5/16	24	Q	32	11/32						
7/16	0.4375	14	U	20	25/64	28	13/32						
1/2	0.5000	13	27/64	20	29/64	28	15/32			12	27/64		
9/16	0.5625	12	31/64	18	33/64	24	33/64			12	31/64		
5/8	0.6250	11	17/32	18	37/64	24	37/64			12	35/64		
11/16	0.6875					24	41/64			12	39/64		
3/4	0.7500	10	21/32	16	11/16	20	45/64			12	43/64	16	11/16
13/16	0.8125					20	49/64			12	47/64	16	3/4
7/8	0.8750	9	49/64	14	13/16	20	53/64			12	51/64	16	13/16
15/16	0.9375					20	57/64			12	55/64	16	7/8
1	1.0000	8	7/8	12	59/64	20	61/64	8	7/8	12	59/64	16	15/16
1 1/16	1.0625					18	1			12	63/64	16	1
1 1/8	1.1250	7	63/64	12	1 3/64	18	1 5/64	8	1	12	1 3/64	16	1 1/16
1 3/16	1.1875					18	1 9/64			12	1 7/64	16	1 1/8
1 1/4	1.2500	7	1 7/64	12	1 11/64	18	1 3/16	8	1 1/8	12	1 11/64	16	1 3/16
1 5/16	1.3125					18	1 17/64			12	1 15/64	16	1 1/4
1 3/8	1.3750	6	1 7/32	12	1 19/64	18	1 5/16	8	1 1/4	12	1 19/64	16	1 5/16
1 7/16	1.4375					18	1 3/8			12	1 23/64	16	1 3/8
1 1/2	1.5000	6	1 11/32	12	1 27/64	18	1 7/16	8	1 3/8	12	1 27/64	16	1 7/16
1 9/16	1.5625					18	1 1/2					16	1 1/2
1 5/8	1.6250					18	1 9/16	8	1 1/2	12	1 35/64	16	1 9/16
1 11/16	1.6875					18	1 5/8					16	1 5/8
1 3/4	1.7500	5	1 9/16					8	1 5/8	12	1 43/64	16	1 11/16
1 7/8	1.8750							8	1 3/4	12	1 51/64	16	1 13/16
2	2.0000	4 1/2	1 25/32					8	1 7/8	12	1 59/64	16	1 15/16
2 1/8	2.1250							8	2	12	2 3/64	16	2 1/16
2 1/4	2.2500	4 1/2	2 1/32					8	2 1/8	12	2 11/64	16	2 3/16
2 3/8	2.3750									12	2 19/64	16	2 5/16
2 1/2	2.5000	4	2 1/4					8	2 3/8	12	2 27/64	16	2 7/16
2 5/8	2.6250									12	2 35/64	16	2 9/16
2 3/4	2.7500	4	2 1/2					8	2 5/8	12	2 43/64	16	2 11/16
2 7/8	2.8750									12	2 51/64	16	2 13/16
3	3.0000	4	2 3/4					8	2 7/8				
3 1/8	3.1250									12	3 3/64	16	3 1/16
3 1/4	3.2500	4	3					8	3 1/8	12	3 11/64	16	3 3/16
3 3/8	3.3750									12	3 19/64	16	3 5/16
3 1/2	3.5000	4	3 1/4					8	3 3/8	12	3 27/64	16	3 7/16
3 5/8	3.6250									12	3 35/64	16	3 9/16
3 3/4	3.7500	4	3 1/2					8	3 5/8	12	3 43/64	16	3 11/16
3 7/8	3.8750									12	3 51/64	16	3 13/16
4	4.0000	4	3 3/4					8	3 7/8	12	3 59/64	16	3 15/16

12. AMERICAN STANDARD BOLTS

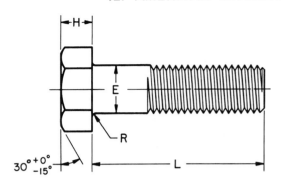
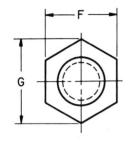

Dimensions of Hex Bolts

Nominal Size or Basic Product Dia	Body Dia E	Width Across Flats F			Width Across Corners G		Height H			Radius of Fillet R
	Max	Basic	Max	Min	Max	Min	Basic	Max	Min	Max
1/4 0.2500	0.260	7/16	0.4375	0.425	0.505	0.484	11/64	0.188	0.150	0.031
5/16 0.3125	0.324	1/2	0.5000	0.484	0.577	0.552	7/32	0.235	0.195	0.031
3/8 0.3750	0.388	9/16	0.5625	0.544	0.650	0.620	1/4	0.268	0.226	0.031
7/16 0.4375	0.452	5/8	0.6250	0.603	0.722	0.687	19/64	0.316	0.272	0.031
1/2 0.5000	0.515	3/4	0.7500	0.725	0.866	0.826	11/32	0.364	0.302	0.031
5/8 0.6250	0.642	15/16	0.9375	0.906	1.083	1.033	27/64	0.444	0.378	0.062
3/4 0.7500	0.768	1 1/8	1.1250	1.088	1.299	1.240	1/2	0.524	0.455	0.062
7/8 0.8750	0.895	1 5/16	1.3125	1.269	1.516	1.447	37/64	0.604	0.531	0.062
1 1.0000	1.022	1 1/2	1.5000	1.450	1.732	1.653	43/64	0.700	0.591	0.093
1 1/8 1.1250	1.149	1 11/16	1.6875	1.631	1.949	1.859	3/4	0.780	0.658	0.093
1 1/4 1.2500	1.277	1 7/8	1.8750	1.812	2.165	2.066	27/32	0.876	0.749	0.093
1 3/8 1.3750	1.404	2 1/16	2.0625	1.994	2.382	2.273	29/32	0.940	0.810	0.093
1 1/2 1.5000	1.531	2 1/4	2.2500	2.175	2.598	2.480	1	1.036	0.902	0.093
1 3/4 1.7500	1.785	2 5/8	2.6250	2.538	3.031	2.893	1 5/32	1.196	1.054	0.125
2 2.0000	2.039	3	3.0000	2.900	3.464	3.306	1 11/32	1.388	1.175	0.125

Dimensions of Heavy Hex Bolts

Nominal Size or Basic Product Dia	Body Dia E	Width Across Flats F			Width Across Corners G		Height H			Radius of Fillet R
	Max	Basic	Max	Min	Max	Min	Basic	Max	Min	Max
1/2 0.5000	0.515	7/8	0.8750	0.850	1.010	0.969	11/32	0.364	0.302	0.031
5/8 0.6250	0.642	1 1/16	1.0625	1.031	1.227	1.175	27/64	0.444	0.378	0.062
3/4 0.7500	0.768	1 1/4	1.2500	1.212	1.443	1.383	1/2	0.524	0.455	0.062
7/8 0.8750	0.895	1 7/16	1.4375	1.394	1.660	1.589	37/64	0.604	0.531	0.062
1 1.0000	1.022	1 5/8	1.6250	1.575	1.876	1.796	43/64	0.700	0.591	0.093
1 1/8 1.1250	1.149	1 13/16	1.8125	1.756	2.093	2.002	3/4	0.780	0.658	0.093
1 1/4 1.2500	1.277	2	2.0000	1.938	2.309	2.209	27/32	0.876	0.749	0.093
1 3/8 1.3750	1.404	2 3/16	2.1875	2.119	2.526	2.416	29/32	0.940	0.810	0.093
1 1/2 1.5000	1.531	2 3/8	2.3750	2.300	2.742	2.622	1	1.036	0.902	0.093
1 3/4 1.7500	1.785	2 3/4	2.7500	2.662	3.175	3.035	1 5/32	1.196	1.054	0.125
2 2.0000	2.039	3 1/8	3.1250	3.025	3.608	3.449	1 11/32	1.388	1.175	0.125

ANSI B18.2.1–1965
Length Increments for Hex. Head Bolts. Available in increments of 1/4 for lengths between 3/4 and 8. Available in increments of 1/2 for lengths between 8 and 20.
UNC, Class 2A
Min. Thread = (2E) + .25 up to 6 long. (2E) + .50 for over 6 long. If too short for formula, thread as close to head as practical.

12. AMERICAN STANDARD BOLTS (Cont'd)

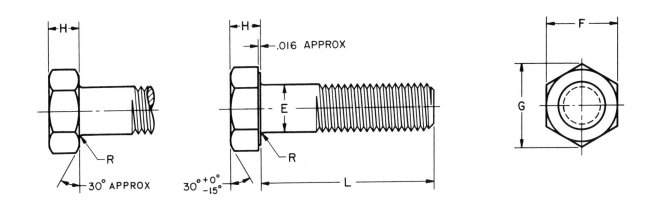

Dimensions of Hex Cap Screws (Finished Hex Bolts)

Nominal Size or Basic Product Dia	Body Dia E		Width Across Flats F			Width Across Corners G		Height H			Radius of Fillet R	
	Max	Min	Basic	Max	Min	Max	Min	Basic	Max	Min	Max	Min
1/4	0.2500	0.2450	7/16	0.4375	0.428	0.505	0.488	5/32	0.163	0.150	0.025	0.015
5/16	0.3125	0.3065	1/2	0.5000	0.489	0.577	0.557	13/64	0.211	0.195	0.025	0.015
3/8	0.3750	0.3690	9/16	0.5625	0.551	0.650	0.628	15/64	0.243	0.226	0.025	0.015
7/16	0.4375	0.4305	5/8	0.6250	0.612	0.722	0.698	9/32	0.291	0.272	0.025	0.015
1/2	0.5000	0.4930	3/4	0.7500	0.736	0.866	0.840	5/16	0.323	0.302	0.025	0.015
9/16	0.5625	0.5545	13/16	0.8125	0.798	0.938	0.910	23/64	0.371	0.348	0.045	0.020
5/8	0.6250	0.6170	15/16	0.9375	0.922	1.083	1.051	25/64	0.403	0.378	0.045	0.020
3/4	0.7500	0.7410	1 1/8	1.1250	1.100	1.299	1.254	15/32	0.483	0.455	0.045	0.020
7/8	0.8750	0.8660	1 5/16	1.3125	1.285	1.516	1.465	35/64	0.563	0.531	0.065	0.040
1	1.0000	0.9900	1 1/2	1.5000	1.469	1.732	1.675	39/64	0.627	0.591	0.095	0.060
1 1/8	1.1250	1.1140	1 11/16	1.6875	1.631	1.949	1.859	11/16	0.718	0.658	0.095	0.060
1 1/4	1.2500	1.2390	1 7/8	1.8750	1.812	2.165	2.066	25/32	0.813	0.749	0.095	0.060
1 3/8	1.3750	1.3630	2 1/16	2.0625	1.994	2.382	2.273	27/32	0.878	0.810	0.095	0.060
1 1/2	1.5000	1.4880	2 1/4	2.2500	2.175	2.598	2.480	15/16	0.974	0.902	0.095	0.060
1 3/4	1.7500	1.7380	2 5/8	2.6250	2.538	3.031	2.893	1 3/32	1.134	1.054	0.095	0.060
2	2.0000	1.9880	3	3.0000	2.900	3.464	3.306	1 7/32	1.263	1.175	0.095	0.060

ANSI B18.2.1—1965
Lengths given on previous table.
UNC, UNF, or 8 UN, Class 2 A.
Min. thread length given on previous table.

13. AMERICAN STANDARD CAP SCREWS

Basic dimensions for drawing.
Compiled from ANSI B18.3–1961.

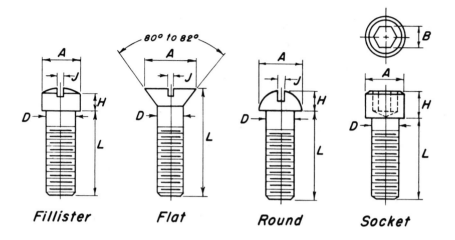

Fillister   Flat   Round   Socket

Diam.	Fillister		Flat	Round		Socket			Slot
	A	H	A	A	H	A	H	B	J
1/4	3/8	11/64	1/2	7/16	.191	3/8	1/4	.190	.070
5/16	7/16	13/64	5/8	9/16	.246	7/16	5/16	.221	.079
3/8	9/16	1/4	3/4	5/8	.273	9/16	3/8	.316	.088
7/16	5/8	19/64	13/16	3/4	.328	5/8	7/16	.316	.098
1/2	3/4	21/64	7/8	13/16	.355	3/4	1/2	.378	.110
9/16	13/16	3/8	1	15/16	.410	13/16	9/16	.378	.123
5/8	7/8	27/64	1 1/8	1	.438	7/8	5/8	.503	.138
3/4	1	1/2	1 3/8	1 1/4	.547	1	3/4	.566	.154
7/8	1 1/8	19/32				1 1/8	7/8	.566	.173
1	1 5/16	21/32				1 5/16	1	.629	.194
1 1/8						1 1/2	1 1/8	.754	
1 1/4						1 3/4	1 1/4	.754	

Length Increments: 1/8" for screw lengths (L) 1/4 to 1;
                       1/4" for screw lengths (L) 1 to 4;
                       1/2" for screw lengths (L) 4 to 6.
Coarse or Fine Thread, class 3 fit.
Thread length = 2D + 1/4". Short screws threaded as close to head as practicable.

APPENDIX

14. AMERICAN STANDARD MACHINE SCREWS
Basic dimensions for drawing.
Compiled from ANSI B18.6.3—1963.

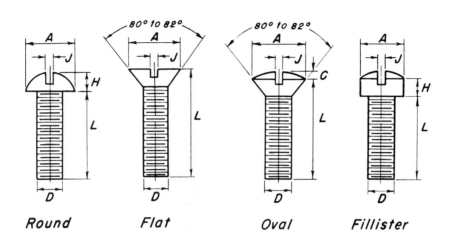

Round   Flat   Oval   Fillister

Size	Diam.	Round		Flat	Oval		Fillister		Slot
		A	H	A	A	C	A	H	J
2	.086	.162	.070	.172	.172	.029	.140	.055	.036
3	.099	.187	.078	.199	.199	.033	.161	.063	.038
4	.112	.211	.086	.225	.225	.037	.183	.072	.040
5	.125	.236	.095	.252	.252	.041	.205	.081	.043
6	.138	.260	.103	.279	.279	.045	.226	.089	.045
8	.164	.309	.119	.332	.332	.053	.270	.106	.050
10	.190	.359	.136	.385	.385	.061	.313	.123	.055
12	.216	.408	.152	.438	.438	.069	.357	.141	.059
1/4	.250	.472	.174	.507	.507	.079	.414	.163	.066
5/16	.3125	.591	.214	.636	.636	.098	.519	.205	.077
3/8	.375	.708	.254	.762	.762	.117	.622	.246	.088

Coarse or Fine Threads, class 2
Thread length: For screws 2″ long or less, full length thread.
                For screws over 2″ long, thread lengths = 1 3/4″.

15. AMERICAN STANDARD WASHERS

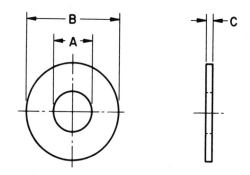

Dimensions of Preferred Sizes of Type A Plain Washers**

Nominal Washer Size***		Inside Diameter A			Outside Diameter B			Thickness C		
		Basic	Tolerance Plus	Tolerance Minus	Basic	Tolerance Plus	Tolerance Minus	Basic	Max	Min
–	–	0.078	0.000	0.005	0.188	0.000	0.005	0.020	0.025	0.016
–	–	0.094	0.000	0.005	0.250	0.000	0.005	0.020	0.025	0.016
–	–	0.125	0.008	0.005	0.312	0.008	0.005	0.032	0.040	0.025
No. 6	0.138	0.156	0.008	0.005	0.375	0.015	0.005	0.049	0.065	0.036
No. 8	0.164	0.188	0.008	0.005	0.438	0.015	0.005	0.049	0.065	0.036
No. 10	0.190	0.219	0.008	0.005	0.500	0.015	0.005	0.049	0.065	0.036
3/16	0.188	0.250	0.015	0.005	0.562	0.015	0.005	0.049	0.065	0.036
No. 12	0.216	0.250	0.015	0.005	0.562	0.015	0.005	0.065	0.080	0.051
1/4	0.250 N	0.281	0.015	0.005	0.625	0.015	0.005	0.065	0.080	0.051
1/4	0.250 W	0.312	0.015	0.005	0.734*	0.015	0.007	0.065	0.080	0.051
5/16	0.312 N	0.344	0.015	0.005	0.688	0.015	0.007	0.065	0.080	0.051
5/16	0.312 W	0.375	0.015	0.005	0.875	0.030	0.007	0.083	0.104	0.064
3/8	0.375 N	0.406	0.015	0.005	0.812	0.015	0.007	0.065	0.080	0.051
3/8	0.375 W	0.438	0.015	0.005	1.000	0.030	0.007	0.083	0.104	0.064
7/16	0.438 N	0.469	0.015	0.005	0.922	0.015	0.007	0.065	0.080	0.051
7/16	0.438 W	0.500	0.015	0.005	1.250	0.030	0.007	0.083	0.104	0.064
1/2	0.500 N	0.531	0.015	0.005	1.062	0.030	0.007	0.095	0.121	0.074
1/2	0.500 W	0.562	0.015	0.005	1.375	0.030	0.007	0.109	0.132	0.086
9/16	0.562 N	0.594	0.015	0.005	1.156*	0.030	0.007	0.095	0.121	0.074
9/16	0.562 W	0.625	0.015	0.005	1.469*	0.030	0.007	0.109	0.132	0.086
5/8	0.625 N	0.656	0.030	0.007	1.312	0.030	0.007	0.095	0.121	0.074
5/8	0.625 W	0.688	0.030	0.007	1.750	0.030	0.007	0.134	0.160	0.108
3/4	0.750 N	0.812	0.030	0.007	1.469	0.030	0.007	0.134	0.160	0.108
3/4	0.750 W	0.812	0.030	0.007	2.000	0.030	0.007	0.148	0.177	0.122
7/8	0.875 N	0.938	0.030	0.007	1.750	0.030	0.007	0.134	0.160	0.108
7/8	0.875 W	0.938	0.030	0.007	2.250	0.030	0.007	0.165	0.192	0.136
1	1.000 N	1.062	0.030	0.007	2.000	0.030	0.007	0.134	0.160	0.108
1	1.000 W	1.062	0.030	0.007	2.500	0.030	0.007	0.165	0.192	0.136
1 1/8	1.125 N	1.250	0.030	0.007	2.250	0.030	0.007	0.134	0.160	0.108
1 1/8	1.125 W	1.250	0.030	0.007	2.750	0.030	0.007	0.165	0.192	0.136
1 1/4	1.250 N	1.375	0.030	0.007	2.500	0.030	0.007	0.165	0.192	0.136
1 1/4	1.250 W	1.375	0.030	0.007	3.000	0.030	0.007	0.165	0.192	0.136
1 3/8	1.375 N	1.500	0.030	0.007	2.750	0.030	0.007	0.165	0.192	0.136
1 3/8	1.375 W	1.500	0.045	0.010	3.250	0.045	0.010	0.180	0.213	0.153
1 1/2	1.500 N	1.625	0.030	0.007	3.000	0.030	0.007	0.165	0.192	0.136
1 1/2	1.500 W	1.625	0.045	0.010	3.500	0.045	0.010	0.180	0.213	0.153
1 5/8	1.625	1.750	0.045	0.010	3.750	0.045	0.010	0.180	0.213	0.153
1 3/4	1.750	1.875	0.045	0.010	4.000	0.045	0.010	0.180	0.213	0.153
1 7/8	1.875	2.000	0.045	0.010	4.250	0.045	0.010	0.180	0.213	0.153
2	2.000	2.125	0.045	0.010	4.500	0.045	0.010	0.180	0.213	0.153
2 1/4	2.250	2.375	0.045	0.010	4.750	0.045	0.010	0.220	0.248	0.193
2 1/2	2.500	2.625	0.045	0.010	5.000	0.045	0.010	0.238	0.280	0.210
2 3/4	2.750	2.875	0.065	0.010	5.250	0.065	0.010	0.259	0.310	0.228
3	3.000	3.125	0.065	0.010	5.500	0.065	0.010	0.284	0.327	0.249

ANSI B 27.2–1965

*The 0.734 in., 1.156 in., and 1.469 in. outside diameters avoid washers which could be used in coin operated devices.

**Preferred sizes are for the most part from series previously designated "Standard Plate" and "SAE." Where common sizes existed in the two series, the SAE size is designated "N" (narrow) and the Standard Plate "W" (wide). These sizes as well as all other sizes of Type A Plain Washers are to be ordered by ID, OD, and thickness dimensions.

***Nominal washer sizes are intended for use with comparable nominal screw or bolt sizes.

16. AMERICAN STANDARD LOCK WASHERS

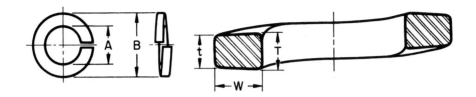

Dimensions of Regular* Helical Spring Lock Washers

Nominal Washer Size		Inside Diameter A		Outside Diameter B	Washer Section	
					Width W	Thickness $\frac{T+t}{2}$
		Min	Max	Max**	Min	Min
No. 2	0.086	0.088	0.094	0.172	0.035	0.020
No. 3	0.099	0.101	0.107	0.195	0.040	0.025
No. 4	0.112	0.115	0.121	0.209	0.040	0.025
No. 5	0.125	0.128	0.134	0.236	0.047	0.031
No. 6	0.138	0.141	0.148	0.250	0.047	0.031
No. 8	0.164	0.168	0.175	0.293	0.055	0.040
No. 10	0.190	0.194	0.202	0.334	0.062	0.047
No. 12	0.216	0.221	0.229	0.377	0.070	0.056
1/4	0.250	0.255	0.263	0.489	0.109	0.062
5/16	0.312	0.318	0.328	0.586	0.125	0.078
3/8	0.375	0.382	0.393	0.683	0.141	0.094
7/16	0.438	0.446	0.459	0.779	0.156	0.109
1/2	0.500	0.509	0.523	0.873	0.171	0.125
9/16	0.562	0.572	0.587	0.971	0.188	0.141
5/8	0.625	0.636	0.653	1.079	0.203	0.156
11/16	0.688	0.700	0.718	1.176	0.219	0.172
3/4	0.750	0.763	0.783	1.271	0.234	0.188
13/16	0.812	0.826	0.847	1.367	0.250	0.203
7/8	0.875	0.890	0.912	1.464	0.266	0.219
15/16	0.938	0.954	0.978	1.560	0.281	0.234
1	1.000	1.017	1.042	1.661	0.297	0.250
1 1/16	1.062	1.080	1.107	1.756	0.312	0.266
1 1/8	1.125	1.144	1.172	1.853	0.328	0.281
1 3/16	1.188	1.208	1.237	1.950	0.344	0.297
1 1/4	1.250	1.271	1.302	2.045	0.359	0.312
1 5/16	1.312	1.334	1.366	2.141	0.375	0.328
1 3/8	1.375	1.398	1.432	2.239	0.391	0.344
1 7/16	1.438	1.462	1.497	2.334	0.406	0.359
1 1/2	1.500	1.525	1.561	2.430	0.422	0.375

ANSI B27.1—1965

*Formerly designated Medium Helical Spring Lock Washers.
**The maximum outside diameters specified allow for the commercial tolerances on cold drawn wire.

17. AMERICAN STANDARD HEX NUTS

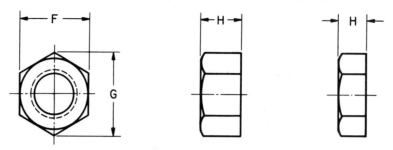

Dimensions of Hex Flat Nuts and Hex Flat Jam Nuts

Nominal Size or Basic Major Dia of Thread		Width Across Flats F			Width Across Corners G		Thickness Hex Flat Nuts H			Thickness Hex Flat Jam Nuts H		
		Basic	Max	Min	Max	Min	Basic	Max	Min	Basic	Max	Min
1 1/8	1.1250	1 11/16	1.6875	1.631	1.949	1.859	1	1.030	0.970	5/8	0.655	0.595
1 1/4	1.2500	1 7/8	1.8750	1.812	2.165	2.066	1 3/32	1.126	1.062	3/4	0.782	0.718
1 3/8	1.3750	2 1/16	2.0625	1.994	2.382	2.273	1 13/64	1.237	1.169	13/16	0.846	0.778
1 1/2	1.5000	2 1/4	2.2500	2.175	2.598	2.480	1 5/16	1.348	1.276	7/8	0.911	0.839

ANSI B18.2.2—1965
UNC—Class 2B

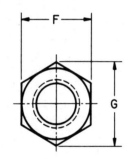 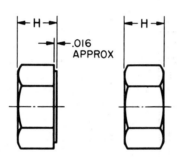 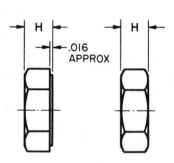

Dimensions of Hex Nuts and Hex Jam Nuts

Nominal Size or Basic Major Dia of Thread		Width Across Flats F			Width Across Corners G		Thickness Hex Nuts H			Thickness Hex Jam Nuts H		
		Basic	Max	Min	Max	Min	Basic	Max	Min	Basic	Max	Min
1/4	0.2500	7/16	0.4375	0.428	0.505	0.488	7/32	0.226	0.212	5/32	0.163	0.150
5/16	0.3125	1/2	0.5000	0.489	0.577	0.557	17/64	0.273	0.258	3/16	0.195	0.180
3/8	0.3750	9/16	0.5625	0.551	0.650	0.628	21/64	0.337	0.320	7/32	0.227	0.210
7/16	0.4375	11/16	0.6875	0.675	0.794	0.768	3/8	0.385	0.365	1/4	0.260	0.240
1/2	0.5000	3/4	0.7500	0.736	0.866	0.840	7/16	0.448	0.427	5/16	0.323	0.302
9/16	0.5625	7/8	0.8750	0.861	1.010	0.982	31/64	0.496	0.473	5/16	0.324	0.301
5/8	0.6250	15/16	0.9375	0.922	1.083	1.051	35/64	0.559	0.535	3/8	0.387	0.363
3/4	0.7500	1 1/8	1.1250	1.088	1.299	1.240	41/64	0.665	0.617	27/64	0.446	0.398
7/8	0.8750	1 5/16	1.3125	1.269	1.516	1.447	3/4	0.776	0.724	31/64	0.510	0.458
1	1.0000	1 1/2	1.5000	1.450	1.732	1.653	55/64	0.887	0.831	35/64	0.575	0.519
1 1/8	1.1250	1 11/16	1.6875	1.631	1.949	1.859	31/32	0.999	0.939	39/64	0.639	0.579
1 1/4	1.2500	1 7/8	1.8750	1.812	2.165	2.066	1 1/16	1.094	1.030	23/32	0.751	0.687
1 3/8	1.3750	2 1/16	2.0625	1.994	2.382	2.273	1 11/64	1.206	1.138	25/32	0.815	0.747
1 1/2	1.5000	2 1/4	2.2500	2.175	2.598	2.480	1 9/32	1.317	1.245	27/32	0.880	0.808

ANSI B18.2.2—1965
UNC, UNF, 8UN—Class 2B

18. AMERICAN STANDARD HEX SLOTTED NUTS

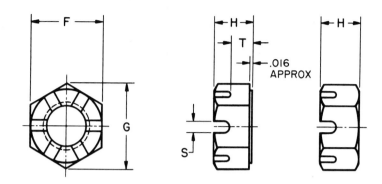

Dimensions of Hex Slotted Nuts

Nominal Size or Basic Major Dia of Thread		Width Across Flats F			Width Across Corners G		Thickness H			Unslotted Thickness T		Width of Slot S	
		Basic	Max	Min	Max	Min	Basic	Max	Min	Max	Min	Max	Min
1/4	0.2500	7/16	0.4375	0.428	0.505	0.488	7/32	0.226	0.212	0.14	0.12	0.10	0.07
5/16	0.3125	1/2	0.5000	0.489	0.577	0.557	17/64	0.273	0.258	0.18	0.16	0.12	0.09
3/8	0.3750	9/16	0.5625	0.551	0.650	0.628	21/64	0.337	0.320	0.21	0.19	0.15	0.12
7/16	0.4375	11/16	0.6875	0.675	0.794	0.768	3/8	0.385	0.365	0.23	0.21	0.15	0.12
1/2	0.5000	3/4	0.7500	0.736	0.866	0.840	7/16	0.448	0.427	0.29	0.27	0.18	0.15
9/16	0.5625	7/8	0.8750	0.861	1.010	0.982	31/64	0.496	0.473	0.31	0.29	0.18	0.15
5/8	0.6250	15/16	0.9375	0.922	1.083	1.051	35/64	0.559	0.535	0.34	0.32	0.24	0.18
3/4	0.7500	1 1/8	1.1250	1.088	1.299	1.240	41/64	0.665	0.617	0.40	0.38	0.24	0.18
7/8	0.8750	1 5/16	1.3125	1.269	1.516	1.447	3/4	0.776	0.724	0.52	0.49	0.24	0.18
1	1.0000	1 1/2	1.5000	1.450	1.732	1.653	55/64	0.887	0.831	0.59	0.56	0.30	0.24
1 1/8	1.1250	1 11/16	1.6875	1.631	1.949	1.859	31/32	0.999	0.939	0.64	0.61	0.33	0.24
1 1/4	1.2500	1 7/8	1.8750	1.812	2.165	2.066	1 1/16	1.094	1.030	0.70	0.67	0.40	0.31
1 3/8	1.3750	2 1/16	2.0625	1.994	2.382	2.273	1 11/64	1.206	1.138	0.82	0.78	0.40	0.31
1 1/2	1.5000	2 1/4	2.2500	2.175	2.598	2.480	1 9/32	1.317	1.245	0.86	0.82	0.46	0.37

ANSI B18.2.2–1965
UNC, UNF, 8UN—Class 2B

19. COTTER PINS

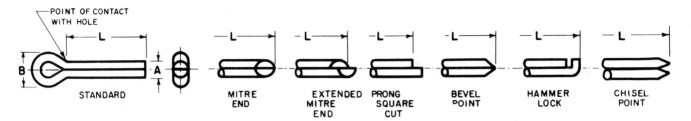

Dimensions of Cotter Pins

Diameter Nominal	Diameter A		Outside Eye Diameter B Min	Hole Sizes Recom- mended
	Max	Min		
0.031	0.032	0.028	1/16	3/64
0.047	0.048	0.044	3/32	1/16
0.062	0.060	0.056	1/8	5/64
0.078	0.076	0.072	5/32	3/32
0.094	0.090	0.086	3/16	7/64
0.109	0.104	0.100	7/32	1/8
0.125	0.120	0.116	1/4	9/64
0.141	0.134	0.130	9/32	5/32
0.156	0.150	0.146	5/16	11/64
0.188	0.176	0.172	3/8	13/64
0.219	0.207	0.202	7/16	15/64
0.250	0.225	0.220	1/2	17/64

ANSI B5.20–1958

20. TAPER PINS

Dimensions of Taper Pins

Number	5/0	4/0	3/0	2/0	0	1	2	3	4
Size (Large End)	0.0940	0.1090	0.1250	0.1410	0.1560	0.1720	0.1930	0.2190	0.2500
Length, L									
0.375									
0.500	X	X	X	X	X				
0.625	X	X	X	X	X				
0.750	X	X	X	X	X	X	X	X	
0.875			X	X	X	X	X	X	
1.000	X	X	X	X	X	X	X	X	X
1.250				X	X	X	X	X	X
1.500					X	X	X	X	X
1.750						X	X	X	X
2.000						X	X	X	X
2.250							X	X	X
2.500							X	X	X
2.750								X	X
3.000								X	X

ANSI B5.20–1958

21. SETSCREWS, HEADS, AND POINTS
Approximated Sizes Used for Drawing

Compiled from ANSI B18.3–1961

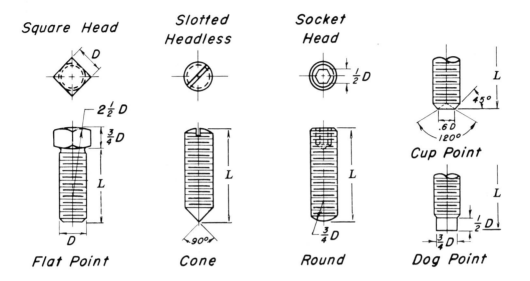

Half Dog Point has point length = D/4

Threads are coarse, fine, or 8 pitch; class 2A, larger sizes in coarse only.

Length increments: 1/16" for diameters of 1/4 to 5/8

1/8" for diameters of 5/8 to 1

1/4" for diameters of 1 to 4

22. ACME THREADS

Compiled from ANSI B1.5–1952

Nominal Diam.	Threads Per Inch
1/4	16
5/16	14
3/8	12
7/16	12
1/2	10
5/8	8
3/4	6
7/8	6
1	5

Nominal Diam.	Threads Per Inch
1 1/8	5
1 1/4	5
1 3/8	4
1 1/2	4
1 3/4	4
2	4
2 1/4	3
2 1/2	3

23. SQUARE THREADS

Nominal Diam.	Threads Per Inch
1/4	10
5/16	9
3/8	8
7/16	7
1/2	6 1/2
9/16	6
5/8	5 1/2
11/16	5
3/4	5

Nominal Diam.	Threads Per Inch
13/16	4 1/2
7/8	4 1/2
15/16	4
1	4
1 1/8	3 1/2
1 1/4	3 1/2
1 3/8	3
1 1/2	3
1 5/8	2 3/4

Nominal Diam.	Threads Per Inch
1 3/4	2 1/2
1 7/8	2 1/2
2	2 1/4
2 1/4	2 1/4
2 1/2	2
2 3/4	2
3	1 3/4
3 1/4	1 3/4
3 1/2	1 5/8

APPENDIX

24. RUNNING AND SLIDING FITS ANSI B4.1–1967

Limits are in thousandths of an inch.
Limits for hole and shaft are applied algebraically to the basic size to obtain the limits of size for the parts.

Nominal Size Range Inches Over — To	Class RC 1 Limits of Clearance	Class RC 1 Standard Limits Hole	Class RC 1 Standard Limits Shaft	Class RC 2 Limits of Clearance	Class RC 2 Standard Limits Hole	Class RC 2 Standard Limits Shaft	Class RC 3 Limits of Clearance	Class RC 3 Standard Limits Hole	Class RC 3 Standard Limits Shaft	Class RC 4 Limits of Clearance	Class RC 4 Standard Limits Hole	Class RC 4 Standard Limits Shaft
0 — 0.12	0.1 / 0.45	+0.2 / 0	−0.1 / −0.25	0.1 / 0.55	+0.25 / 0	−0.1 / −0.3	0.3 / 0.95	+0.4 / 0	−0.3 / −0.55	0.3 / 1.3	+0.6 / 0	−0.3 / −0.7
0.12 — 0.24	0.15 / 0.5	+0.2 / 0	−0.15 / −0.3	0.15 / 0.65	+0.3 / 0	−0.15 / −0.35	0.4 / 1.12	+0.5 / 0	−0.4 / −0.7	0.4 / 1.6	+0.7 / 0	−0.4 / −0.9
0.24 — 0.40	0.2 / 0.6	0.25 / 0	−0.2 / −0.35	0.2 / 0.85	+0.4 / 0	−0.2 / −0.45	0.5 / 1.5	+0.6 / 0	−0.5 / −0.9	0.5 / 2.0	+0.9 / 0	−0.5 / −1.1
0.40 — 0.71	0.25 / 0.75	+0.3 / 0	−0.25 / −0.45	0.25 / 0.95	+0.4 / 0	−0.25 / −0.55	0.6 / 1.7	+0.7 / 0	−0.6 / −1.0	0.6 / 2.3	+1.0 / 0	−0.6 / −1.3
0.71 — 1.19	0.3 / 0.95	+0.4 / 0	−0.3 / −0.55	0.3 / 1.2	+0.5 / 0	−0.3 / −0.7	0.8 / 2.1	+0.8 / 0	−0.8 / −1.3	0.8 / 2.8	+1.2 / 0	−0.8 / −1.6
1.19 — 1.97	0.4 / 1.1	+0.4 / 0	−0.4 / −0.7	0.4 / 1.4	+0.6 / 0	−0.4 / −0.8	1.0 / 2.6	+1.0 / 0	−1.0 / −1.6	1.0 / 3.6	+1.6 / 0	−1.0 / −2.0
1.97 — 3.15	0.4 / 1.2	+0.5 / 0	−0.4 / −0.7	0.4 / 1.6	+0.7 / 0	−0.4 / −0.9	1.2 / 3.1	+1.2 / 0	−1.2 / −1.9	1.2 / 4.2	+1.8 / 0	−1.2 / −2.4
3.15 — 4.73	0.5 / 1.5	+0.6 / 0	−0.5 / −0.9	0.5 / 2.0	+0.9 / 0	−0.5 / −1.1	1.4 / 3.7	+1.4 / 0	−1.4 / −2.3	1.4 / 5.0	+2.2 / 0	−1.4 / −2.8
	Close Sliding Fit. No Perceptible Play.			Free Sliding Fit. For Constant Temperature.			Precision Running Fit. Slow Speeds, and Light Loads.			Close Running Fit. Medium Speeds and Loads.		

Nominal Size Range Inches Over — To	Class RC 5 Limits of Clearance	Class RC 5 Standard Limits Hole	Class RC 5 Standard Limits Shaft	Class RC 6 Limits of Clearance	Class RC 6 Standard Limits Hole	Class RC 6 Standard Limits Shaft	Class RC 7 Limits of Clearance	Class RC 7 Standard Limits Hole	Class RC 7 Standard Limits Shaft	Class RC 8 Limits of Clearance	Class RC 8 Standard Limits Hole	Class RC 8 Standard Limits Shaft	Class RC 9 Limits of Clearance	Class RC 9 Standard Limits Hole	Class RC 9 Standard Limits Shaft
0 — 0.12	0.6 / 1.6	+0.6 / −0	−0.6 / −1.0	0.6 / 2.2	+1.0 / −0	−0.6 / −1.2	1.0 / 2.6	+1.0 / 0	−1.0 / −1.6	2.5 / 5.1	+1.6 / 0	−2.5 / −3.5	4.0 / 8.1	+2.5 / 0	−4.0 / −5.6
0.12 — 0.24	0.8 / 2.0	+0.7 / −0	−0.8 / −1.3	0.8 / 2.7	+1.2 / −0	−0.8 / −1.5	1.2 / 3.1	+1.2 / 0	−1.2 / −1.9	2.8 / 5.8	+1.8 / 0	−2.8 / −4.0	4.5 / 9.0	+3.0 / 0	−4.5 / −6.0
0.24 — 0.40	1.0 / 2.5	+0.9 / −0	−1.0 / −1.6	1.0 / 3.3	+1.4 / −0	−1.0 / −1.9	1.6 / 3.9	+1.4 / 0	−1.6 / −2.5	3.0 / 6.6	+2.2 / 0	−3.0 / −4.4	5.0 / 10.7	+3.5 / 0	−5.0 / −7.2
0.40 — 0.71	1.2 / 2.9	+1.0 / −0	−1.2 / −1.9	1.2 / 3.8	+1.6 / −0	−1.2 / −2.2	2.0 / 4.6	+1.6 / 0	−2.0 / −3.0	3.5 / 7.9	+2.8 / 0	−3.5 / −5.1	6.0 / 12.8	+4.0 / 0	−6.0 / −8.8
0.71 — 1.19	1.6 / 3.6	+1.2 / −0	−1.6 / −2.4	1.6 / 4.8	+2.0 / −0	−1.6 / −2.8	2.5 / 5.7	+2.0 / 0	−2.5 / −3.7	4.5 / 10.0	+3.5 / 0	−4.5 / −6.5	7.0 / 15.5	+5.0 / 0	−7.0 / −10.5
1.19 — 1.97	2.0 / 4.6	+1.6 / −0	−2.0 / −3.0	2.0 / 6.1	+2.5 / −0	−2.0 / −3.6	3.0 / 7.1	+2.5 / 0	−3.0 / −4.6	5.0 / 11.5	+4.0 / 0	−5.0 / −7.5	8.0 / 18.0	+6.0 / 0	−8.0 / −12.0
1.97 — 3.15	2.5 / 5.5	+1.8 / −0	−2.5 / −3.7	2.5 / 7.3	+3.0 / −0	−2.5 / −4.3	4.0 / 8.8	+3.0 / 0	−4.0 / −5.8	6.0 / 13.5	+4.5 / 0	−6.0 / −9.0	9.0 / 20.5	+7.0 / 0	−9.0 / −13.5
3.15 — 4.73	3.0 / 6.6	+2.2 / −0	−3.0 / −4.4	3.0 / 8.7	+3.5 / −0	−3.0 / −5.2	5.0 / 10.7	+3.5 / 0	−5.0 / −7.2	7.0 / 15.5	+5.0 / 0	−7.0 / −10.5	10.0 / 24.0	+9.0 / 0	−10.0 / −15.0
	Medium Running Fits. High Speeds and Heavy Loads.						Free Running Fit. For Varying Temperature.			Loose Fits. Large Allowances and Tolerances.					

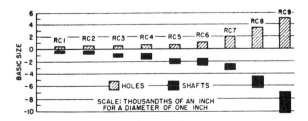

GRAPHICAL REPRESENTATION OF STANDARD
RUNNING OR SLIDING CLEARANCE FITS

25. LOCATIONAL CLEARANCE FITS ANSI B4.1–1967

Limits are in thousandths of an inch.
Limits for hole and shaft are applied algebraically to the basic size to obtain the limits of size for the parts

Nominal Size Range Inches Over — To	Class LC 1 Limits of Clearance	Class LC 1 Standard Limits Hole	Class LC 1 Standard Limits Shaft	Class LC 2 Limits of Clearance	Class LC 2 Standard Limits Hole	Class LC 2 Standard Limits Shaft	Class LC 3 Limits of Clearance	Class LC 3 Standard Limits Hole	Class LC 3 Standard Limits Shaft	Class LC 4 Limits of Clearance	Class LC 4 Standard Limits Hole	Class LC 4 Standard Limits Shaft	Class LC 5 Limits of Clearance	Class LC 5 Standard Limits Hole	Class LC 5 Standard Limits Shaft
0 — 0.12	0 / 0.45	+0.25 / −0	+0 / −0.2	0 / 0.65	+0.4 / −0	+0 / −0.25	0 / 1	+0.6 / −0	+0 / −0.4	0 / 2.6	+1.6 / −0	+0 / −1.0	0.1 / 0.75	+0.4 / −0	−0.1 / −0.35
0.12 — 0.24	0 / 0.5	+0.3 / −0	+0 / −0.2	0 / 0.8	+0.5 / −0	+0 / −0.3	0 / 1.2	+0.7 / −0	+0 / −0.5	0 / 3.0	+1.8 / −0	+0 / −1.2	0.15 / 0.95	+0.5 / −0	−0.15 / −0.45
0.24 — 0.40	0 / 0.65	+0.4 / −0	+0 / −0.25	0 / 1.0	+0.6 / −0	+0 / −0.4	0 / 1.5	+0.9 / −0	+0 / −0.6	0 / 3.6	+2.2 / −0	+0 / −1.4	0.2 / 1.2	+0.6 / −0	−0.2 / −0.6
0.40 — 0.71	0 / 0.7	+0.4 / −0	+0 / −0.3	0 / 1.1	+0.7 / −0	+0 / −0.4	0 / 1.7	+1.0 / −0	+0 / −0.7	0 / 4.4	+2.8 / −0	+0 / −1.6	0.25 / 1.35	+0.7 / −0	−0.25 / −0.65
0.71 — 1.19	0 / 0.9	+0.5 / −0	+0 / −0.4	0 / 1.3	+0.8 / −0	+0 / −0.5	0 / 2	+1.2 / −0	+0 / −0.8	0 / 5.5	+3.5 / −0	+0 / −2.0	0.3 / 1.6	+0.8 / −0	−0.3 / −0.8
1.19 — 1.97	0 / 1.0	+0.6 / −0	+0 / −0.4	0 / 1.6	+1.0 / −0	+0 / −0.6	0 / 2.6	+1.6 / −0	+0 / −1	0 / 6.5	+4.0 / −0	+0 / −2.5	0.4 / 2.0	+1.0 / −0	−0.4 / −1.0
1.97 — 3.15	0 / 1.2	+0.7 / −0	+0 / −0.5	0 / 1.9	+1.2 / −0	+0 / −0.7	0 / 3	+1.8 / −0	+0 / −1.2	0 / 7.5	+4.5 / −0	+0 / −3	0.4 / 2.3	+1.2 / −0	−0.4 / −1.1
3.15 — 4.73	0 / 1.5	+0.9 / −0	+0 / −0.6	0 / 2.3	+1.4 / −0	+0 / −0.9	0 / 3.6	+2.2 / −0	+0 / −1.4	0 / 8.5	+5.0 / −0	+0 / −3.5	0.5 / 2.8	+1.4 / −0	−0.5 / −1.4

Nominal Size Range Inches Over — To	Class LC 6 Limits of Clearance	Class LC 6 Standard Limits Hole	Class LC 6 Standard Limits Shaft	Class LC 7 Limits of Clearance	Class LC 7 Standard Limits Hole	Class LC 7 Standard Limits Shaft	Class LC 8 Limits of Clearance	Class LC 8 Standard Limits Hole	Class LC 8 Standard Limits Shaft	Class LC 9 Limits of Clearance	Class LC 9 Standard Limits Hole	Class LC 9 Standard Limits Shaft	Class LC 10 Limits of Clearance	Class LC 10 Standard Limits Hole	Class LC 10 Standard Limits Shaft	Class LC 11 Limits of Clearance	Class LC 11 Standard Limits Hole	Class LC 11 Standard Limits Shaft
0 — 0.12	0.3 / 1.9	+1.0 / 0	−0.3 / −0.9	0.6 / 3.2	+1.6 / 0	−0.6 / −1.6	1.0 / 3.6	+1.6 / −0	−1.0 / −2.0	2.5 / 6.6	+2.5 / −0	−2.5 / −4.1	4 / 12	+4 / −0	−4 / −8	5 / 17	+6 / −0	−5 / −11
0.12 — 0.24	0.4 / 2.3	+1.2 / 0	−0.4 / −1.1	0.8 / 3.8	+1.8 / 0	−0.8 / −2.0	1.2 / 4.2	+1.8 / −0	−1.2 / −2.4	2.8 / 7.6	+3.0 / −0	−2.8 / −4.6	4.5 / 14.5	+5 / −0	−4.5 / −9.5	6 / 20	+7 / −0	−6 / −13
0.24 — 0.40	0.5 / 2.8	+1.4 / 0	−0.5 / −1.4	1.0 / 4.6	+2.2 / 0	−1.0 / −2.4	1.6 / 5.2	+2.2 / −0	−1.6 / −3.0	3.0 / 8.7	+3.5 / −0	−3.0 / −5.2	5 / 17	+6 / −0	−5 / −11	7 / 25	+9 / −0	−7 / −16
0.40 — 0.71	0.6 / 3.2	+1.6 / 0	−0.6 / −1.6	1.2 / 5.6	+2.8 / 0	−1.2 / −2.8	2.0 / 6.4	+2.8 / −0	−2.0 / −3.6	3.5 / 10.3	+4.0 / −0	−3.5 / −6.3	6 / 20	+7 / −0	−6 / −13	8 / 28	+10 / −0	−8 / −18
0.71 — 1.19	0.8 / 4.0	+2.0 / 0	−0.8 / −2.0	1.6 / 7.1	+3.5 / 0	−1.6 / −3.6	2.5 / 8.0	+3.5 / −0	−2.5 / −4.5	4.5 / 13.0	+5.0 / −0	−4.5 / −8.0	7 / 23	+8 / −0	−7 / −15	10 / 34	+12 / −0	−10 / −22
1.19 — 1.97	1.0 / 5.1	+2.5 / 0	−1.0 / −2.6	2.0 / 8.5	+4.0 / 0	−2.0 / −4.5	3.0 / 9.5	+4.0 / −0	−3.0 / −5.5	5 / 15	+6 / −0	−5 / −9	8 / 28	+10 / −0	−8 / −18	12 / 44	+16 / −0	−12 / −28
1.97 — 3.15	1.2 / 6.0	+3.0 / 0	−1.2 / −3.0	2.5 / 10.0	+4.5 / 0	−2.5 / −5.5	4.0 / 11.5	+4.5 / −0	−4.0 / −7.0	6 / 17.5	+7 / −0	−6 / −10.5	10 / 34	+12 / −0	−10 / −22	14 / 50	+18 / −0	−14 / −32
3.15 — 4.73	1.4 / 7.1	+3.5 / 0	−1.4 / −3.6	3.0 / 11.5	+5.0 / 0	−3.0 / −6.5	5.0 / 13.5	+5.0 / −0	−5.0 / −8.5	7 / 21	+9 / −0	−7 / −12	11 / 39	+14 / −0	−11 / −25	16 / 60	+22 / −0	−16 / −38

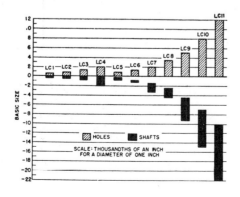

GRAPHICAL REPRESENTATION OF STANDARD LOCATIONAL CLEARANCE FITS

APPENDIX

26. LOCATIONAL TRANSITION FITS ANSI B4.1–1967

Limits are in thousandths of an inch.
Limits for hole and shaft are applied algebraically to the basic size to obtain the limits of size for the mating parts.

Nominal Size Range Inches Over To	Class LT 1			Class LT 2			Class LT 3			Class LT 4			Class LT 5			Class LT 6		
	Fit	Standard Limits		Fit	Standard Limits		Fit	Standard Limits		Fit	Standard Limits		Fit	Standard Limits		Fit	Standard Limits	
		Hole	Shaft		Hole	Shaft		Hole	Shaft		Hole	Shaft		Hole	Shaft		Hole	Shaft
0 – 0.12	−0.10 +0.50	+0.4 −0	+0.10 −0.10	−0.2 +0.8	+0.6 −0	+0.2 −0.2							−0.5 +0.15	+0.4 −0	+0.5 +0.25	−0.65 +0.15	+0.4 −0	−0.65 +0.25
0.12 – 0.24	−0.15 +0.65	+0.5 −0	+0.15 −0.15	−0.25 +0.95	+0.7 −0	+0.25 −0.25							−0.6 +0.2	+0.5 −0	+0.6 +0.3	−0.8 +0.2	+0.5 −0	+0.8 +0.3
0.24 – 0.40	−0.2 +0.8	+0.6 −0	+0.2 −0.2	−0.3 +1.2	+0.9 −0	+0.3 −0.3	−0.5 +0.5	+0.6 −0	+0.5 +0.1	−0.7 +0.8	+0.9 −0	+0.7 +0.1	−0.8 +0.2	+0.6 −0	+0.8 +0.4	−1.0 +0.2	+0.6 −0	+1.0 +0.4
0.40 – 0.71	−0.2 +0.9	+0.7 −0	+0.2 −0.2	−0.35 +1.35	+1.0 −0	+0.35 −0.35	−0.5 +0.6	+0.7 −0	+0.5 +0.1	−0.8 +0.9	+1.0 −0	+0.8 +0.1	−0.9 +0.2	+0.7 −0	+0.9 +0.5	−1.2 +0.2	+0.7 −0	+1.2 +0.5
0.71 – 1.19	−0.25 +1.05	+0.8 −0	+0.25 −0.25	−0.4 +1.6	+1.2 −0	+0.4 −0.4	−0.6 +0.7	+0.8 −0	+0.6 +0.1	−0.9 +1.1	+1.2 −0	+0.9 +0.1	−1.1 +0.2	+0.8 −0	+1.1 +0.6	−1.4 +0.2	+0.8 −0	+1.4 +0.6
1.19 – 1.97	−0.3 +1.3	+1.0 −0	+0.3 −0.3	−0.5 +2.1	+1.6 −0	+0.5 −0.5	−0.7 +0.9	+1.0 −0	+0.7 +0.1	−1.1 +1.5	+1.6 −0	+1.1 +0.1	−1.3 +0.3	+1.0 −0	+1.3 +0.7	−1.7 +0.3	+1.0 −0	+1.7 +0.7
1.97 – 3.15	−0.3 +1.5	+1.2 −0	+0.3 −0.3	−0.6 +2.4	+1.8 −0	+0.6 −0.6	−0.8 +1.1	+1.2 −0	+0.8 +0.1	−1.3 +1.7	+1.8 −0	+1.3 +0.1	−1.5 +0.4	+1.2 −0	+1.5 +0.8	−2.0 +0.4	+1.2 −0	+2.0 +0.8
3.15 – 4.73	−0.4 +1.8	+1.4 −0	+0.4 −0.4	−0.7 +2.9	+2.2 −0	+0.7 −0.7	−1.0 +1.3	+1.4 −0	+1.0 +0.1	−1.5 +2.1	+2.2 −0	+1.5 +0.1	−1.9 +0.4	+1.4 −0	+1.9 +1.0	−2.4 +0.4	+1.4 −0	+2.4 +1.0

27. LOCATIONAL INTERFERENCE FITS

Limits are in thousandths of an inch.
Limits for hole and shaft are applied algebraically to the basic size to obtain the limits of size for the parts.

Nominal Size Range Inches Over To	Class LN 1			Class LN 2			Class LN 3		
	Limits of Interference	Standard Limits		Limits of Interference	Standard Limits		Limits of Interference	Standard Limits	
		Hole	Shaft		Hole	Shaft		Hole	Shaft
0 – 0.12	0 0.45	+0.25 −0	+0.45 +0.25	0 0.65	+0.4 −0	+0.65 +0.4	0.1 0.75	+0.4 −0	+0.75 +0.5
0.12 – 0.24	0 0.5	+0.3 −0	+0.5 +0.3	0 0.8	+0.5 −0	+0.8 +0.5	0.1 0.9	+0.5 0	+0.9 +0.6
0.24 – 0.40	0 0.65	+0.4 −0	+0.65 +0.4	0 1.0	+0.6 −0	+1.0 +0.6	0.2 1.2	+0.6 −0	+1.2 +0.8
0.40 – 0.71	0 0.8	+0.4 −0	+0.8 +0.4	0 1.1	+0.7 −0	+1.1 +0.7	0.3 1.4	+0.7 −0	+1.4 +1.0
0.71 – 1.19	0 1.0	+0.5 −0	+1.0 +0.5	0 1.3	+0.8 −0	+1.3 +0.8	0.4 1.7	+0.8 −0	+1.7 +1.2
1.19 – 1.97	0 1.1	+0.6 −0	+1.1 +0.6	0 1.6	+1.0 −0	+1.6 +1.0	0.4 2.0	+1.0 −0	+2.0 +1.4
1.97 – 3.15	0.1 1.3	+0.7 −0	+1.3 +0.7	0.2 2.1	+1.2 −0	+2.1 +1.4	0.4 2.3	+1.2 −0	+2.3 +1.6
3.15 – 4.73	0.1 1.6	+0.9 −0	+1.6 +1.0	0.2 2.5	+1.4 −0	+2.5 +1.6	0.6 2.9	+1.4 −0	+2.9 +2.0

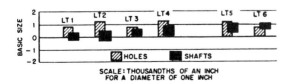

GRAPHICAL REPRESENTATION OF STANDARD LOCATIONAL TRANSITION FITS

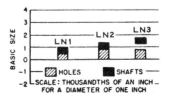

GRAPHICAL REPRESENTATION OF STANDARD LOCATIONAL INTERFERENCE FITS

28. FORCE AND SHRINK FITS ANSI B4.1–1967

Limits are in thousandths of an inch.
Limits for hole and shaft are applied algebraically to the basic size to obtain the limits of size for the parts.

Nominal Size Range Inches Over To	Class FN 1 Limits of Interference	Class FN 1 Standard Limits Hole	Class FN 1 Standard Limits Shaft	Class FN 2 Limits of Interference	Class FN 2 Standard Limits Hole	Class FN 2 Standard Limits Shaft	Class FN 3 Limits of Interference	Class FN 3 Standard Limits Hole	Class FN 3 Standard Limits Shaft	Class FN 4 Limits of Interference	Class FN 4 Standard Limits Hole	Class FN 4 Standard Limits Shaft	Class FN 5 Limits of Interference	Class FN 5 Standard Limits Hole	Class FN 5 Standard Limits Shaft
0 – 0.12	0.05 / 0.5	+0.25 / −0	+0.5 / +0.3	0.2 / 0.85	+0.4 / −0	+0.85 / +0.6				0.3 / 0.95	+0.4 / −0	+0.95 / +0.7	0.3 / 1.3	+0.6 / −0	+1.3 / +0.9
0.12 – 0.24	0.1 / 0.6	+0.3 / −0	+0.6 / +0.4	0.2 / 1.0	+0.5 / −0	+1.0 / +0.7				0.4 / 1.2	+0.5 / −0	+1.2 / +0.9	0.5 / 1.7	+0.7 / −0	+1.7 / +1.2
0.24 – 0.40	0.1 / 0.75	+0.4 / −0	+0.75 / +0.5	0.4 / 1.4	+0.6 / −0	+1.4 / +1.0				0.6 / 1.6	+0.6 / −0	+1.6 / +1.2	0.5 / 2.0	+0.9 / −0	+2.0 / +1.4
0.40 – 0.56	0.1 / 0.8	−0.4 / −0	+0.8 / +0.5	0.5 / 1.6	+0.7 / −0	+1.6 / +1.2				0.7 / 1.8	+0.7 / −0	+1.8 / +1.4	0.6 / 2.3	+1.0 / −0	+2.3 / +1.6
0.56 – 0.71	0.2 / 0.9	+0.4 / −0	+0.9 / +0.6	0.5 / 1.6	+0.7 / −0	+1.6 / +1.2				0.7 / 1.8	+0.7 / −0	+1.8 / +1.4	0.8 / 2.5	+1.0 / −0	+2.5 / +1.8
0.71 – 0.95	0.2 / 1.1	+0.5 / −0	+1.1 / +0.7	0.6 / 1.9	+0.8 / −0	+1.9 / +1.4				0.8 / 2.1	+0.8 / −0	+2.1 / +1.6	1.0 / 3.0	+1.2 / −0	+3.0 / +2.2
0.95 – 1.19	0.3 / 1.2	+0.5 / −0	+1.2 / +0.8	0.6 / 1.9	+0.8 / −0	+1.9 / +1.4	0.8 / 2.1	+0.8 / −0	+2.1 / +1.6	1.0 / 2.3	+0.8 / −0	+2.3 / +1.8	1.3 / 3.3	+1.2 / −0	+3.3 / +2.5
1.19 – 1.58	0.3 / 1.3	+0.6 / −0	+1.3 / +0.9	0.8 / 2.4	+1.0 / −0	+2.4 / +1.8	1.0 / 2.6	+1.0 / −0	+2.6 / +2.0	1.5 / 3.1	+1.0 / −0	+3.1 / +2.5	1.4 / 4.0	+1.6 / −0	+4.0 / +3.0
1.58 – 1.97	0.4 / 1.4	+0.6 / −0	+1.4 / +1.0	0.8 / 2.4	+1.0 / −0	+2.4 / +1.8	1.2 / 2.8	+1.0 / −0	+2.8 / +2.2	1.8 / 3.4	+1.0 / −0	+3.4 / +2.8	2.4 / 5.0	+1.6 / −0	+5.0 / +4.0
1.97 – 2.56	0.6 / 1.8	+0.7 / −0	+1.8 / +1.3	0.8 / 2.7	+1.2 / −0	+2.7 / +2.0	1.3 / 3.2	+1.2 / −0	+3.2 / +2.5	2.3 / 4.2	+1.2 / −0	+4.2 / +3.5	3.2 / 6.2	+1.8 / −0	+6.2 / +5.0
2.56 – 3.15	0.7 / 1.9	+0.7 / −0	+1.9 / +1.4	1.0 / 2.9	+1.2 / −0	+2.9 / +2.2	1.8 / 3.7	+1.2 / −0	+3.7 / +3.0	2.8 / 4.7	+1.2 / −0	+4.7 / +4.0	4.2 / 7.2	+1.8 / −0	+7.2 / +6.0
3.15 – 3.94	0.9 / 2.4	+0.9 / −0	+2.4 / +1.8	1.4 / 3.7	+1.4 / −0	+3.7 / +2.8	2.1 / 4.4	+1.4 / −0	+4.4 / +3.5	3.6 / 5.9	+1.4 / −0	+5.9 / +5.0	4.8 / 8.4	+2.2 / −0	+8.4 / +7.0
3.94 – 4.73	1.1 / 2.6	+0.9 / −0	+2.6 / +2.0	1.6 / 3.9	+1.4 / −0	+3.9 / +3.0	2.6 / 4.9	+1.4 / −0	+4.9 / +4.0	4.6 / 6.9	+1.4 / −0	+6.9 / +6.0	5.8 / 9.4	+2.2 / −0	+9.4 / +8.0
	Light Drive Fit. Light pressures, thin sections, long engagements, or cast-iron housings			Medium Drive Fit. For thin steel or high grade cast-iron housings			Heavy Drive Fit For Medium Steel Sections			Force Fits. For highly stressed and shrink fits.					

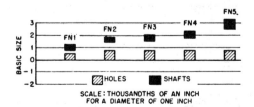

GRAPHICAL REPRESENTATION OF STANDARD FORCE OR SHRINK FITS

29. GRAPHICAL SYMBOLS

Pipe Fittings and Valves.
ANSI Y32.4–1955.

	Flanged	Screwed	Bell and Spigot	Welded	Soldered
Joint					
Elbow—90 degree					
Elbow—45 degree					
Elbow—Turned Up					
Elbow—Turned Down					
Elbow—Long Radius					
Tee—Outlet Up					
Tee—Outlet Down					
Tee					
Reducer, Concentric					
Reducer, Eccentric					
Lateral					
Cross					
Reducing Elbow					
Gate Valve, Elevation					
Globe Valve, Elevation					
Stop Cock					
Safety Valve					
Check Valve					

30. AMERICAN STANDARD WROUGHT IRON AND STEEL PIPE AND TAPER PIPE THREADS

Compiled from ANSI B36.10–1959, ANSI B36.19–1965, ANSI B2.1–1960 and Manufacturers Catalogs

Nominal Pipe Size	Outside Dia (O.D.)	Threads Per Inch	Thread Data Thread Engagement Tight	Thread Data Tap Drill Sizes	Schedule 40 I.D.	Schedule 40 Wt, lb/ft	Schedule 80 I.D.	Schedule 80 Wt, lb/ft	Schedule 160 I.D.	Schedule 160 Wt, lb/ft
1/8	0.405	27	1/4	11/32	.269	.244 as	.215	.314		
1/4	0.540	18	3/8	7/16	.364	.424 pas	.302	.535 p		
3/8	0.675	18	3/8	37/64	.493	.567 pas	.423	.738 ps		
1/2	0.840	14	1/2	23/32	.622	.850 pas	.546	1.087 ps	.466	1.300
3/4	1.050	14	9/16	59/64	.824	1.130 pas	.742	1.473 ps	.614	1.940
1	1.315	11 1/2	11/16	1 5/32	1.049	1.678 pas	.957	2.171 pas	.815	2.840
1 1/4	1.660	11 1/2	11/16	1 1/2	1.380	2.272 pas	1.278	2.996 pas	1.160	3.764
1 1/2	1.990	11 1/2	11/16	1 47/64	1.610	2.717 pas	1.500	3.631 pas	1.338	4.862
2	2.375	11 1/2	3/4	2 7/32	2.067	3.652 pas	1.939	5.022 pas	1.689	7.440
2 1/2	2.875	8	15/16	2 5/8	2.469	5.79 pas	2.323	7.66 pas	2.125	10.01
3	3.500	8	1	3 1/4	3.068	7.58 pas	2.900	10.25 pas	2.624	14.32
3 1/2	4.000	8	1 1/16	3 3/4	3.548	9.11 pas	3.364	12.51 as		
4	4.500	8	1 1/8	4 1/4	4.026	10.79 pas	3.826	14.98 pas	3.438	22.51
5	5.563	8	1 1/4	5 5/16	5.047	14.62 pas	4.813	20.78 pas	4.313	32.96
6	6.625	8	1 5/16	6 5/16	6.065	18.97 pas	5.761	28.57 pas	5.189	45.30
8	8.625	8	1 7/16		7.981	28.55 pas	7.625	43.39 pas	6.813	74.69
10	10.750	8	1 5/8		10.020	40.48 pas	9.564	64.33 p	8.500	115.65
12	12.750	8			11.938	53.53 pa	11.376	88.51	10.126	160.27
14	14.00	8			13.124	63.37	12.500	106.13	11.188	189.12
16	16.00	8			15.000	82.77	14.314	136.46	12.814	245.11
18	18.00	8			16.876	104.75	16.126	170.75	14.438	308.51
20	20.00	8			18.814	122.91	17.938	208.87	16.064	379.01
24	24.00	8			22.626	171.17	21.564	296.36	19.314	541.94

Notes:
p: Available in rigid plastic: weight reduction factor, f ≅ 0.18
a: Available in aluminum: f ≅ .35
s: Available in stainless steel: f ≅ 1.00

APPENDIX

31. SCREWED AND FLANGED FITTINGS AND VALVE SIZES

Compiled from Manufacturer's Catalogs.

Screwed Fitting	Nominal Dia.	Outside Length	
		Screwed	Flanged
Unions (Mal. Iron)	1/2	1 7/8	
	3/4	2 1/8	
	1	2 3/8	
	1 1/4	2 5/8	
	1 1/2	2 15/16	
	2	3 1/4	
	2 1/2	3 9/16	
Globe Valves (Brass)	1/2	2 11/16	
	3/4	3 3/16	
	1	3 3/4	4 3/8
	1 1/4	4 1/4	4 13/16
	1 1/2	4 3/4	5 1/2
	2	5 3/4	6 1/2
	2 1/2	6 3/4	7 1/2
Gate Valves (Brass)	1/2	2 1/8	
	3/4	2 3/8	
	1	2 7/8	3 3/8
	1 1/4	3 3/16	3 7/8
	1 1/2	3 7/16	4 3/8
	2	3 7/8	5 1/2
	2 1/2	4 1/2	6 1/2
Reducers (Mal. Iron) R = Smaller Size	1/2 X R	1 1/4	
	3/4 X R	1 7/16	
	1 X R	1 11/16	
	1 1/4 X R	2 1/16	
	1 1/2 X R	2 5/16	
	2 X R	2 13/16	5
	2 1/2 X R	3 1/4	5 1/2
Reducing Tees (Mal. Iron)		X Y Z	
	1 1/4 X 1 X 3/4	1 7/16 1 5/16 1 9/16	Sizes correspond to sizes for largest opening. (Table 33)
	1 X 3/4 X 1/2	1 1/4 1 3/16 1 5/16	
	3/4 X 1/2 X 1/2	1 3/16 1 1/8 1 1/4	
	1/2 X 1/2 X 3/4	1 1/4 1 1/4 1 3/16	
Nipples*	1/2	Close 1 1/8 Short 1 1/2	
	3/4	Close 1 3/8 Short 2	
	1	Close 1 1/2 Short 2	
	1 1/4	Close 1 5/8 Short 2 1/2	
	1 1/2	Close 1 3/4 Short 2 1/2	
	2	Close 2 Short 2 1/2	
	2 1/2	Close 2 1/2 Short 3	

* Long Nipples: Short to 6", in 1/2" increments of length
                6" to 12", in 1" increments of length
                12" to 24", in 2" increments of length

32. MALLEABLE IRON SCREWED FITTINGS—150 LB
Compiled from ANSI B16.3—1958.

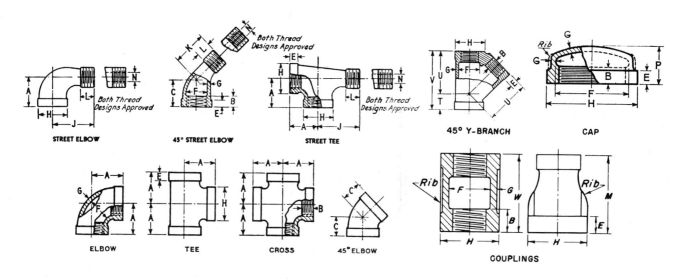

Nom. Pipe Size	A	B	C	E	F	G	H	J	K	L	M	N	P	T	U	V	W
1/8	0.69	0.25		0.200	0.405	0.090	0.693	1.00		0.2638		0.20	0.53				0.96
1/4	0.81	0.32	0.73	0.215	0.540	0.095	0.844	1.19	0.94	0.4018	1.00	0.26	0.63				1.06
3/8	0.95	0.36	0.80	0.230	0.675	0.100	1.015	1.44	1.03	0.4078	1.13	0.37	0.74	0.50	1.43	1.93	1.16
1/2	1.12	0.43	0.88	0.249	0.840	0.105	1.197	1.63	1.15	0.5337	1.25	0.51	0.87	0.61	1.71	2.32	1.34
3/4	1.31	0.50	0.98	0.273	1.050	0.120	1.458	1.89	1.29	0.5457	1.44	0.69	0.97	0.72	2.05	2.77	1.52
1	1.50	0.58	1.12	0.302	1.315	0.134	1.771	2.14	1.47	0.6828	1.69	0.91	1.16	0.85	2.43	3.28	1.67
1 1/4	1.75	0.67	1.29	0.341	1.660	0.145	2.153	2.45	1.71	0.7068	2.06	1.19	1.28	1.02	2.92	3.94	1.93
1 1/2	1.94	0.70	1.43	0.368	1.900	0.155	2.427	2.69	1.88	0.7235	2.31	1.39	1.33	1.10	3.28	4.38	2.15
2	2.25	0.75	1.68	0.422	2.375	0.173	2.963	3.26	2.22	0.7565	2.81	1.79	1.45	1.24	3.93	5.17	2.53
2 1/2	2.70	0.92	1.95	0.478	2.875	0.210	3.589	3.86	2.57	1.1375	3.25	2.20	1.70	1.52	4.73	6.25	2.88
3	3.08	0.98	2.17	0.548	3.500	0.231	4.285	4.51	3.00	1.2000	3.69	2.78	1.80	1.71	5.55	7.26	3.18
3 1/2	3.42	1.03	2.39	0.604	4.000	0.248	4.843						1.90				
4	3.79	1.08	2.61	0.661	4.500	0.265	5.401	5.69	3.70	1.3000	4.38	3.70	2.08	2.01	6.97	8.98	3.69
5	4.50	1.18	3.05	0.780	5.563	0.300	6.583	6.86		1.4063		4.69	2.32				
6	5.13	1.28	3.46	0.900	6.625	0.336	7.767	8.03		1.5125		5.67	2.55				

33. CLASS 125 CAST IRON FLANGED FITTINGS

Compiled from ANSI B16.1–1960.

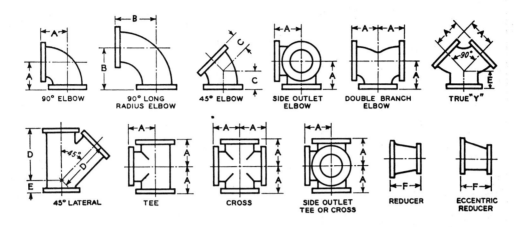

Nom. Pipe Size	Inside Diam. Fitting	A	B	C	D	E	F	Flange Diam.	Flange Thickness	Wall Thickness
1	1	3 1/2	5	1 3/4	5 3/4	1 3/4		4 1/4	7/16	5/16
1 1/4	1 1/4	3 3/4	5 1/2	2	6 1/4	1 3/4		4 5/8	1/2	5/16
1 1/2	1 1/2	4	6	2 1/4	7	2		5	9/16	5/16
2	2	4 1/2	6 1/2	2 1/2	8	2 1/2	5	6	5/8	5/16
2 1/2	2 1/2	5	7	3	9 1/2	2 1/2	5 1/2	7	11/16	5/16
3	3	5 1/2	7 3/4	3	10	3	6	7 1/2	3/4	3/8
3 1/2	3 1/2	6	8 1/2	3 1/2	11 1/2	3	6 1/2	8 1/2	13/16	7/16
4	4	6 1/2	9	4	12	3	7	9	15/16	1/2
5	5	7 1/2	10 1/4	4 1/2	13 1/2	3 1/2	8	10	15/16	1/2
6	6	8	11 1/2	5	14 1/2	3 1/2	9	11	1	9/16
8	8	9	14	5 1/2	17 1/2	4 1/2	11	13 1/2	1 1/8	5/8
10	10	11	16 1/2	6 1/2	20 1/2	5	12	16	1 3/16	3/4
12	12	12	19	7 1/2	24 1/2	5 1/2	14	19	1 1/4	13/16
14 OD	14	14	21 1/2	7 1/2	27	6	16	21	1 3/8	7/8
16 OD	16	15	24	8	30	6 1/2	18	23 1/2	1 7/16	1

34. TYPICAL DRAWN THINWALL TUBING
Compiled and Condensed from Warehousers' Catalogs

O.D.	Wall	I.D.	Wt./Ft.=Lb. Aluminum	Wt./Ft.=Lb. C.S. & S.S.
.250	.028	.194	.023	.066
	.035	.180	.028	.080
.3125	.035	.242	.036	.104
	.049	.214	.048	.138
.375	.035	.305	.044	.127
	.049	.277	.059	.171
.4375	.035	.367	.052	.151
	.049	.339	.070	.204
.500	.028	.444	.049	.141
	.049	.402	.082	.236
	.065	.370	.104	.302
.625	.028	.569	.062	.178
	.049	.527	.104	.301
	.065	.495	.134	.389
.750	.035	.680	.092	.267
	.065	.620	.164	.476
	.083	.584	.205	.591
.875	.035	.805	.109	.314
	.049	.777	.150	.432
	.065	.745	.194	.562
1.000	.035	.930	.125	.361
	.058	.884	.202	
	.065	.870	.225	.649
	.083	.834	.281	.813
1.250	.035	1.180	.157	.454
	.058	1.134	.255	
	.065	1.120	.285	.823
	.083	1.084	.358	1.034
1.500	.035	1.430	.189	.548
	.058	1.384	.309	
	.065	1.370	.345	.996
	.083	1.334	.435	1.256

35. TYPICAL SQUARE AND RECTANGULAR TUBING
Compiled and Condensed from Warehousers' Catalogs

Outside Dimensions	Wall	Wt./Ft.=Lb. Steel	Wt./Ft.=Lb. Aluminum
1 x 1	.125		.526
	.120	1.436	
1.25 x 1.25	.125		.696
	.120	1.8444	
1.50 x 1.50	.125		.826
	.120	2.252	
1.75 x 1.75	.125		.974
	.120	2.660	
2 x 2	.125		1.126
	.120	3.068	
4 x 4	.250	12.02	
6 x 6	.250	18.82	
8 x 8	.250	36.83	
1.75 x 3	.125		1.350
x 4	.125		1.650
x 5	.125		1.950
1 x 2	.083	1.60	
	.120	3.88	
2 x 3	.125		1.426
	.188	8.14	
2 x 5	.125		2.026

36. AMERICAN STANDARD SQUARE AND FLAT, PLAIN TAPER, AND GIB HEAD KEYS
Compiled from ANSI B17.1–1967.

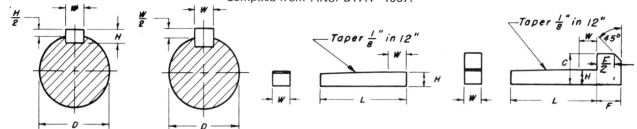

Shaft Size D	Square W	Flat W × H	Gib Head Square C	F	Flat C	F
1/2 - 9/16	1/8	1/8 × 3/32	1/4	1/4	3/16	1/8
5/8 - 7/8	3/16	3/16 × 1/8	5/16	5/16	1/4	1/4
15/16-1 1/4	1/4	1/4 × 3/16	7/16	3/8	5/16	5/16
1 5/16-1 3/8	5/16	5/16 × 1/4	1/2	7/16	7/16	3/8
1 7/16-1 3/4	3/8	3/8 × 1/4	5/8	1/2	7/16	3/8
1 13/16-2 1/4	1/2	1/2 × 1/4	7/8	5/8	5/8	1/2
2 5/16-2 3/4	5/8	5/8 × 7/16	1	3/4	3/4	9/16
2 7/8 -3 1/4	3/4	3/4 × 1/2	1 1/4	7/8	7/8	5/8

Increments of L = 2W   Maximum L = 16W   Minimum L = 4W

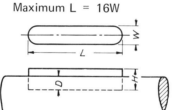

37. PRATT AND WHITNEY KEYS

Key No.	L	W	H	D	Key No.	L	W	H	D
1	1/2	1/16	3/32	1/16	C	1 1/8	5/16	15/32	5/16
2	1/2	3/32	9/64	3/32	19	1 1/4	3/16	9/32	3/16
3	1/2	1/8	3/16	1/8	20	1 1/4	7/32	21/64	7/32
4	5/8	3/32	9/64	3/32	21	1 1/4	1/4	3/8	1/4
5	5/8	1/8	3/16	1/8	D	1 1/4	5/16	15/32	5/16
6	5/8	5/32	15/64	5/32	E	1 1/4	3/8	9/16	3/8
7	3/4	1/8	3/16	1/8	22	1 3/8	1/4	3/8	1/4
8	3/4	5/32	15/64	5/32	23	1 3/8	5/16	15/32	5/16
9	3/4	3/16	9/32	3/16	F	1 3/8	3/8	9/16	3/8
10	7/8	5/32	15/64	5/32	24	1 1/2	1/4	3/8	1/4
11	7/8	3/16	9/32	3/16	25	1 1/2	5/16	15/32	5/16
12	7/8	7/32	21/64	7/32	G	1 1/2	3/8	9/16	3/8
A	7/8	1/4	3/8	1/4	51	1 3/4	1/4	3/8	1/4
13	1	3/16	9/32	3/16	52	1 3/4	5/16	15/32	5/16
14	1	7/32	21/64	7/32	53	1 3/4	3/8	9/16	3/8
15	1	1/4	3/8	1/4	26	2	3/16	9/32	3/16
B	1	5/16	15/32	5/16	27	2	1/4	3/8	1/4
16	1 1/8	3/16	9/32	3/16	28	2	5/16	15/32	5/16
17	1 1/8	7/32	21/64	7/32	29	2	3/8	9/16	3/8
18	1 1/8	1/4	3/8	1/4	54	2 1/4	1/4	3/8	1/4

Length must equal at least 2W.

38. WOODRUFF KEYS
Compiled from ANSI B17.2–1967.

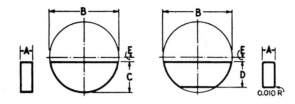

Key Number	Nominal Key Size A X B	Height of Key C Max.	C Min.	D Max.	D Min.	Distance Below Center E	Depth of Key Slot in Shaft
204	1/16 X 1/2	0.203	0.198	0.194	0.188	3/64	.1668
304	3/32 X 1/2	.203	.198	.194	.188	3/64	.1511
305	3/32 X 5/8	.250	.245	.240	.234	1/16	.1981
404	1/8 X 1/2	.203	.198	.194	.188	3/64	.1355
405	1/8 X 5/8	.250	.245	.240	.234	1/16	.1825
406	1/8 X 3/4	.313	.308	.303	.297	1/16	.2455
505	5/32 X 5/8	.250	.245	.240	.234	1/16	.1669
506	5/32 X 3/4	.313	.308	.303	.297	1/16	.2299
507	5/32 X 7/8	.375	.370	.365	.359	1/16	.2919
606	3/16 X 3/4	.313	.308	.303	.297	1/16	.2143
607	3/16 X 7/8	.375	.370	.365	.359	1/16	.2763
608	3/16 X 1	.438	.433	.428	.422	1/16	.3393
609	3/16 X 1 1/8	.484	.479	.475	.469	5/64	.3853
807	1/4 X 7/8	.375	.370	.365	.359	1/16	.2450
808	1/4 X 1	.438	.433	.428	.422	1/16	.3080
809	1/4 X 1 1/8	.484	.479	.475	.469	5/64	.3540
810	1/4 X 1 1/4	.547	.542	.537	.531	5/64	.4170
811	1/4 X 1 3/8	.594	.589	.584	.578	3/32	.4640
812	1/4 X 1 1/2	.641	.636	.631	.625	7/64	.5110
1008	5/16 X 1	.438	.433	.428	.422	1/16	.2768
1009	5/16 X 1 1/8	.484	.479	.475	.469	5/64	.3228
1010	5/16 X 1 1/4	.547	.542	.537	.531	5/64	.3858
1011	5/16 X 1 3/8	.594	.589	.584	.578	3/32	.4328
1012	5/16 X 1 1/2	.641	.636	.631	.625	7/64	.4798
1210	3/8 X 1 1/4	.547	.542	.537	.531	5/64	.3545
1211	3/8 X 1 3/8	.594	.589	.584	.578	3/32	.4015
1212	3/8 X 1 1/2	.641	.636	.631	.625	7/64	.4485

Key numbers indicate the nominal key dimensions. The last two digits give the nominal diameter (B) in eighths of an inch and the digits preceding the last two give the nominal width (A) in thirty-seconds of an inch.

39. BIBLIOGRAPHY

A selected list of references related to Engineering Graphics supplements this text which is limited to brief explanations on some material.

STANDARDS

American National Standards Institute, 345 East 47th St., New York, N.Y. 10017, for complete standards listing, index and price list.

I.S.O. = International Organization for Standardization.

Abbreviations

Abbreviations for Scientific and Engineering Terms, Z10.1-1941
Abbreviations for Use on Drawings, Z32.13-1950

Bolts, Nuts, Rivets, and Screws

Hexagonal Head Cap Screws, Slotted Head Cap Screws, Square Head Setscrews, and Slotted Headless Setscrews, B18.6.2-1956
Slotted and Recessed Head Wood Screws, B18.6.1-1961
Hexagon Cap Screws, Slotted Head Cap Screws, Square Head Set Screws, Slotted Headless Set Screws, B18.6.2-1956
Socket Cap, Shoulder and Set Screws, B18.3-1961
Machine Screw Nuts, B18.6.3-1962 (I.S.O. R272)
Square and Hexagonal Bolts and Screws, Including Hex Cap Screws and Lag Screws, B18.2.1-1965
Square and Hex Nuts, B18.2.2-1965

Charts and Symbols

Illustrations for Publication and Projection, Y15.1-1959
Graphical Symbols for Electrical and Electronics Diagrams, Y32.2-1967
Graphical Symbols for Logic Diagrams, Y32.14-1962
Reference Designations for Electrical and Electronic Parts and Equipment, Y32.16-1968
Graphical Symbols for Heating, Ventilating, and Air Conditioning, Z32.2.4-1953
Graphical Symbols for Plumbing, Y32.4-1955
Graphical Symbols for Welding, Y32.3-1959

Dimensioning and Specifications

Preferred Limits and Fits for Cylindrical Parts, B4.1-1967 (I.S.O. 286)
Surface Texture, B46.1-1962

Drafting Standards

Sec. 1, Size and Format, Y14.1-1957
Sec. 2. Line Conventions, Sectioning and Lettering, Y14.2-19
Sec. 3. Multiview Drawing, Y14.3-19
Sec. 4. Pictorial Drawing, Y14.4-1957
Sec. 5. Dimensioning and Tolerances for Engineering Drawings, Y14.5-1966
Sec. 6, Screw Threads, Y14.6-1957
Sec. 7, Gears, Splines, and Serrations, Y14.7-1958

Threads

Nomenclature, Definitions, and Letter Symbols for Screw Threads, B1.7-1965
Pipe Threads, B2.1-1968
Unified Screw Threads, B1.1-1960

Miscellaneous

Lock Washers, B27.1-1965
Plain Washers, B27.2-1965
Keys and Keyseats, B17.1-1967
Woodruff Keys and Keyseats, B17.2-1967
Many other Standards are available and others are in preparation.

REFERENCE BOOKS

Aeronautical

Anderson, N. H., *Aircraft Layout and Detail Design*. McGraw-Hill
Katz, H. H., *Aircraft Drafting*. Macmillan
Meadowcroft, N., *Aircraft Detail Drafting*. McGraw-Hill
SAE Aeronautical Drafting Manual. Society of Automotive Engineers, Inc., 485 Lexington Ave., New York, N.Y. 10017

Descriptive Geometry

Grant, H. E., *Practical Descriptive Geometry*. McGraw-Hill
Hood, Palmerlee, and Baer, *Geometry of Engineering Drawing*. McGraw-Hill
Paré, Loving, and Hill, *Descriptive Geometry*. Macmillan
Slaby, S. M., *Fundamentals of Three-Dimensioned Descriptive Geometry*. Harcourt, Brace, and World
Warner and McNeary, *Applied Descriptive Geometry*. McGraw-Hill
Wellman, B. L., *Technical Descriptive Geometry*. McGraw-Hill

Engineering Drawing

French, T. E. and Vierck, C. J., *Engineering Drawing*. McGraw-Hill
Giesecke, Mitchell, Spencer, Hill, and Loving, *Technical Drawing*. Macmillan
Hoelscher, R. P., and Springer, C. H., *Engineering Drawing and Geometry*. Wiley
Luzadder, W. J., *Fundamentals of Engineering Drawing*. Prentice-Hall
Zozzora, F., *Engineering Drawing*. McGraw-Hill

Engineering Graphics and Design

Arnold, J. N., *Introductory Graphics*. McGraw-Hill
Earle, J. H., *Engineering Design Graphics*. Addison-Wesley
French, T. E. and Vierck, C. J., *Graphic Science*. McGraw-Hill
Hammond, Buck, Rogers, Walsh, Ackert, *Engineering Graphics*. Ronald
Hoelscher, Springer, and Dobrovolny, *Graphics for Engineers*. Wiley
Levens, A. S., *Graphics*. Wiley
Luzadder, W. J., *Basic Graphics*. Prentice-Hall
Paré, Francis, and Kimbrell, *Introduction to Engineering Design*. Holt, Rinehart, and Winston

Rising, J. S. and Almfeldt, M. W., *Engineering Graphics*. Wm. C. Brown Company
Rule, J. T. and Coons, S. A., *Graphics*. McGraw-Hill
Springer, Palmer, Klienhenz, and Bullen, *Basic Graphics*. Allyn and Bacon
Svensen, C. L. and Street, W. E., *Engineering Graphics*. Van Nostrand

Graphical Computations

Hoelscher, Arnold, and Pierce, *Graphic Aids in Engineering Computations*. McGraw-Hill
Karsten, K. G., *Charts and Graphs*. Prentice-Hall
Lipka, J., *Graphical and Mechanical Computation*. Wiley
Mackey, C. O., *Graphic Solutions*. Wiley
Malcolm, C. W., *Graphic Statics*. McGraw-Hill

Illustration

Farmer, Hoecker, and Vavrin, *Illustrating for Tomorrow's Production*. Macmillan
Hoelscher, Springer, and Pohle, *Industrial Production Illustration*. McGraw-Hill
Treacy, J., *Production Illustration*. Wiley
McCartney, T. O., *Precision Perspective Drawing*. McGraw-Hill

Nomography

Davis, D. S., *Empirical Equations and Nomography*. McGraw-Hill
Douglass, and Adams, *Elements of Nomography*. McGraw-Hill
Johnson, L. H., *Nomography and Empirical Equations*. Wiley
Levens, A. S., *Nomography*. Wiley

Structural Drafting

Bishop, C. T., *Structural Drafting*. Wiley
Steel Construction Manual. American Institute of Steel Construction, 101 Park Ave., New York, N.Y. 10017
Structural Steel Detailing. American Institute of Steel Construction, 101 Park Ave., New York, N.Y. 10017

Welding

Procedure Handbook of Arc Welding Design and Practice. Lincoln Electric Co., Cleveland, Ohio

GREEK ALPHABET

A, α Alpha	H, η Eta	N, ν Nu	T, τ Tau
B, β Beta	Θ, θ Theta	Ξ, ξ Xi	Υ, υ Upsilon
Γ, γ Gamma	I, ι Iota	O, o Omicron	Φ, ϕ Phi
Δ, δ Delta	K, κ Kappa	Π, π Pi	X, χ Chi
E, ϵ Epsilon	Λ, λ Lambda	P, ρ Rho	Ψ, ψ Psi
Z, ζ Zeta	M, μ Mu	Σ, σ, ς Sigma	Ω, ω Omega

Index

A
Acme thread table, 200, 362
Adhesives, 208
Adjacent chart
  construction of, 311
Adjacent views, defined, 48
Algebra, graphical, 305, 307
Aligned views, 47
Alignment charts, 308
  conversion charts, 311
  forms of, 319
  functional scales, 309, 310
  N charts, 316-319
  parallel scale charts, 312-316
  problems, 320-322
Allowance, defined, 231
  preferred, 233
American threads, 351
Analysis, 29
Angle between line and plane, 148, 150
Angular perspective, 101
Appendix, 339-374
  table list, 338
Architect's scale, 36
Area chart, 292
Area law, 329
Area, plane figures, 345, 346
Arithmetic scale, 309
Automation, 223
Auxiliary views, 50-52
  elevation, 50
  inclined, 50, 51
  oblique, 51
  problems, 53, 54
Axonometric, 90-95
  defined, 90
  four center ellipse, 94
  problems, 102, 103
  types, 90

B
Bar chart, 292
Basic hole system, 231
Basic shaft system, 231
Basic size, defined, 230
Basic sizes, preferred, 233
Bearing, lines, 58
Bibliography, 228, 265, 375, 376
Blueprint, 286
Bolts, 204
  nuts, 204
  standard table, 352, 353
Box of projection, 47, 48, 49
Broach, 221

C
Cabinet projection, 96, 97
Calculus, 323-325
  applied problem, 332
  area law, 329
  constant of integration, 331
  derivative curves, 324, 325
  differentiation, 323-328
  differentiation problems, 333, 334
  Durand's rule, 331
  function, 323
  integration, 328-333
  integration problems, 334, 335
  mechanical integration, 331
  pole distance, 331
  rate of change, 323, 324
  ray polygon, 329
  scale calibration, 325, 330
  semi-graphical, 331
  Simpson's rule, 331
  slope law, 324
  string polygon, 328
  summary, 332
  trapezoidal rule, 331
  Weddle's rule, 331
Calipers, 225
Cap screws, 202
  table, 353, 354
Castings, 221
  centrifugal, 223
  continuous, 223
  die, 223
  sand, 221

Cavalier projection, 96, 97
Center lines, 81
Charts
  adjacent, 311
  alignment, 308
  area, 292
  bar, 292
  classification of, 292
  column, 292
  conversion, 311
  distribution, 294
  flow, 293
  forms of, 319, 320
  N charts, 316-319
  network, 305
  organization, 294
  parallel scale, 311-316
  pictorial, 294
  pie diagram, 293
  polar, 296
  progress, 294
  rank, 294
  scale design, 309
  semilogarithmic, 299
  trilinear, 295
Clearance
  point and line, 134
  skew lines, 135, 136
Communication, 2
Component, vector, 156
Computer design, 5, 261-276
Cone, 168
  development, 189
  intersection, 179
Conic sections, 168
Constructions, geometrical, 340, 341
Contour rule, 126
Control
  computer, 275
  contouring, 272
  numerical, 272
  point to point, 272
Conventional practices, 109-114
  conventional breaks, 112

definition, 109
fillets, 113
half-view drawings, 112
intersections, 112
ribs, 109
rotated features, 112
rounds, 113
spokes, 109
thin plates, 112
Convolutes
tangent line, 169
tangent plane, 169, 191, 192
Cotter pins, 208
table, 360
Counterbore, 122
Countersink, 122
Creative design, 2, 5
Crest, 199
Critical point, 176
Curve fitting, 39, 300
averages, 300
selected points, 300
Curves
derived, 324, 325
drawing, 297
exponential, 301
fitting, 297
irregular, 39
identification, 297
isometric, 91
power, 302
straight-line, 301
titles, 297
Cutting plane, 105, 144, 174
Cylinders, 168
development, 187
intersecting, 177

D

Datum, defined, 118
Decimal equivalents, 342
Derivative curve scale, 325
Design
cathode ray tube, 275
computer aided, 5, 261-276
conclusions, 276
data processing, 267
parameters, 272
Design, creative
case study, 10
conceptualization, 8
creative, 6
defined, 6
drawings, 8
evaluation, 8, 9
finalization, 8
goals, 7
ideas, 13
ideation, 8
needs & goals, 7
parameters, 7
presentation, 10
process, 6, 12
projects, 13
reports, 10
research, 7
sketches, 8
steps, 7, 12
Design size, defined, 231
Detail drawing, 280
Developability, 166, 185

Development, 185-197
cone, 189, 190
cylinder, 187
defined, 185
practices, 185
prism, 187
problems, 196, 197
pyramid, 187
tangent plane convolute, 191, 192
transition pieces, 191, 192
Diazo, 286
Die casting, 223
Differentiation, graphical, 323-328
applied problem, 332
area law, 329
broken line, 327
chord method, 325
problems, 333, 334
scale, 330
slope law, 324
Dihedral angle, 146, 148
Dimensioning, basic, 117-132
aligned, 119
arcs, 123
base line, 126
chain, 126
contour rule, 126
criteria, 117
cylinders, 123
datum, 118
decimal, 118
definition, 118
features, 118
fractions, 118
geometrical shapes, 121
good practices, 128
holes, 123
leaders, 119
lines, 119
location, 120
measurement, units of, 118
overall, 126
problems, 130
procedure, 127
reference (REF), 126, 127
rules, 118
size, 120
standards, 117
unidirectional, 119
Dimensioning, production, 230-243
allowance selection, 232
ANSI tables, use of, 236, 237
definitions, 230
fits, classification, 233
limits, computing, 235
limits, manufacturing, 235
notes, 241
problems, 241-243
surface quality, 240
tolerances, 232, 237, 238, 240
Dimetric projection, 90
Dip, plane, 70
Director, 163
Directrix, 163
Distribution charts, 294
Dividers, 32
Double curved surfaces, 169
Dowel pins, 208
Drawing, assembly, 279
checking, 280
detail, 280

layout, 279
parts list, 280, 282
reproduction, 284
simplified, 284
titles, 280
Drawing boards, 33
Drawing equipment, 31-42
instruments, 32
list of, 31
problems, 42
scales, 36-38
special, 41
Drill
size table, 350
Drill press, 221
Durand's rule, 331

E

Electronic diagrams, 261-266
block, 261
connection, 262
diagrams, 262
grouping, 263
interconnection, 262
layout, 263
logic, 264
problems, 266
references, 265
schematic, 264
single line, 262
standards, 263
symbols, 349
wiring, 264
Electrical discharge machining, 221
Electro-chemical machining, 221
Electroforming, 223
Electron beam machining, 221
Element of a surface, 163
Ellipsoid
oblate, 170
prolate, 170
Empirical equations, 300-303
determination of, 300
exponential curves, 301
power curves, 302
straight-line curves, 301
Engineering graphics
defined, 1
divisions of, 2
Engineering illustrations, 281-284
shading, 281
types of, 281
use of, 281
Engineer's scale, 36
Equilibrant, 155
Equilibrium
definition, 155
Equivalents, table of, 348
Erasers, 40
Exponential curves, 301
Extrusion, 217

F

Fasteners, 198-216
permanent, 206, 208-214
problems, 205, 215, 216
removable, 206-208
threaded, 198-205
Fillets and rounds, 113
Fit
classification, 233

clearance, 232
interference, 232
transition, 232
Fits, ANSI B4.1
   clearance locational, 364
   force and shrink, 366
   interference locational, 365
   running and sliding, 363
   transition locational, 365
Fittings (see piping), 247-249
Flanged fittings table, 369, 371
Flow charts, 293
Force, defined, 155
   resolution of, 156
Forging, 219
Four center ellipse, 94, 97
Fractions, 26
Freehand drawing (sketching), 14-21
   application to design, 19
   circles, 17
   ellipses, 17
   equipment, 14
   objectives, 14
   pictorial, 15, 19
   problems, 20, 21
   technique, 15
   types, 14
Functional scale, 309
Funicular diagram, 328

G

Gages, 225-228
General oblique, 96
Geometric scale, 309
Geometrical constructions, 340, 341
Generatrix, 163
Grade connector, 138
Grade of lines, 59
Graphical algebra, 305
   (see Graphical Mathematics)
Graphical differentiation, 323-327
Graphical integration, 328-335
   applied problem, 332
   problems, 334, 335
   scale, 330
   string polygon, 328
   summary, 332
Graphical language, divisions of, 2
Graphical mathematics, 305-307
   network charts, 305
   powers and roots, 305
   simultaneous equations, 306
   word problems, 307
Graphical representation of design data, 292-303
   area charts, 292
   axes, location of, 296
   bar charts, 292
   column charts, 292
   curve fitting, 300
   curve identification, 297
   distribution charts, 294
   empirical equations, 300
   exponential curves, 301
   flow charts, 293
   general information charts, 293, 294
   graphing procedure, 296
   grid selection, 296
   logarithmic grids, 300
   organization charts, 294
   pictorial charts, 294
   pie diagrams, 292
   plotting, 297
   polar charts, 296
   power curves, 302
   problems, 303, 304
   progress charts, 294
   ranking charts, 294
   reproduction, 298
   scale identification, 297
   scale units, 297
   semilog curves, 299
   straight line, 301
   technical charts, 295-303
   titles, 297
   trilinear charts, 295
Graphical symbols
   electrical, 349
   fittings and valves, 367
Graphing procedure, 296, 297
Grid paper, 295, 296
   logarithmic, 300
Grinder, 219
Guide lines, 23

H

Half-view drawings, 112
Hectograph, 286
Helicoid, 166
Hidden lines, 81
Hinge lines, 48, 49
Hobber, 221
Horizontal connector, 137, 138
Hyperbolic paraboloid, 165
Hyperboloid, two nappes, 170
Hyperboloid of revolution, 166

I

Illustration, engineering, 281-284
   shading, 281
   types of, 281
   uses of, 281
Inclination-Declination
   defined, 59
   by rotation, 61
Integral scale, 330
   calibration, 330
   pole distance, 331
Integration, graphical, 328-335
   applied problem, 332
   constant of integration, 331
   problems, 334, 335
   scale, 330
   string polygon, 328
   summary, 333
Intermediate points, 177
Intersections, 174-184
   cylinder and cone, 179
   cylinder and prism, 176
   cylinder and pyramid, 178
   general analyses, 174
   numbering system, 174, 177
   other intersections, 179
   problems, 183, 184
   sphere and line, 180
   sphere and plane, 181
   two cylinders, 177
   two prisms, 174, 175
Isometric drawing, 91
   advantages, 95
   axes, 41
   curves, 93, 94
   dimensioning, 95
   problems, 102, 103
   shading, 95
Isometric projection, 90

J

Job analysis, 223
Johannsson blocks, 225

K

Keys
   gib head, 207, 373
   keyways, 207
   Pratt & Whitney, 206, 373
   square and flat, 206, 373
   tables, 373, 374
   Woodruff, 206, 374

L

Laser machining, 221
Lathe, 219
Layout, 185
Lead, 199
Leonardo da Vinci, 1
Leroy, lettering, 28
Lettering, 22-30
   chart, 25
   composition, 27
   device, 23
   equipment, 23
   exercises, 29, 30
   guide lines, 23
   history, 22
   ink, 28, 29
   instruments, 27-29
   operation, 24
   slope, 27
   spacing, 26
   special alphabets, 29
   vertical capital, 24
   vertical lower-case, 26
Letterpress, 286
Limiting planes, 177
Limits, 230
   computations, 235
   definitions, 230
   expressing, 232
   tables, 363-366
Line
   standards and conventions, 81, 82
Lines, in space, 55-63
   as points, 61
   bearing, 58
   classified, 55
   clearance between, 135, 136
   defined, 55
   grade connector, 138
   horizontal connector, 137, 138
   problems, 62, 63
   slope, grade, 59
   specifications, 56
   true-length, 57
   true length by rotation, 57
   vertical connector, 139
Lines and planes, 142
   angle between, 148, 150
   piercing point, 142, 145
   problems, 151, 154
   visibility, 142
Lithography, 286
Logarithmic grids, 300

Logarithmic scale, 310
Logarithms, 343
Logic diagrams, 264, 265
Lunar excursion module, 44

M

Machine screws, 203, 355
Manufacturing limits, 235
Manufacturing measurements
  historical, 224
  linear devices, 225
  special devices, 228
Manufacturing processes
  job analysis, 223
  machining, 217, 219-221
  molding, 217, 221
  shaping, 217
  welding, 217, 210-214
Maximum material limit, 231
Mechanical integration, 331
Method of averages, 300
Method of selected points, 300
Microfilm, 286
Micrometer, 225
Milling machine, 221
Mimeograph, 286
Mongé, 1
Multiview projection, 43, 44

N

N charts, 316-319
  construction of, 319
  diagonal, graduation, 316
  mathematical graduation, 318
  problems, 322
  temporary scale method, 326
Nominal size, defined, 230
Nonuniform scale, 309
Numerals, 25
Nuts,
  hex, 204, 358
  jam, 205, 358
  slotted, 205, 359
  square, 204

O

Oblique projection,
  advantages and disadvantages, 98
  axes, 95, 96
  cabinet, 96, 97
  cavalier, 96, 97
  curves in, 97
  dimensioning, 98
  general, 96
  position of object, 97
  problems, 103
One-point perspective, 99
Optical comparitor, 225
Order of drawing, 82
Organization charts, 294
Orthographic projection,
  defined, 43
  lines, 55-63
  planes, 64-77
  points, 43-54
  solids, 78-89

P

Paper, drawing, 41
  bristol board, 41
  detail, 41
  film, 41
  tracing, 41
  tracing cloth, 41
  white, 41
Paraboloid, 170
Parallel perspective, 99, 100
Parallel scale chart
  construction of, 312-316
  four or more variables, 316
  theory of, 311
Parallelogram law, 156
Pattern, 185
Pencils, 39
Perspective drawing, 98-101
  advantages and disadvantages, 101
  angular or two-point, 101
  axis of vision, 98
  cone of vision, 99
  definite perspective, 99
  definition, 98, 99
  ground line, 98
  ground plane, 98
  horizon, 99
  indefinite perspective, 99
  initial point, 99
  parallel or one-point, 99, 100
  picture plane, 98
  principles of, 98
  problems, 103, 104
  sight point, 98
  sketching, 101
  vanishing point, 99
Photostat, 286
Pictorial charts, 294
Pictorial systems 90-104
  axonometric, 90-95
  dimetric, 90
  isometric drawing, 91
  isometric projection, 90
  oblique, 95-98
  trimetric, 90
Pie diagram, 292
Piercing point, 142, 145, 180
Piping, 244-250
  cast iron, 245
  computing lengths, 249
  copper and brass, 245
  fittings, 247-249
  joints, 245
  problems, 258
  representation, 246, 247
  sizes, 244
  tables, 367-371
  thread, 245
  tubing, 244
  valves, 248
  wrought iron and steel, 244
Pitch, 199
Pitch diameter, 199
Plane surface, 163
Planer, 219
Planes, 64-77
  classification of, 64
  definitions, 64, 65
  dihedral angle, 146-148
  dip, 70
  edge view, 67
  line of intersection, 143, 145
  line piercing, 142, 145
  location of solid upon, 150
  point and line in, 66
  problems, 72-77
  projection of a point on, 147
  specifications, 64
  strike, 69
  true size, 68, 69
  true size by rotation, 69
Planimeters, 332
Plotting data, 297
Point
  auxiliary views, 50-52
  coordinate system, 49
  location, 45
  principal projections, 45, 46
  problems, 53, 54
  projection onto a plane, 147
  relationship of views, 47
  standards of projection, 48, 49
Points and lines, 134
  problems, 139-141
Polar charts, 296
Polygon method, 156
Powder metallurgy, 223
Power curves, 302
  empirical equations, 302
Powers and roots, 306
  network chart, 305
Principal planes of projection, 45, 46
Principle of perpendicularity, 132, 147
Prisms, development of, 187
  intersecting, 174, 175
Problems at ends of units
  alignment charts, 330-332
  angle between line and plane, 152
  auxiliary views, 53, 54
  axonometric, 102, 103
  connectors, 139-141
  developments, 196, 197
  dihedral angle, 151
  dimensioning, basic, 130-132
  dimensioning, production, 241-243
  electrical, 266
  fasteners, 205, 215
  graphical differentiation, 333, 334
  graphical integration, 334, 335
  graphical mathematics, 320, 321
  graphical representation of data, 303, 304
  isometric, 102
  lettering, 29, 30
  line of intersection, 151
  line piercing a plane, 151
  lines in space, 62, 63
  oblique, 103
  perspective, 103, 104
  piping, 258
  plane vectors, 161
  planes, 72-77
  points, 53, 54
  processing methods, 229
  production illustration, 291
  projection of points and lines, 152
  reproduction, 290
  scales, 42
  sections, 115
  simplified drawing, 290
  sketching, 20, 21
  solids, orthographic, 87-89
  space vectors, 162
  structural, 258-260
  surface intersections, 183, 184
  surfaces, 170-173

INDEX

vectors, 161, 162
welding, 215, 216
worded mathematics, 320, 321
working drawings, 287-290
Processing, 217-224
   problems, 229
Proctor and Gamble model, 45
Production dimensioning, 230-243
Production illustration, 281-284
Progress charts, 294
Protractor, 39
Pyramid
   development of, 187
   intersections, 178

R

Ranking charts, 294
Rate of change
   nonuniform, 324
   uniform, 323
Reading a drawing, 82
Ream, 122
Related views, defined, 67
Reproduction, 286
   problems, 290
Resolution, 156, 157
Resultant,
   definition, 155
Ribs, 109
Right section, 189
Right triangle method, 189
Rivets, 208, 253
   gage, pitch, edge distances, 251, 252
   symbols, 253
Root, 199
Rotation
   double, 69
   edge view of plane, 69
   line about an axis, 57
   surfaces of, 169
   true size, 69
Rounds and fillets, 113
Ruled surfaces, 163
Ruling pens, 33

S

Sand casting, 221
Scalar quantity, definition, 155
Scales, 309
   adjacent, 311
   architect's, 36
   calibration, 325, 330
   construction, 311
   definition, 309
   design, 309
   distance, 310, 311
   engineers (civil), 36
   functional, 309, 310
   graphical integration, 330
   identification, 297
   length, 309
   mechanical engineers, 37
   metric, 37
   nonuniform, 309
   parallel, 311-316
   uniform, 309
   units, 297
Screw threads, 193
Screwed fittings tables, 369, 370
Section lining, 108
Sections, 105-109
   broken out, 107
   cutting plane, 105
   defined, 105
   detail, 107
   full, 105
   half, 105
   offset, 106
   partial, 107
   phantom, 107
   problems, 115
   removed, 107
   revolved, 107
   right, 187
Semigraphical integration, 331
Setscrews, 203, 361
Shaper, 219
Shortest distance
   point and line, 134
   skew lines, 135, 136
Shortest grade connector, 138
Shortest horizontal connector, 137, 138
Simplified drafting, 284
Simpson's rule, 331
Simultaneous equations, 306
Single-curved surfaces, 168
Sketching, see freehand drawing
Slope, defined, 59
Slope angle, lines, 59
Slope law, 324
Solids, 78-89
   auxiliary views, 83-85
   choice of views, 78
   necessary views, 80
   oblique views, 86
   order of drawing, 82
   partial views, 86
   principal views, 78
   problems, 87-89
Space analysis, defined, 56
Space vectors, 158-160
Specifications
   basic, 117-132
   production, 230-243
Sphere
   development of, 193
   line piercing, 180
   plane intersecting, 181
   point on surface, 180
Splines, 207
Spotface, 122
Springs, 208
Square thread, 199, 362
Standards of projection, 48
Straight-line curves, 301
   empirical equations, 301
Stretchout, 185
Strike, 69
String polygon, 328
Structural drawing, 250-258
   definitions, 251
   design drawings, 253, 254
   dimensioning, 257
   dimensioning standards, 258
   gusset plate, 255
   high strength bolts, 254
   joints, 255
   problems, 259, 260
   rivet symbols, 253
   scale, detail, 256
   scale, working, 256
   shop details, 256
   standard designations, 252
   steel shapes, 251
   terms defined, 251, 253
   trusses, 256
   types of drawing, 255
Stud, 203
Surface of revolution, 169
Surface quality, 240
Surfaces
   classification chart, 164
   definition of, 163
   developments, 185-197
   double-curved, 169
   intersections, 174-184
   plane, 163
   problems, 170-173
   ruled, 163
   single-curved, 168
   warped, 163
Symbols, electronic, 349

T

Tables (see Appendix), 339
Tangent line construction, 324
Tangent line convolute, 169
Tangent plane convolute, 169
   development of, 191, 192
Taper pin sizes, 360
Thermofax, 286
Thread forms, 199-202
   Acme, 200
   American National, 199, 201
   buttress, 200
   classes, 202
   knuckle, 200
   multiple, 200
   pipe, 245
   representation, 200
   sharp V, 199
   specifications, 202
   square, 167, 199
   unified, 199, 201
Threaded fasteners, 198-205
   bolts, 202, 204
   cap screws, 202
   inserts, 203
   jam nuts, 205
   machine screws, 203
   miscellaneous, 204
   multiple thread, 200
   nuts, 204
   problems, 205
   right and left hand, 200
   setscrews, 203
   studs, 203
Tolerance
   bilateral, 232
   control symbols, 237
   cumulative, 237
   defined, 231
   expression of, 232
   general notes, 241
   geometrical, 240
   positional, 238
   preferred, 233
   unilateral, 232
Torus, annular, 170
Tracing cloth, 41
Tracing paper, 41
Transition pieces, 191, 192
Trapezoidal rule, 331

Triangle method, vectors, 156
Triangles, 34
  use of, 34-36
Trigonometric functions, 344
Trilinear charts, 295
Trimetric projection, 90
Tripod
  nonvertical load, 158
  vertical load, 160
True length
  auxiliary view, 57
  right triangle method, 189
  rotation, 57
T-square, 33
Tubing (see piping), 244, 372
Twist drill sizes, 350
Two-point perspective, 101

U

Unified thread, 199, 201, 351
Uniform scale, 309
Units of measurement, 118

V

Valves (see piping), 248
Van Dyke, 286
Vector
  component, 156
  composition, 155
  defined, 155
  resolution, 156, 157
Vector chain, defined, 155
Vector quantity, defined, 155
Vectors
  collinear, 156
  concurrent, 156
  coplanar, 156
  definitions, 155
  equilibrant, 155
  nonconcurrent, 156
  noncoplanar, 156-158
  plane, 156
  problems, coplanar, 161
  problems, space, 162
  resultant, 155, 157
Verifax, 287
Vernier, 225
Vertical connector, 139
Visibility, 142, 176

W

Warped surfaces, 163
  problems, 170, 171
Washers
  lock, 205
  plain, 204
  tables, 356, 357
Weddle's rule, 331
Welding, 210-214, 217
  defined, 210
  designations, 212, 213
  joints, 210
  problems, 215, 216
  processes, 210
  symbols, 214
  types of welds, 210
Wire gages, 350
Working drawings, 277
  problems, 287-291
Wrico (lettering), 28

X

Xerography, 286

Z

Z charts, 316-319
  construction of, 319
  diagonal, graduation, 316
  mathematical graduation, 318
  problems, 322
  temporary scale method, 316